INDUSTRIAL ELECTRONICS AND INSTRUMENTATION

For
Engineering Students

By
Sunil Kumar
Instrumentation and Control Engg. Deptt.
P.D.M. College of Engineering
Bahadurgarh

S.K. KATARIA & SONS®
Publisher of Engineering Books
4885/109, Prakash Mahal, Dr. Subhash Bhargav Lane,
Opposite Delhi Medical Association, Daryaganj, New Delhi-110002 (INDIA)
Phone: +91-11-43551243; Mobile: +91-9871775858, 7678425209
e-mail: katariabook@gmail.com, katariabooks@yahoo.com
Website : www.skkatariaandsons.com
(e-books, subscriptions and bulk buying also available on our website)

Published by :

S.K. Kataria & Sons®

4885/109, Prakash Mahal, Dr. Subhash Bhargav Lane,
Opposite Delhi Medical Association, Daryaganj, New Delhi-110002
Phone: +91-11-43551243; Mobile: +91-9871775858, 7678425209
e-mail: katariabook@gmail.com, katariabooks@yahoo.com

Price : ₹ 395/-

Printed at : S.N. Computers

PREFACE TO THE FOURTH EDITION

This book was first written while keeping in view the requirements of the students. I feel very glad that it has been successful in meeting their demands. The material has been presented in a way as to give clear understanding of the subject to the readers.

The whole text has been divided into twelve units. The first unit is dedicated to thyristor while second unit is the extension of the first one and covers the methods to protect the device against excessive heat dissipation. The SCR has various applications in today's life, few of which have been covered in third unit. Fourth, fifth, sixth and seventh units provide further use of thyristor for different applications. Heating of any specimen is required in various applications. Some of the methods of heating are presented in unit eight.

Next section of the book is dedicated to instrumentation which comprises units nine through eleven. Unit nine gives the introduction to instrumentation. Some new topics have been added to it. The next unit describes transducer which is an important component in instrumentation. Eleventh unit is just the extension for processing the signal obtained from a transducer. The last unit describes some devices used for different purposes in instrumentation.

For better understanding of the subject, numerical problems have been given at the end of the chapters. The solutions to those problems have been given at the end of the book. In addition, ten experiments prescribed have been covered and at last the subject review questions are given for quick and better understanding of the matter.

AUTHOR

ACKNOWLEDGEMENT

I am thankful to the readers who have taken interest in this text and always stimulated me towards the betterment of it. I am also grateful to my family and friends who have always encouraged me directly or indirectly for improving it.

I am also thankful to **Mr. Sanjeev Kumar Kataria** and staff of **S.K. Kataria & Sons** Publishers to provide me opportunity and also to encourage me improve the text to the present status.

Finally the criticism and suggestion for the improvements of the text are always welcome.

AUTHOR

CONTENTS

1

Thyristors

1.0. INTRODUCTION

Before the semiconductor devices were fabricated input power delivered from the source to the load was controlled by some crude methods such as rheostat etc. With advancement in semiconductor devices, these devices are now used for interaction between source and load. Various semiconductor devices such as MOSFETs, Transistors, GTO's, SCR etc. are now-a-days used as switches between load and source, in different power ranges. Earlier the power losses across the control elements would be the significant fraction of the total power. Such a loss has been decreased using the semiconductor devices, considerably. For control of power supplied from the source to load or vice versa these devices operate in switching mode which means that (1) In the OFF state the current through the device is zero and voltage across it has maximum value and (2) in the ON state the current through the device is maximum and voltage is ideally zero. Thyristor is a very important element for power control applications. A brief description of this device is presented in this chapter.

1.1. THYRISTOR RATINGS AND GATE RATINGS

"Thyristor is a four layer three terminal and three junction device". The thyristor acts like a controlled switch in forward mode of operation while in reverse mode it acts like a simple *p-n* junction diode. The construction, symbol and static characteristics of the thyristor are shown in Fig. 1.1. In normal working of the device, the junctions J_1 and J_3 remain forward biased while junction J_2 reverse biased. Three terminals of the thyristor are anode, cathode and gate designated respectively as A, K and G as shown in the figure. From the V-I characteristics of the SCR it can be seen that the anode to cathode voltage required to trigger the device decreases as the gate current increases. This is due to the fact that at higher gate current most of the charge at the junction forming the barrier potential is removed. As a consequence of it the energy required by the charge carrier to overcome the barrier reduces which can be supplied by smaller values of anode to cathode voltage. When the gate circuit is open the voltage required to turn the device

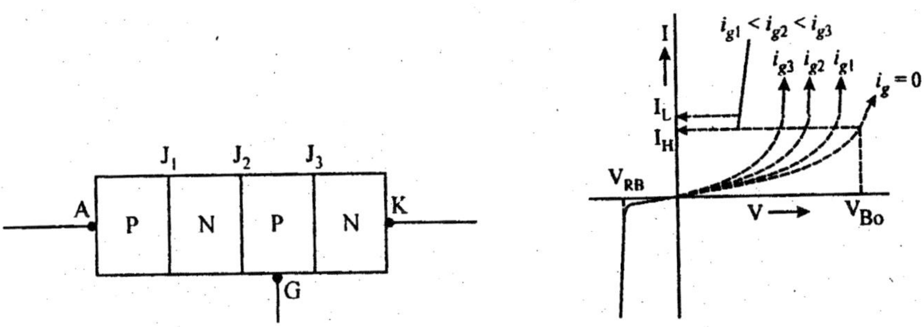

(a) (b)

Fig. 1.1. SCR (a) construction (b) characteristics

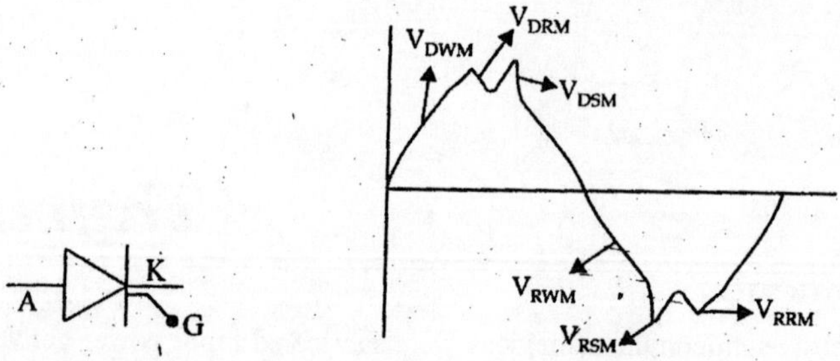

(c) (d)

Fig. 1.1. SCR (c) symbol (d) voltage ratings

ON is, obviously, high. This value of voltage is called breakover voltage. It is designated as V_{BO}. In reverse mode the SCR acts like a simple p-n junction diode. In this mode the reverse voltage at which device starts conducting in reverse direction is called breakdown voltage. Various ratings for the thyristor are defined below:

1.1.1. Voltage Ratings

For proper working of the device it is to be protected against any damage to it. The voltage capability of the thyristor should not exceed even for very short duration. The voltage rating of the device should be high enough so that it can withstand anticipated voltage transients. Various ratings are described below :

1. **Repetitive peak-off state forward voltage (V_{DRM})** : This is repetitive peak transient voltage that a thyristor can withstand in off state, in the forward direction.
2. **Non repetitive peak-off state forward voltage (V_{DSM})** : This is instantaneous non-repetitive value that occurs across thyristor in off state.
3. **Working peak forward voltage (V_{DWM})**: This is the maximum voltage which the device can withstand in forward blocking state.
4. **Working peak reverse voltage (V_{RWM})** : This is maximum reverse voltage that occur across thyristor in reverse mode.
5. **Repetitive peak reverse voltage (V_{RRM})** : This is maximum transient voltage across thyristor which occurs repeatedly in reverse direction at specified temperature. It varies with temperature.
6. **Non repetitive peak reverse voltage (V_{RSM})** : It is maximum transient reverse voltage that a thyristor can block in reverse direction.
7. **ON state voltage (V_T)** : It is the maximum voltage across thyristor in ON state. It lies between 1 - 1.5 volts.
8. **Gate Trigger voltage** : This is the maximum gate voltage to produce the minimum gate current necessary for firing the thyristor.
9. **Voltage safety factor (V_f)** : To avoid damage to thyristor in normal operation, voltage is kept below peak inverse voltage (V_{RSM}) value of the device. The normal operating voltage and V_{RSM} are related by the voltage safety factor V_f defined by

$$V_f = \frac{V_{RSM}\ (\text{or PIV})}{\sqrt{2} \times \text{RMS value of input voltage}}$$

This factor has value between 2 and 2.5.

10. **Forward $\frac{dV}{dt}$ rating :** This indicates the maximum rate of rise of anode voltage across thyristor that will not trigger the device. Various voltage ratings are shown in Fig. 1.1 (*d*).

1.1.2. Current Ratings

Current ratings of thyristor are determined at junction temperature of the device. Various current ratings pertaining to thyristor are given below :

1. Average on state current ($I_{T(av)}$): The average current through the thyristor depends upon the conduction angle if the input is AC. To avoid any damage to the thyristor the junction temperature should not exceed the rated value. The average current through the thyristor decreases with a decrease in conduction angle because the power loss increases with decrease in θ for given average current as shown in Fig. 1.2.

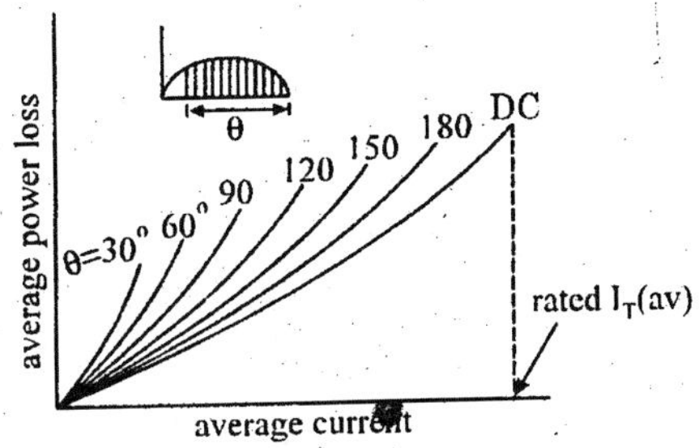

Fig. 1.2. Thyristor ratings.

2. RMS on state current : This is important when thyristor is subjected to current of high peak value. This current is of repetitive type.

3. Non-repetitive, peak on state current or surge current (I_{TSM}) : This is the maximum possible non-repetitive current which the device can withstand. It is defined for the device for half input cycle at frequency of 50 Hz, i.e., for a time of 10 ms. This non-repetitive surge current depends upon the duration of the input cycle as given by relation:

$$I_{sub}^2 t = I^2 T$$

or

$$I_{sub} = I\sqrt{\frac{T}{t}}$$

where T = time for one half cycle of input (10 m sec)

I = surge current for one cycle

t = duration of subcycle surge

I_{sub} = subcycle surge current rating.

The surge current which occur due to faults or short circuit increases with a decrease in the duration of the surge. For example I_{TSM} = 3000 A for 1/2 cycle, I_{TSM} = 1730 A for 3 cycles etc.

4. $I^2 t$ rating : This rating is the measure of thermal energy that the device can absorb before the fault is cleared by the fuse. It puts limit on the fuse used to protect the device. I^2t rating of the device should be more than that of the fuse so that the fuse may burn out before the deivce is damaged.

5. $\frac{di}{dt}$ rating : This is maximum rate of rise of anode current at which device will not turn on. If $\frac{di}{dt}$ rating is higher than the spreading velocity of carriers at junction J_2, a hot spot heating occurs which may damage the device.

6. Holding current (I_H) : This is the minimum value of forward anode current below which the device stops conducting.

7. Latching current (I_L) : This is the minimum ON state current required to maintain the device in ON state even if the gate pulse is removed.

8. Gate current ($I_{g\ min}$ and $I_{g\ max}$) : Gate current is supplied to the gate circuit for control purpose. Minimum value of gate current $I_{g\ min}$ is the current required through the gate for triggering the device. Maximum value of gate current $I_{g\ max}$ is the gate current that may be applied to the gate in the safe mode.

1.1.3. Thermal Rating

Thermal ratings of thyristor are the following :

1. Ambient temperature
2. Junction temperature
3. Transient thermal resistance

1. Ambient temperature : It defines the range of external temperature to which the device may be subjected without affecting its performance.

2. Junction temperature (T_j) : The junction temperature of the device puts limit on voltage and current rating of the device. It is the temperature above which the device is turned on even if no gate current has been supplied to the device. Typical value of junction temperature is 125°C. Above this temperature the device damages due to excessive heating. It may, in some cases, be as high as 200°C.

3. Transient thermal impedance (T_{Th}) : This is the thermal resistance between the junction of a thyristor and its cooling surface. For larger thyristor the value of thermal resistance will be smaller.

1.1.4. Power Ratings

Power generated in junction region of thyristor consists of the following components:

1. Forward conduction loss
2. Turn on switching loss
3. Turn off loss
4. Forward and reverse blocking losses
5. Gate pulse triggering loss

1. Forward conduction loss : Average anode current multiplied by voltage drop across thyristor in the ON state is the average power dissipated in the thyristor. Power dissipated leads to a rise in junction temperature. Average power loss can be controlled by changing the firing angle as shown in Fig. 1.2. By changing firing angle average current in a cycle can be controlled to a value so that power loss may remain within rated value. The conduction loss across SCR increases with a decrease in frequency.

2. Turn on switching loss : Switching process takes a finite time as all the charges at middle junction take non zero time to take part in conduction. During this period high voltage appears across the device while the current still flows through it. As a result power dissipated in the device generates heat, called switching loss. Switching loss can be minimized by decreasing switching time of the device.

3. Turn off loss : This loss arises during decay of reverse current. It is the product of reverse current and reverse voltage and may reach very high value. During reverse recovery period, the reverse current flows through the device as the charges are swept away from the junction. As a result a voltage given by Ldi/dt is developed across the SCR. The product of the voltage and reverse current is responsible for the generation of heat in it. It is called local heating and generates what is called hot spot. It can be controlled by decreasing rate of decay of reverse current, *i.e.*, by increasing reverse recovery time.

4. Forward and reverse blocking losses : In forward off state the leakage current flows through the device. The product of leakage current and off state voltage across the device is the power loss in the device. Similar type of power loss occurs in reverse blocking state.

5. Gate pulse triggering loss : This loss is due to gate current through gate circuit.

1.2. TURN ON METHODS USING R-FIRING CIRCUIT

In forward mode the thyristor acts like open switch unless it is turned ON in which case it acts like a closed switch. Various methods are used to turn it ON. Some of the methods are given below :

1. Voltage turn ON method
2. Thermal turn ON method
3. *dV*/*dt* turn ON method
4. Light turn ON
5. Gate turn ON.

The last method viz. gate turn on method will be discussed in this chapter.

We refer to Fig. 1.1 which shows the construction of the thyristor. In forward mode junctions J_1 and J_3 are forward biased while the junction J_2 is reverse biased. As a result the charge carriers crossing the junctions J_1 and J_3 are accumulated around junction J_2. If by any means an amount of energy is supplied to these charges such that the barrier potential is overcome, these charge will take part into conduction and the device is said to be fired or turned ON. One of such means is by applying positive pulse at the gate terminal. Various gate turn ON methods are described below :

1.2.1. DC Gate Turn ON Method

In this method of triggering a DC voltage of proper magnitude and polarity is applied between gate and cathode terminals of the device as shown in Fig. 1.3. By changing the resistance R_g in the gate circuit the gate current necessary to trigger the device is supplied. One drawback of this method is that gate current flows continuously which leads to high power loss. There is no isolation between anode and gate circuits.

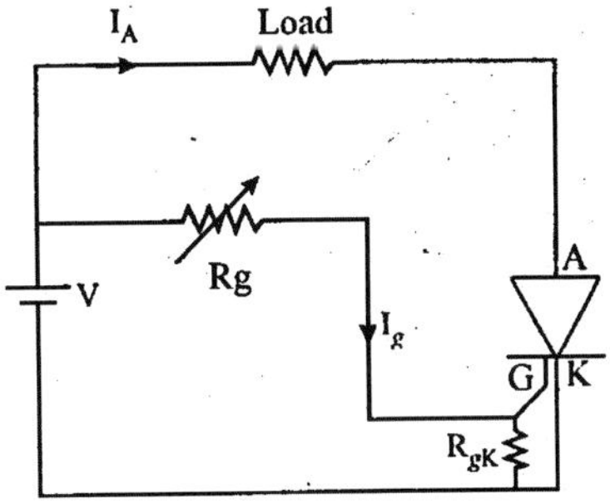

Fig. 1.3. DC gate turn ON

1.2.2. AC Gate Turn ON Method using R-firing circuit

In this method of triggering, AC voltage is applied between gate and cathode terminal as shown in Fig. 1.4 *(a)*. In the Fig., by changing the value of variable resistance R in the gate circuit the minimum gate current required to trigger the device is obtained at different values of the instantaneous voltages which correspond to different phase angles. The angle at which the device is triggered is called the firing angle or delay angle. The voltage across the thyristor in conduction state is ideally zero. The angle representing the conduction

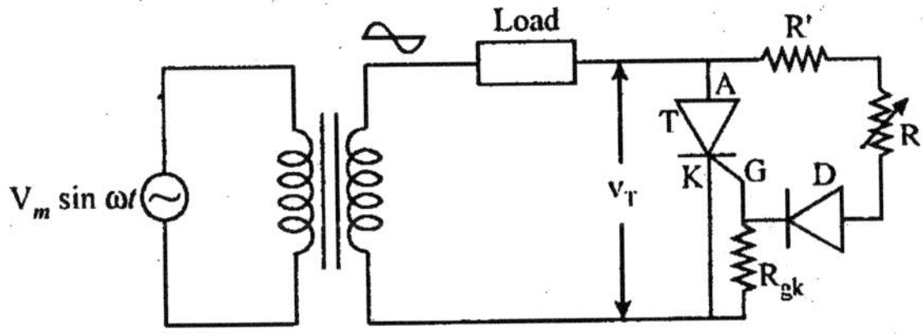

Fig. 1.4. AC gate turn ON (a) Circuit

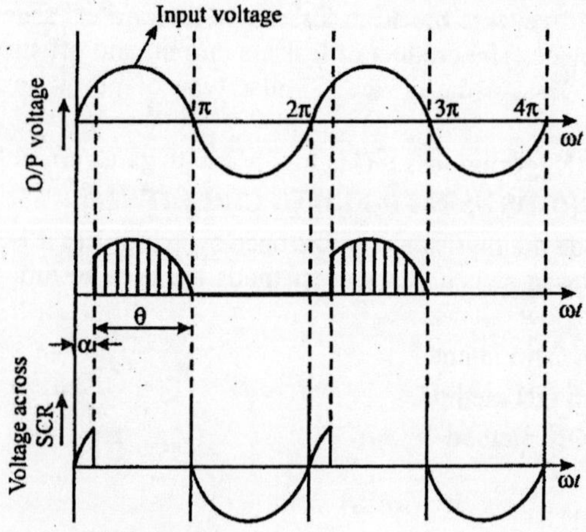

Fig. 1.4. AC gate turn ON (b) Wave form

of the device is called conduction angle θ. The waveforms are shown in Fig. 1.4 *(b)*. Operation of the circuit is given below :

Operation : In the OFF state input voltage appears across the SCR and voltage across the load is zero as no current flows through it. Voltage across the SCR is the instantaneous value of supply voltage hence it acts as power source for gate circuit. For the thyristors to turn ON minimum gate current, I_{gmin}, should flow through the gate circuit. By changing the value of potentiometer R, the firing angle can be controlled as for different values of resistance R the instantaneous voltage required to supply minimum gate current I_{gmin} to turn ON the device will have different values which means different firing angles. Diode D is used to protect the gate circuit against negative cycle of the input supply. Resistance R_{gK} maintains minimum voltage required between gate and cathode and improves $\frac{dv}{dt}$ capability of the device. Its value is given by the expression $\frac{V_{gk}}{R_{gk}} \geq \frac{V_m - V_{gk}}{R + R'}$.

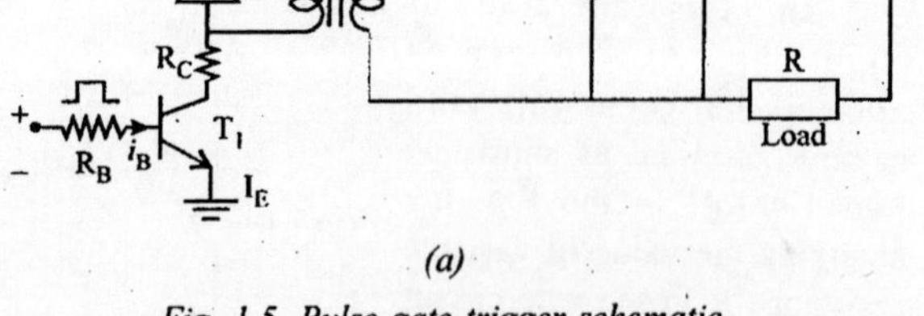

(a)

Fig. 1.5. Pulse gate trigger schematic.

Resistance R′ is connected to ensure that gate current does not exceed maximum value I_{gmax}.

1.2.3. Pulse Gate Triggering Circuit

The drawback of continuous gate triggering is that after the device is triggered, the gate current leads only to the dissipation of power in the gate circuit. This fault is overcome by using pulses of small duration at the gate terminal. The circuit is as shown in Fig. 1.5 *(a)*. Pulse transformer is used for isolation between gate and anode circuits. The transistor switch T_1 is set into conduction as the pulse appears at the base terminal. The current flowing through the primary of the transformer induces an e.m.f. in the secondary which triggers the thyristor. The waveforms are shown in Fig. 1.5 *(b)*. We observe that firing of the SCR has no dependence on input signal. Whenever the gate pulse is applied the SCR is turned ON. It is naturally turned OFF at zero crossing and during negative cycle remains OFF due to reverse biasing. When T_1 is turned OFF, transformer primary releases its energy through freewheeling diode D_{FW}.

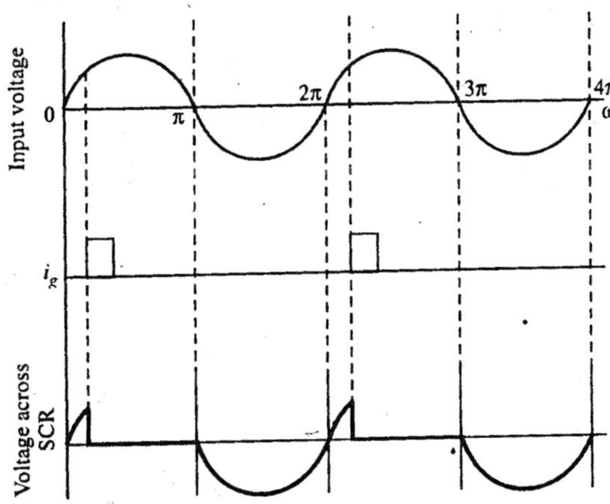

(b)
Fig. 1.5. Waveforms.

1.2.4. R-C Firing Circuit

The limitation of AC turn ON method using R-firing circuit is that the firing angle ranges from 0 to 90° only. This limitation is eliminated in R-C firing circuit whose firing angle ranges from 0 to 180° which enhances the range of control of load voltage and current. The circuit is shown in Fig. 1.6 *(a)*.

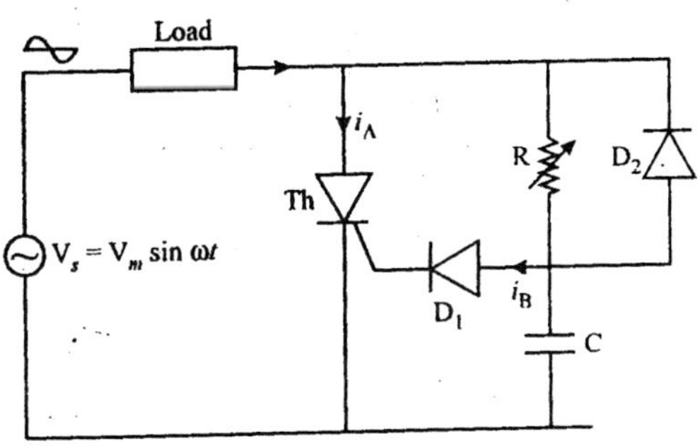

Fig. 1.6. (a) RC firing circuit.

Operation : During negative half cycle, the capacitor C is charged to peak value V_m through diode D_2 with its lower plate positive positive at angle ωt = 270°. Beyond this angle it starts discharging through the path C-R-load-source-C exponentially. Its discharge rate is controlled by potentiometer resistance R. When $\omega t = 2\pi$, the capacitor voltage is still negative and is discharged completely at any angle between 2π and 3π. The discharge rate is determined

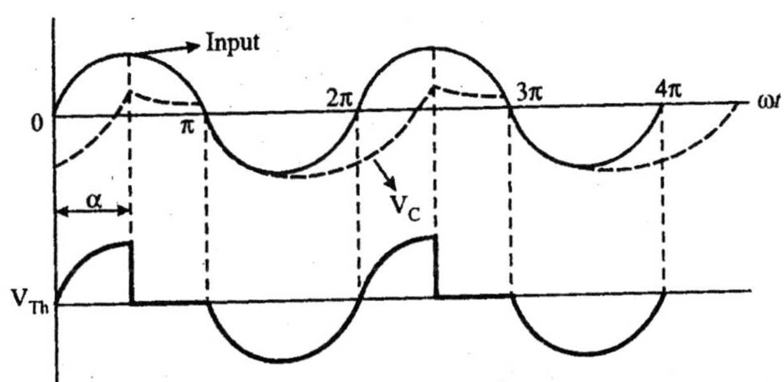

Fig. 1.6. (b) Waveforms.

by time constant RC. Now it is charged with upper plate positive. When the voltage across it reaches the value $V_g + V_{D1}$ the thyristor is turned ON and conduction starts. At beginning of negative half cycle again the thyristor is turned OFF due to line commutation while capacitor starts charging in reverse direction. This process continues. The waveforms are shown in Fig. 1.6 *(b)*.

The empirical relation for product RC is

$$RC \geq 0.65\,T \qquad \text{...(1)}$$

T is time period of input voltage.

Since $T = \dfrac{2\pi}{\omega}$, we obtain

$$RC \geq 0.65\frac{2\pi}{\omega} = \frac{4}{\omega}$$

For thyristor to trigger $V_c = V_g + V_{D1}$. The instantaneous value of V_s at the instant of triggering is

$$V_s \geq I_{g\,(min)}\,R + V_g + V_{D1}$$

giving

$$R \leq \frac{V_s - V_g - V_{D1}}{I_{g(min)}} \qquad \text{...(2)}$$

where V_g is minimum value of voltage between gate and cathode so that minimum current $I_{g\,(min)}$ required for triggering flows through the gate circuit.

Example 1.1. A voltage given by V = 300 sin 100 πt is used to supply power to the load through RC firing circuit showin in Fig. 1.6 (*a*). If firing angle is 30° and minimum gate current is 3 mA, calculate maximum value of resistance R.

Solution. The instantaneous value of source voltage

$$v_s = 300 \sin 30° = 150° \text{ volts}$$

Using expression (2), we get

$$R = \frac{150 - 0.7 - 0.7}{3 \times 10^{-3}} = \frac{148.6}{3} \times 10^3\Omega = 49.5 \times 10^3\ \Omega = 49.5\ k\Omega. \textbf{ Ans.}$$

1.3. TURN OFF METHODS

The thyristor does not turn OFF once it is triggered, unless the anode current is reduced below the value of holding current (I_H). Thyristor can be turned OFF in two ways which are

1. Natural turn OFF.
2. Forced turn OFF

The methods are discussed in what follows :

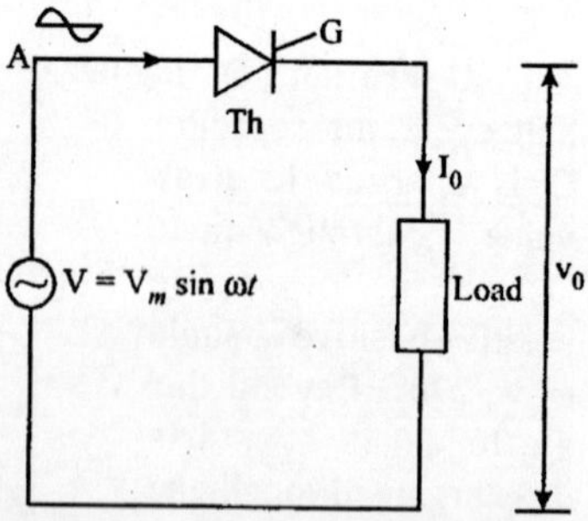

Fig. 1.7 (a) Natural turn OFF

1. Natural turn OFF : We refer to Fig. 1.7 *(a)*. AC voltage source is used to deliver power to the load. Let the thyristor be fired at angle α. It will conduct for positive half cycle as it is forward biased. As the voltage falls to zero at $\omega t = \pi$, the anode current falls to zero value which is less than the holding current. Consequently the thyristor is turned OFF. We observe that the thyristor is turned OFF due to input AC current which falls to zero. This is called line turn OFF or natural turn OFF. The waveforms are shown in Fig. 1.7 *(b)*.

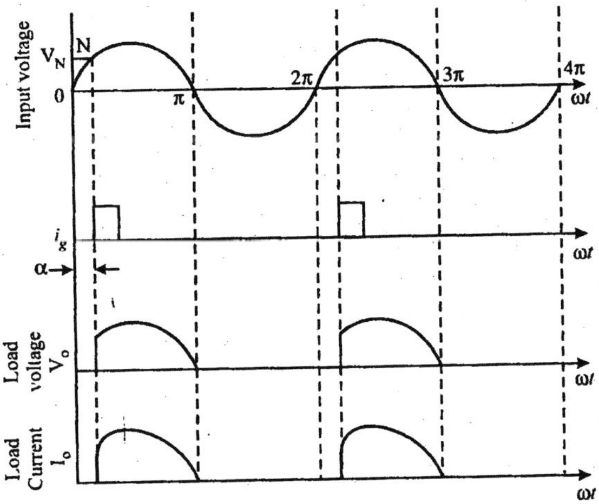

Fig. 1.7 (b) Waveforms.

Working : Thyristor is in forward mode when point A in Fig. 1.7 *(a)* is positive. A pulse is applied at the gate of thyristor at angle α. At this instant the instantaneous value of voltage at anode of the thyristor is V_N as shown by point N in Fig. 1.7 *(b)*. The thyristor is turned ON. The load voltage follows the source voltage as the current starts flowing through the load. At $\omega t = \pi$, the anode current falls to zero, thereby turning it OFF naturally. The load voltage falls to zero, the load current also falls to zero for purely resistive load. During next cycle the device can be turned ON by application of gate pulse as shown in Fig. 1.7 *(b)*.

2. Forced turn OFF : For DC power source, thyristor is forcibly turned OFF by applying reverse voltage or reverse current across it. Some of the methods are discussed below:

Complementary voltage commutation : The circuit is shown in Fig. 1.8 *(a)*. The operation of the circuit is explained as below :

We assume at instant $t = 0$ that voltage across the capacitor is zero when thyristor Th_1 is triggered. The capacitor starts charging exponentially with left plate positive. As the voltage across capacitor reaches V_{dc} it blocks the current. The current now flows only through load 1, *i.e.*, $i_{L_1} = I_{T_1}$. If we trigger thyristor Th_2 now the capacitor will try to discharge through the path C-Th_2-Th_1-C making cathode voltage of Th_1 equal to $+V_{dc}$. As a result the thyristor Th_1 is turned off as it becomes reverse biased. The capacitor discharges through the path C, Th_2, source, load, C. When it is charged to voltage $-V_{dc}$ the current is blocked through it. The thyristor current I_{T_2} is now equal to load current i_{L_2}. In complementary voltage commutation, thus, one thyristor turns off the other and vice versa. One thyristor conducts during the period other is OFF. Thus these are complementary to each other. The waveforms are shown in Fig. 1.8 *(b)*.

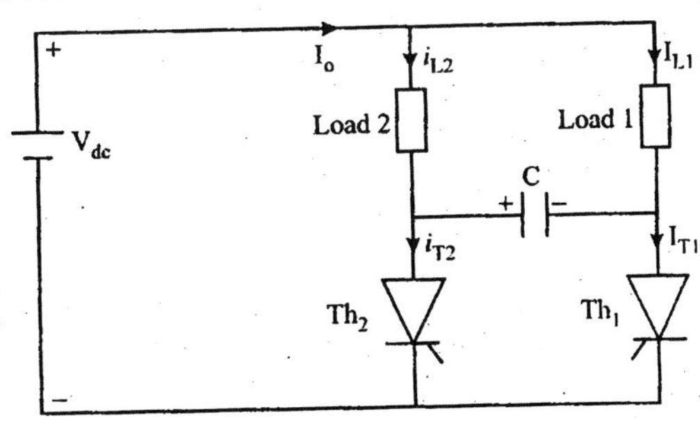

Fig. 1.8 (a) Complementary voltage commutation.

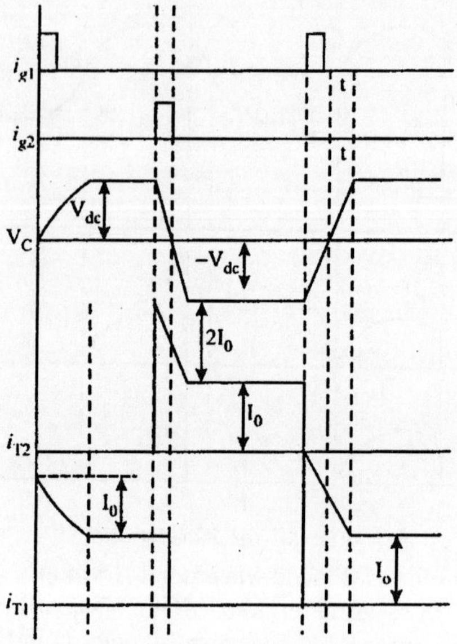

Fig. 1.8 (b) Waveforms.

Example 1.2. In Fig. 1.8 *(a)*, $V_{dc} = 100$ V, C = 47 μF and load$_1$ and load$_2$ both are resistive and have value 50 Ω each. Let thyristor Th_2 is fired at time $t = 0$. What will be the time taken by capacitor to discharge to 1/3 of the total voltage across it.

Solution : The capacitor will discharge through load$_1$ and source using eqn.

$$v_c = 2V_{dc}\left(e^{-\frac{t}{RC}}\right) - V_{dc}$$

Factor 2 is due to the presence of voltage source in the path of capacitor. Using condition given in the question, we get

$$\frac{1}{3}V_{dc} = 2V_{dc}\left(e^{-\frac{t}{47\times10^{-6}\times50}}\right) - V_{dc}$$

$$\frac{t}{2e^{235\times10^{-5}}} = 4/3$$

Taking ln on both sides, we get

$$\frac{t}{235\times10^{-5}} = \ln 2/3$$

or $$t = 2.35\times10^{-3}\ln 2/3.$$

$$= 4.28 \text{ m sec. } \textbf{Ans.}$$

Current commutation or resonance commutation : In this method of commutation a current flowing opposite to the direction of SCR current is used to turn OFF the thyristor. The circuit is shown in Fig. 1.9 *(a)*. First the capacitor is charged by some auxiliary means with the polarity shown. Switch S is closed when thyristor Th is in ON state. The capacitor will try to discharge through the path C- L-Th-C when the switch S is closed. The capacitor current i_C and thyristor current i_T are opposite to each other. Under the condition $i_C = I_T$ the thyristor is turned off by forced commutation. The waveforms are shown in Fig. 1.9 *(b)*. Capacitor current rises sinusoidally and its maximum value is given by $I_{cm} = \frac{V}{\sqrt{L/C}}$. At time $t = t_1$ the capacitor current i_C reaches the value of load current. Hence the thyristor is turned off as is shown by the value of thyristor current I_T at $t = t_1$.

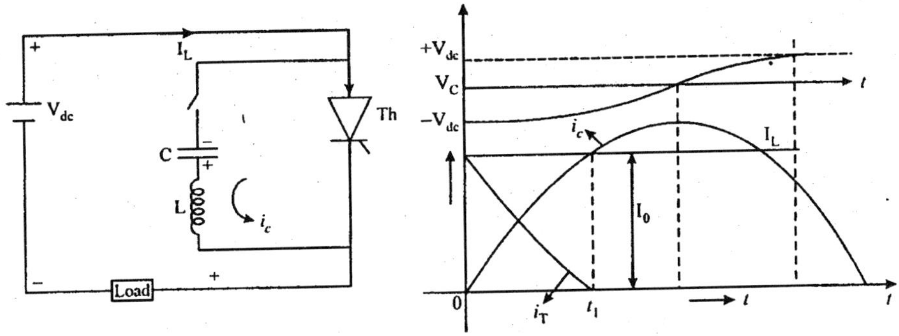

(a) Circuit *(b) Waveform.*

Fig. 1.9. Resonance commutation.

External pulse communication : The circuit is shown in Fig. 1.10. As the pulse is applied at the primary of the transformer, induced e.m.f. in secondary reverse biases the thyristor, turning it OFF finally. The load current is now carried by the capacitor C.

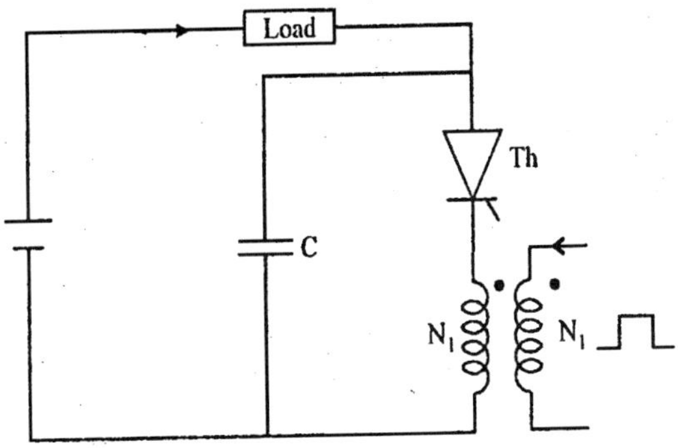

Fig. 1.10. External pulse communication circuit.

Load Side Commutation: In load side commutation, the commutating capacitor remains in series with the load. The circuit is shown in Fig. 1.11. In the figure, the commutating capacitor C charges and discharges through the load. When SCR T_2 is turned ON, the capacitor charges through the path L-T_2-load-V_{dc}-C. When SCR T_3 is triggered the capacitor voltage changes the polarity through the path consisting of C-T_3-L in the same sense of current flow. It discharges through T_2 again after turning OFF main SCR T_1.

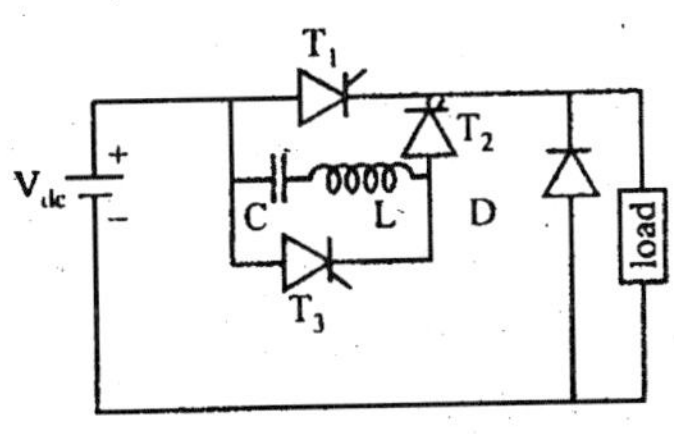

Fig. 1.11: Load Side Commutation.

SUMMARY

Thyristor or SCR is a switching device. For proper working of this switch various ratings such as current rating, voltage rating etc. are defined. Once SCR is turned ON it is to be turned OFF. The circuits used to turn it OFF are called commutation circuits. Such circuits also are described in this chapter.

EXERCISES

1. What are various types of current ratings for a thyristor ? Signify i^2t rating. .
2. Describe AC gate turn on method. What is advantage of pulse triggering over continuous triggering.
3. Explain R-C trigger circuit.
4. What is difference between natural and forced turn off method. Describe complementary voltage commutation circuit.
5. Define the following :
   (a) Latching current
   (b) Repetitive peak reverse voltage
   (c) Gate trigger voltage
   (d) Gate current
   (e) Voltage safety factor.
6. Describe the power rating of an SCR.
7. Describe DC gate turn ON method.
8. Define firing angle and conduction angle of an SCR.
9. What are natural and forced turn OFF methods. Give examples.
10. Draw the current commutation circuit. Describe it.
11. Draw external pulse commutation circuit and explain its working.
12. What is meant by load side commutation? Draw the circuit

NUMERICAL PROBLEMS

1. Voltage safety factor for an SCR is 2.1. The rated peak value of input votlage in normal working condition is 70 V. What is the peak inverse voltage that can be applied across the device ?
2. A conducting SCR is turned OFF by applying rated voltage of 50 V in reverse mode. A current of 3 amperes was flowing through the device when it was conducting. It reaches zero value in 10 μs, calculate the approximate power dissipated across the device.
3. Refer to Fig. 1.6 *(a)* showing RC firing circuit. If R = 6 Ω and C_1 = 10 μF, calculate the time period of input cycle.
4. Again in Fig. 1.6 *(a)* if instantaneous value of input voltage is 2.0 volt and minimum gate current required for triggering ($I_{g\ min}$) is 6 mA, calculate potentiometer resistance R.
5. In resonance commutation method source voltage is 100 V and L = 20 mH, C = 47 μF, calculate the peak value of capacitor current.

OBJECTIVE TYPE QUESTIONS

1. Thyristor is a
   (a) three terminal, three layer device
   (b) Three terminal, three junction four layer device
   (c) simple diode in forward as well as reverse mode
   (d) two junction, three terminal, four layer diode

2. In ON state the voltage across SCR is
   (a) zero (b) 1-1.5 volts
   (c) very high voltage (d) not specified
3. Breakover voltage of thyristor is defined as
   (a) a voltage at which SCR breaksdown
   (b) a voltage at which thyristor starts conducting with the given conditions
   (c) maximum voltage which a thyristor can sustain in OFF state
   (d) the maximum voltage in the reverse mode.
4. Breakover voltage
   (a) is constant for any SCR (b) depends upon the gate current
   (c) depends upon anode current
   (d) should be less than peak repetive forward voltage in OFF state.
5. Latching current is
   (a) minimum value of anode current to maintain the SCR in ON state in the abscence of gate pulse
   (b) maximum value of anode current in the ON state
   (c) dependent on thyristor type
6. Holding current
   (a) is the same as latching current
   (b) is minimum value of anode below which SCR is turned OF
   (c) puts the limit on max current through SCR
   (d) is maximum value of anode current above which it is turned ON.
7. Turn OFF loss in an SCR is due to
   (a) finite time required by reverse current to reach zero value
   (b) voltage across the thyristor in OFF state
   (c) leakage current (d) none of the above
8. Forward blocking loss in SCR is due to
   (a) leakage current through SCR in forward blocking state
   (b) defects in SCR
   (c) forward voltage and forward current
   (d) incorrect connections
9. dV/dt turn ON method is
   (a) the turning ON of SCR by changing across it
   (b) the generation of high current given by dV/dt across middle junction
   (c) used when ac power source is used (d) used in reverse mode only.
10. DC turned ON method is less favourable then pulse gate triggering method because
   (a) dc power is expensive (b) dc method leads to wastage of power
   (c) in dc turn ON method high gate current is required
   (d) in pulse triggering height of pulse is more then constant
11. In R-C firing circuit, the firing angle range from
   (a) 0 to 180° (b) 0 to 90°
   (c) 0 to 360° (d) 3° to 77°
12. In R-firing circuit firing angle ranges from
   (a) 0 to 180° (b) 0 to 360°
   (c) 0 to 90° (d) 3° to 77°

13. In complementary voltage commutation method

(a) two thyristors are required (b) one thyristor is required

(c) one thyrisor and one diode is required

(d) two diodes are required.

14. In resonance commutation method, capacitor peak current is equal to

(a) $\dfrac{V}{\sqrt{LC}}$ (b) $\dfrac{V}{\omega L}$

(c) $\dfrac{V\sqrt{C}}{\sqrt{L}}$ (d) both (b) and (c)

15. In thyristor under normal operation

(a) All the junction remain reverse biased

(b) J_1, T_2 reverse biased while J_2 forward biased

(c) J_1, J_3 forward while J_2 remain reverse biased

(d) J_1, J_3 remain reverse biased while J_2 remain forward biased.

ANSWERS

1. (b)	2. (b)	3. (b)	4. (b)	5. (a)	6. (b)
7. (a)	8. (a)	9. (b)	10. (b)	11. (a)	12. (c)
13. (a)	14. (d)	15. (c).			

❑❑❑

2
Heat Sinks

2.0. INTERNAL POWER DISSIPATION AND HEAT SINKS

As we have discussed in Chapter 1 that a semiconductor device when in the circuit dissipates power. In practical circuit design, this power is taken into account. Power dissipation in the device is due to many factors as discussed in Chapter 1. For good thermal design of the device it is to ensure that its temperature should not exceed the rated value. The junction temperature is deterministic of the thermal considerations of the device. Thyristor current ratings are determined by keeping in view the thermal characteristics of the device. Typical junction temperature value of the conventional thyristor is 125°C. At higher temperature the device may be permanently damaged.

The electrical losses in the device produce thermal energy, which must be dissipated into the surroundings in order to avoid overheating of the device. This is achieved by mounting the device on metal plate called heat sink. Heat generated at the junction is transferred to the surrounding through device case and the heat sink. The heat is removed from the heat sink predominantly by convection. The effectiveness of the heat sink to conduct heat is enhanced by fitting it with protruding fins to enlarge the surface area.

The heat dissipated by the heat sink is primarily a function of the volume contained within the surrounding of the heat sink. For excessively large cooling system a more compact heat sink can be achieved by employing forced aircooling. In high power installations, where thyristor losses are large, water cooling is very much promising. In what follows we shall discuss the parameters of heat sinks.

2.1. THERMAL RESISTANCE OF HEAT SINKS

Electric current through a conductor flows from higher potential to lower potential. Thermal energy flows from higher temperature to lower temperature region. Thus there is an analogy between thermal power flow and current flow. Thermal power and temperature are analogous, respectively, to electric current and electric potential. Analogous to electric resistance we consider a quantity called thermal resistance denoted by θ.

By using a law analogus to Ohm's law the temperature difference ($T_1 - T_2$) when average thermal powrer P_{av} flows through a medium having thermal resistance θ is given by

$$T_1 - T_2 = P_{av} \times \theta \text{ °C watt}$$

or

$$\theta = \frac{T_1 - T_2}{P_{av}} \text{ °C/watt} \quad ...(1)$$

In a semiconductor heat flows from semiconductor junction to ambient through device case and heat sink which can be represented by thermal equivalent circuit shown in Fig. 2.1. Here average thermal power P_{av} at junction is analogous to constant current source.

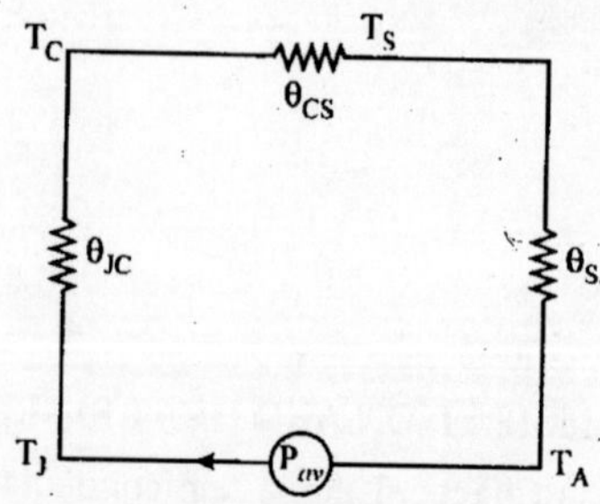

Fig. 2.1. Thermal equivalent circuit.

In the circuit θ_{jc} is the resistance between junction and device case, θ_{cs} between case and heat sink while θ_{sA} is the thermal resistance between heat sink and the ambient. By definition, thermal resistance between junction and the ambient, θ_{jA} will be given by

$$\theta_{jA} = \frac{T_j - T_A}{P_{av}} \quad ...(2)$$

By using eqn. (1) or (2), we can find the resistance between any two media. For example the thermal resistance between heat sink and ambient will be given by

$$\theta_{sA} = \frac{T_s - T_A}{P_{av}} \quad ...(3)$$

Example 2.1. In a thyristor the junction temperature is 60°C. The thermal power flow between junction and heat sink is 4 watt. What will be thermal resistance between junction and heat sink if later is at a temperature of 20°C.

Solution : Here we are given

Junction temperature $T_j = 60°$

Heat sink temperature $T_s = 20°$

Average power $P_{av} = 4$ watt

Using relationship, $\theta_{js} = \frac{T_j - T_s}{P_{av}}$, we get

$$\theta_{js} = \frac{60° - 20°}{4} \text{ °C/watt}$$

$$= \frac{40}{4}$$

$$\theta_{js} = 10 \text{ °C/watt} \quad \textbf{Ans.}$$

2.2. HEAT DISSIPATION AREA OF HEAT SINK

Heat dissipating capability of heat sink depends on its surface area which is exposed to surroundings. The area of contact between sink and ambient can be increased by bolting fins with the heat sink. The thermal resistance θ_{sA} between sink and the surrounding depends on parameters such as heat sink material, surface area and finish volume occupied etc. Heat sink efficiency can be greatly changed by changing the heat dissipation area of heat sink.

2.3. THYRISTOR MOUNTING TECHNIQUES

Internal power loss in thyristor gives rise to high thermal stresses which in turn give rise to mechanical forces which may disrupt the device. Therefore, the thyristor must be braced to withstand these forces. Also the thyristor mounting should ensure flow of heat from junction to case. We have following five methods of mounting thyristor.

1. Lead mounting : Lead mounted thyristor is shown in Fig. 2.2 *(a)*. This mounting is suitable for current rating upto 1 A. No heat sink is required for lead mounted thyristors.

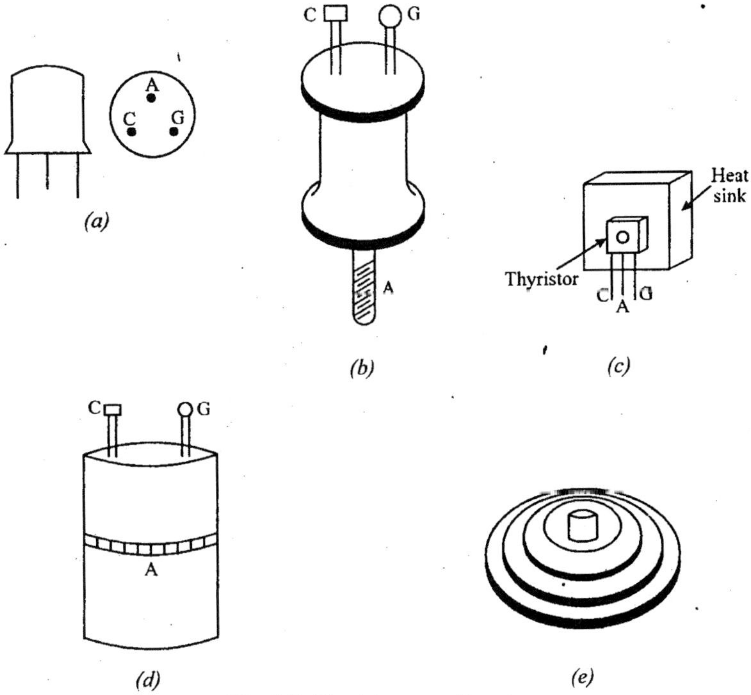

Fig. 2.2. Thyristor mountings.

2. Stud mounting : Stud mounting is shown in Fig. 2.2 *(b)*. It is most commonly used mounting. It uses a copper or aluminium stud which acts as anode also. The anode is also used for exchanging heat. Thus anode and heat sink are electrically in contact. It is used for small and medium sized device.

3. Bolted mounting : The bolted mounting is shown in Fig. 2.2 *(c)*. It is also called bolt down mounting. The device is attached to the mounting with bolts. The device and heat sink are electrically insulated from each other.

4. Press-fit mounting : It is shown in Fig. 2.2 *(d)*. The device is pressed into the hole made in the heat sink. This type of SCR has a lower thermal drop, than the stud mounting. It is employed for thyristor of large rating.

5. Pressure mounting : Pressure mounting is shown in Fig. 2.2 *(e)*. It is also known as disc type mounting due to its shape. The device is clamped between two heat sinks using external force. This type of thyristors are used for very high current rating.

2.4. TYPES OF HEAT SINKS

Two types of heat sinks are generally used which are :

1. Flat heat sink : The construction of flat type heat sink is shown in Fig. 2.3 *(a)*. It is bolted to the device.

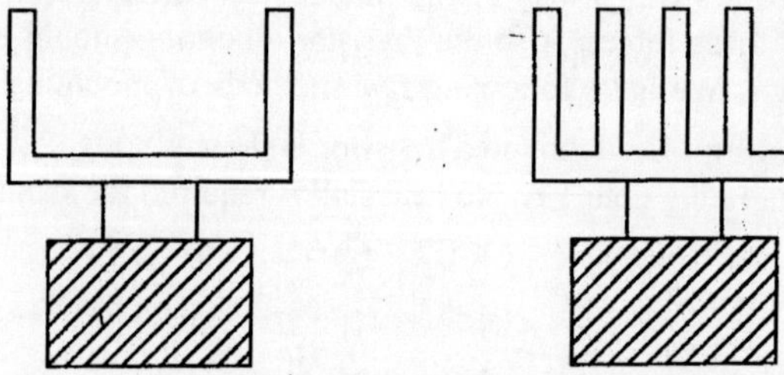

(a) Flat type *(b) Block type*

Fig. 2.3. Types of heat sinks.

2. Block type heat sink : This type of heat sink is used to dissipate higher power to the ambient. Instead of flat metallic plates, it has total surface area divided into a number of small strips as shown in Fig. 2.3 *(b)*. Heat dissipation capacity of this heat sink is higher.

2.5. HEAT SINK TRANSFER COEFFICIENT

The temperature of heat sink can be lowered by increasing its surface area. However beyond a particular limit the temperature can not be lowered by increasing the surface area alone. The cooling of the heat sink can be made more effective by forced convection with an optimum area of the heat sink. The transfer of the heat to the cooling medium can be discribed by heat transfer coefficient which is given by

$$h = 0.15\,(V/L)^{1/2}\ \text{W/m}^2\,^{\circ}\text{C} \quad \text{...(4)}$$

where h = heat transfer coefficient

V = linear cooling air velocity across the fin surface in metre per second

L = length of fin parallel to air flow in metres

Eqn. (4) can also be written in FPS system, as

$$h = 11.2\,(V/L)^{1/2} \times 10^{-4}\ \text{W/inch}^2\,^{\circ}\text{C} \quad \text{...(5)}$$

Variation of heat dissipated with the area of heat sink is given in Fig. 2.4. As the area of heat sink is increased the power dissipation by heat sink becomes almost constant at large area. Thus increase in surface area of heat sink above optimum value does not lead to improvement in its performance.

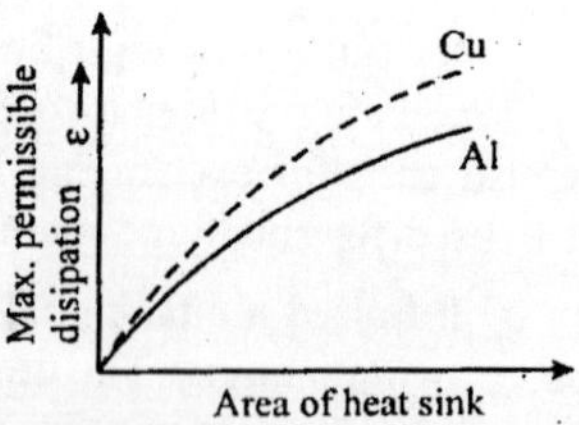

Fig. 2.4. Heat sink transfer coefficient

2.6. HEAT SINK EFFICIENCY

The efficiency of heat sink is given by

$$\eta_H = \frac{\text{Power dissipated to ambinet from heat sink}}{\text{Power received by the heat sink}}$$

Referring to Sec. 2.1, the power dissipated to ambient

$$P_{sA} = \frac{T_s - T_A}{\theta_{sA}}$$

Power delivered to the heat sink = P_{av}

Hence the efficiency of heat sink in terms of average power is

$$\eta_H = \frac{T_s - T_A}{\theta_{sA} P_{av}} \qquad ...(6)$$

The efficiency of heat sink should be high so that maximum heat generated in the device may be transferred to the ambient.

Example 2.2 The average power flow from a heat sink to ambient is 10 watt. The heat sink efficiency is 70% . If the resistance between junction and heat sink has a value of 3° C/watt. Calculate the junction temperature. Given heat sink temperature = 60° C.

Solution : Heat sink efficiency is given as

$$\eta H = \frac{P_{out}}{P_{av}}$$

or $$0.7 = \frac{10}{P_{av}}$$

or $$P_{av} = 14.28 \text{ watt}$$

Now $$P_{av} \times \theta_{js} = T_j - T_s$$

or $$T_j = T_s + P_{av} \times \theta_{js}$$

$$= 60 + 14.28 \times 3 = 60 + 42.84$$

$$= 102.84° \text{C.} \quad \textbf{Ans.}$$

SUMMARY

In this chapter we have discussed how to protect SCR against high temperature rise. The heat developed in the device is transferred to the surrounding through what is called heat sink. How much heat out of the developed in the device is transferred is given in terms of heat sink efficiency. All these terms have been covered in this chapter.

EXERCISES

1. What is the need of heat sink in thyristor ? How does the dissipated power by heat sink depends on its area ?
2. Define heat sink efficiency. What is the significance of the efficiency ?
3. Estabilish relationship between thermal resistance and average power dissipated at the thyristor junction.
4. What are the types of thyristors.
5. Describe in detail the thyristor mounting techniques.
6. Maximum junction temperature of a thyristor is 125°C. Average power dissipated at junction is 10 W. If heat sink is at temperature of 55°C find the thermal resistance between junction and the heat sink.
7. What is heat transfer coefficient ? Write down its expression for heat sink.

NUMERICAL PROBLEMS

1. Junction temperature of a thyristor is 100°C and heat sink is at 50°C. If the heat dissipated to ambient by junction is 3 watt what is the thermal resistance between heat sink and Junction?
2. In the above problem if device case temperature is 60°C what is the total resistance of the circuit ?

3. Power flow between thyristor junction and heat sink is 10 watt. If heat sink temperature is 40°C and thermal resistance between heat sink and ambient at room temperature is 35 °C/watt, calculate the heat sink efficiency.
4. Length of a heat sink parallel to cooling air moving at speed of 5 m/sec is 22 cm. Calculate heat transfer coefficient.

OBJECTIVE TYPE QUESTIONS

1. Analogous to electric current is electric circuit the quantity in thermal equivalent circuit i
   (*a*) average thermal power (*b*) resistance
   (*c*) temperature (*d*) none of the above.
2. A quantity in thermal equivalent circuit which is analogous to electric potential is
   (*a*) thermal power (*b*) thermal resistance
   (*c*) temperature (*d*) all of the above.
3. Typical maximum value of junction temperature in a SCR is
   (*a*) 500°C (*b*) 350°C
   (*c*) 125°C (*d*) 23°C
4. Heat dissipation capacity of a heat sink is affected by
   (*a*) surface area
   (*b*) temperature and velocity of cooling medium
   (*c*) both (*a*) and (*b*) (*d*) none of the above.
5. Unit of heat transfer coefficient is
   (*a*) w/m^2 (*b*) w/°C
   (*c*) w/m^2 °C (*d*) wm^2
6. For high power rating thyristors
   (*a*) pressure mounting is used (*b*) stud mounting is used
   (*c*) lead mounting is used (*d*) bolted mounting is used.
7. Unit of thermal resistance is
   (*a*) ohm (*b*) watt/°C
   (*c*) °C/watt (*d*) °C watt.
8. The temperature is analogous to electrical quantity
   (*a*) current (*b*) electric field
   (*c*) electric potential (*d*) electrical power
9. Thermal power is analogous to
   (*a*) electric power (*b*) electric current
   (*c*) electric resistance (*d*) potential
10. Electrical resistance is analogous to thermal
    (*a*) resistance (*b*) power
    (*c*) temperature (*d*) heat.

ANSWERS

1. (*a*)	2. (*c*)	3. (*c*)	4. (*c*)	5. (*c*)	6. (*a*)
7. (*c*)	8. (*c*)	9. (*b*)	10. (*a*).		

❑❑❑

3
Applications of Thyristors

3.0. INTRODUCTION

SCR and many other semiconductor devices are used for power control Switching property of the SCR and other devices are used for various applications. Few of the applications of SCRs are described in the sections which follow :

3.1. AUTOMATIC BATTERY CHARGER

Battery charger circuit using an SCR is shown in Fig. 3.1. It consists of a step down transformer whose output is rectified by using centre tapped rectifier which employs diodes D_1 and D_2. The output of the rectifier is used to charge the battery E which is connected between the terminals (1, 1′) as shown in the figure. A zener diode D_3 is used to maintain the minimum gate voltage to trigger the SCR. By varying potentiometer resistance R required gate current can be supplied to the SCR. The circuit operation is as follows :

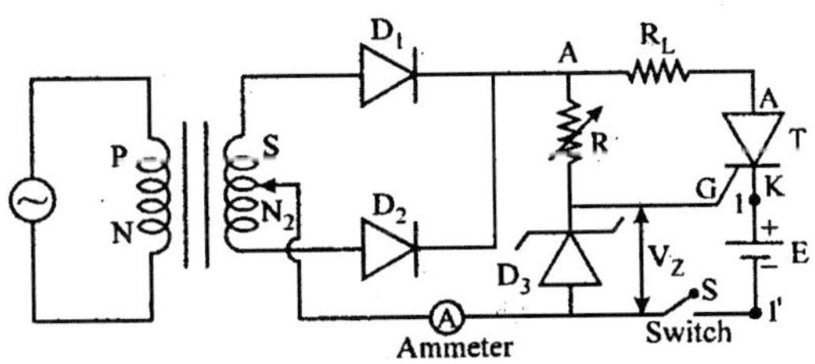

Fig. 3.1. Automatic battery charger.

Operation : Step down transformer converts 230 V AC to E volt AC which is rectified by using two diodes D_1 and D_2. The rectifier output is used to charge the battery E which has zero initial voltage between its plates. The point A in the circuit is at E volt DC which is connected to anode of the thyristor. The gate of the thyristor is at constant voltage V_Z determined by zenor diode D_3. The gate current for firing is controlled by the potentiometer R. After the thyristor T is fired the battery starts charging when switch S is closed. When battery voltage reaches the value near to that at point A viz. E V, the thyristor T_1 is turned off and the charging of the battery stops which is indicated by zero reading of the ammeter connected in the circuit.

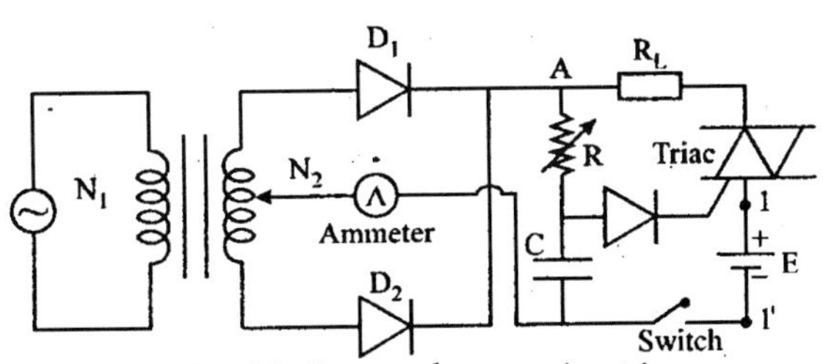

Fig. 3.2. Battery charger using triac.

In the above circuit battery charger can be designed by using triac also instead of SCR. The circuit employing triac is shown in Fig. 3.2. The operation of the circuit is as follows :

When input cycle is positive, the diode D_1 conducts and point A is at transformer secondary voltage. The pulse at the gate terminal triggers the triac which is used to charge the battery E, which is initially at zero voltage, *i.e.,* it is discharged initially. During negative half cycle diode D_2 conducts. As the battery is charged to rated voltage, the triac is commutated.

3.2. VOLTAGE REGULATOR OR VOLTAGE CONTROLLER

A regulator or a controller circuit is the one which is used to feed desired output to the load. As the thyristor is a controlled switch, *i.e.,* desired part of the input can be allowed to pass through the thyristor by controlling the firing angle, desired output can be obtained across the load. In the following lines we shall explain the regulatory action of the SCR.

The basic circuit of a voltage regulator is shown in Fig. 3.3 *(a)* and *(b)* where we have used thyristors and triac, respectively for voltage regulation. The load in the circuit may either a purely resistive load or a R-L load. We shall consider two cases separately.

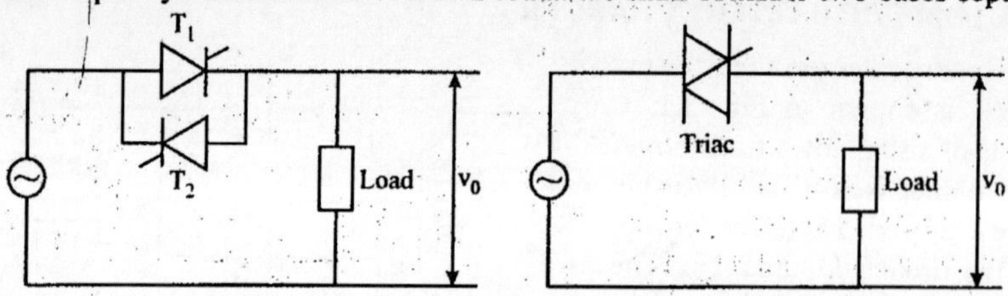

(a) *(b)*

Fig. 3.3. Voltage regulator circuits.

1. AC voltage regulator with R-load : The circuit of AC voltage regulator with pure resistive load is shown in Fig. 3.4 *(a)*. The thyristor T_1 is used for positive half cycle while thyristor T_2 works during negative half cycle. Since in this circuit the controlled power is fed to the load in both the half cycles, it is called full wave AC regulator. For a purely resistive load the voltage and current are in the same phase, hence the output waveforms will be as shown in Fig. 3.4 *(b)*.

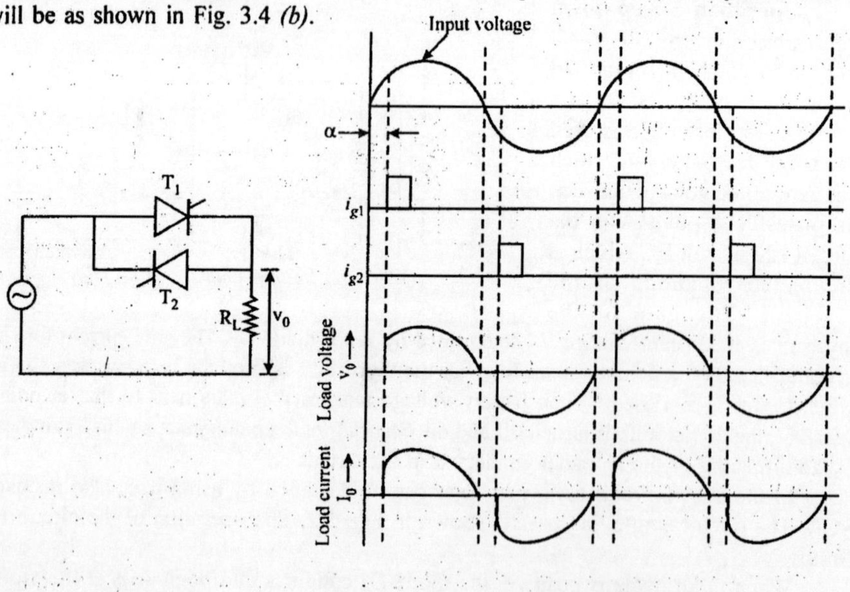

(a) Circuit *(b) Waveforms.*

Fig. 3.4. Voltag regulator.

Let α be the firing angle for the thyristor T_1. At angle α, T_1 is set into conduction and the load voltage follows the input voltage. At $\omega t = \alpha$, the current through the load starts rising above instantaneous value at this angle. At zero crossing T_1 is turned off, the voltage across the load falls to zero. The current also falls to zero at $\omega t = \pi$.

At $\omega t = \pi + \alpha$, thyristor T_2 is turned on. The voltage and current across the load are in negative direction. Finally at $\omega t = 2\pi$, load voltage and load current fall to zero.

For converting full wave AC regulator into half wave AC regulator one of the thyristors is replaced by a diode. The circuit of a half wave AC regulator is shown in Fig. 3.4 *(c)* and corresponding waveforms in Fig. 3.4 *(d)*.

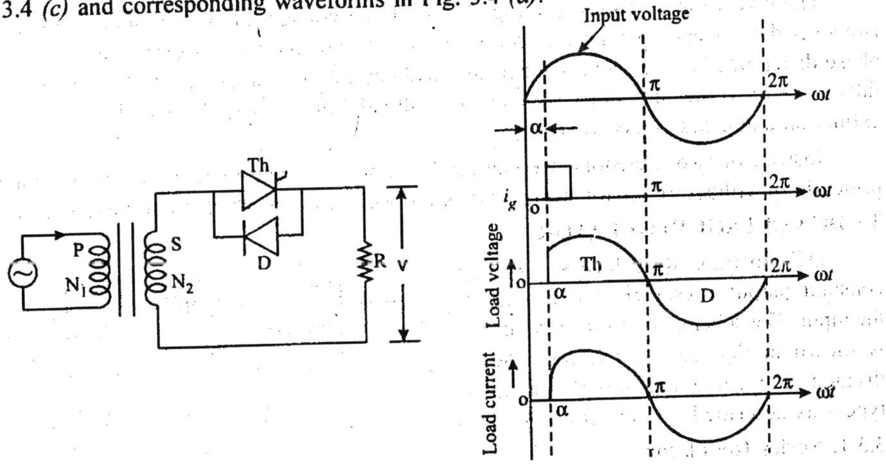

(c) Circuit *(d) Waveforms.*

Fig. 3.4. Half wave AC regulator.

2. AC voltage regulator with R-L load : The circuit of AC voltage regulator with R-L load is shown in Fig. 3.5 *(a)*:

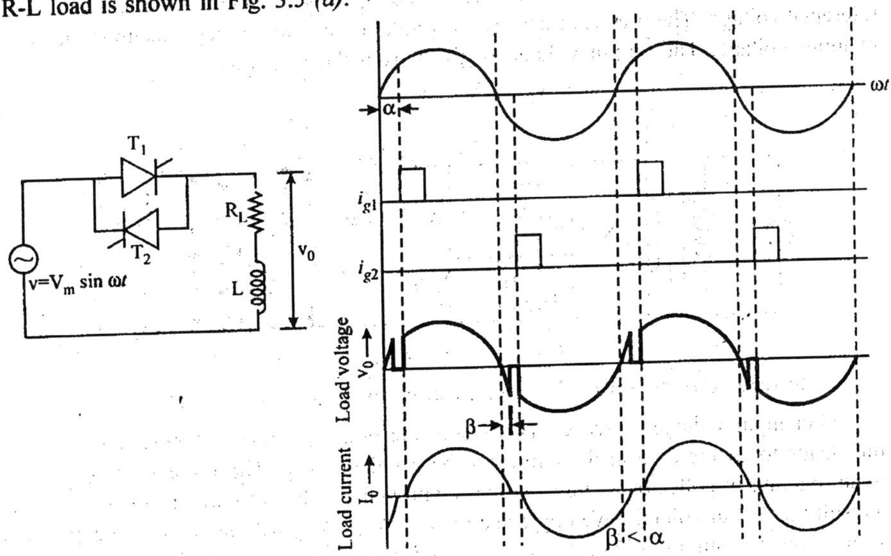

(a) Circuit *(b) Waveform for discontinuous mode.*

Fig. 3.5. Full wave ac regulator using R-L load.

Across the resistive part R_L of the load the current and voltage are in the same phase while across inductor the current lags the voltage by 90°. We can say that voltage and current across R-L load are not in the same phase. The phase angle between voltage and current is given from Fig. 3.5 *(c)* by

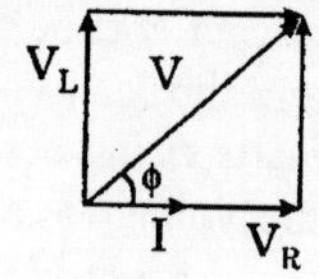

Fig. 3.5 *(c)*

$$\tan \phi = \frac{\omega L}{R}$$

where ω is the frequency of input.

The waveforms are shown in Fig. 3.5 *(b)*. Load voltage follows the source voltage. However the current does not fall to zero when load voltage falls to zero. This is due to phase difference between load voltage and load current. The angle at which load current falls to zero value and device conducting already turns to OFF state is known as extinction angle designated β (Beta).

Instead of two thyristors, we can use a triac for voltage regulation. The working principle of voltage regulator using triac is the same as one which is discussed above.

3.3. DC VOLTAGE REGULATOR

DC voltage Regulators produce constant output irrespective of change in the input. The schematic of such a regulator is shown in Fig. 3.6. A DC regulator is divided into categories-series and shunt type - as described in what follows.

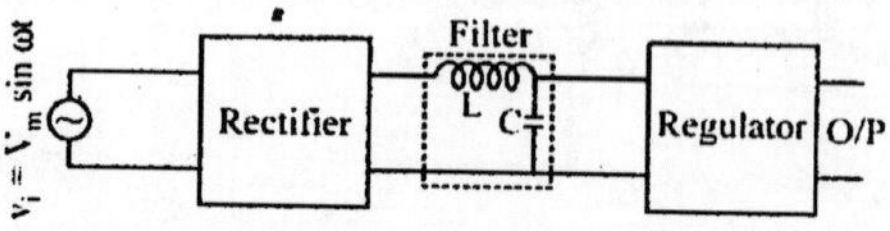

Fig. 3.6: Schematic of DC Regulator

3.3.1. Series Regulator

In this category the pass element is in series with load and the supply as shown in Fig. 3.7. As shown in the figure, the unregulated DC input voltage is applied at the collector terminal of NPN transistor (pass element) whose collector and emitter are connected in series with the load. A zener diode D_z is used which provides the constant reference voltage. The transistor acts as a variable resistor which depends upon the value of input voltage. The output voltage is given from the figure as

$$V_0 = V_z + V_{EB}$$

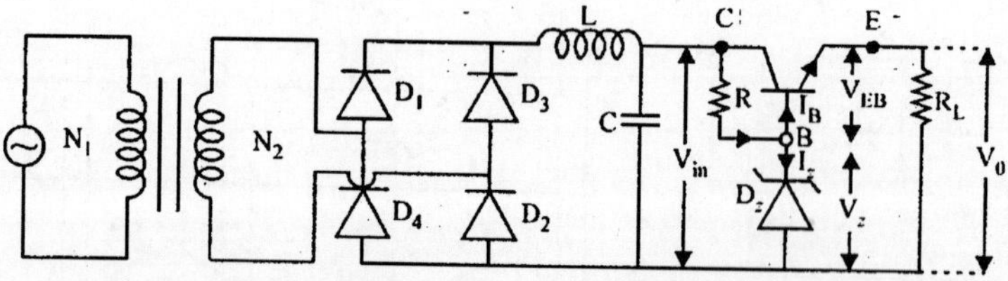

Fig. 3.7: Series Regulator Circuit

The regulatory action of the circuit can be explained as follows:

Let input voltage increase. As a result, the output voltage V_0 increases which in turn leads to an increase in the emitter to base voltage V_{EB}. The increase in V_{EB} leads to an increase in collector-emitter resistance and hence collector emitter voltage V_{CE}. As a result the output voltage given by equation $V_0 = V_{in} - V_{CE}$ will decrease to the previous value. Thus the output voltage remains constant. In a similar manner, the response of the circuit to decrease in the input supply voltage can be explained.

Second, the regulator's response to the variation in load current is to be examined. Consider that the output current increases with a decrease in R_L. Under such a case the O/P voltage V_0 tends to fall. As a result the conduction level of the transistor increases which leads to a decrease in collector to emitter voltage. Hence the output voltage increases. In this way the decrease in load voltage is compensated by the decrease in collector to emitter voltage. Hence the output is regulated.

3.3.2. Shunt Regulator

In shunt regulator, the pass transistor is connected in parallel with the load as shown in Fig. 3.8. From the figure we can write

$$V_0 = V_z + V_{BE} \quad ...(3.2)$$

and

$$V_0 = V_{in} - I_R \quad ...(3.3)$$

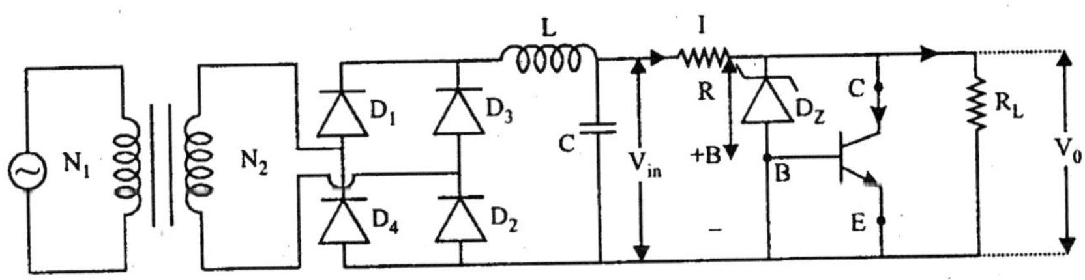

Fig. 3.8: Shunt Voltage Regulator

The regulatory action of the circuit can be explained as follows:

Let us consider that the supply voltage increases. As a result, the output voltage V_0 increases which, in turn, leads to an increases in the base to emitter voltage V_{BE} as per equation (3.2) with an increase in V_{BE}, the base and hence the collector current of the transistor also increases which flows through the series resistance R. As a consequence of it, the voltage drop across R increases which compensates the increase in the supply voltage. Hence the output voltage remains constant. Similarly the regulating action of the circuit can be explained when the input voltage decreases.

3.3.3. IC Regulator

Regulator are now available in the form of integrated circuits in different ranges. A very common IC regulator is 7805 which produces constant output of 5 volts. It is a three pin device which produces the fixed output at pin 3 when input is applied at pin 1-pin 2 is used for ground. The pin diagram is shown in Fig. 3.9.

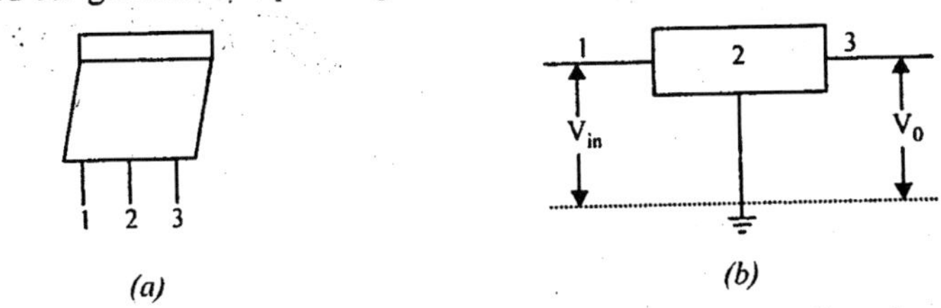

Fig. 3.9: Voltage Regulator IC 7805 (a) strcuture (b) schematic

3.4. EMERGENCY LIGHT

The emergency light is used in the event of AC power failure. It works on low DC voltage. In the presence of AC power DC source gets trickle charged and the lamp does not glow.

Fig. 3.10 shows the circuit of emergency light. The mains supply is stepped down to required value say 6 V in the present case. Four diodes D_1, D_2, D_3, D_4 with the protective resistance are used to increase the current rating. When AC power is on, the capacitor is charged with upper plate positive. The gate of the SCR is connected with negative plate of the capacitor. Hence it remains in the off state and the bulb in series with SCR does not glow. The AC power charges the battery through 5 Ω, 10 W resistance. When AC power is not available, the capacitor C discharges through diodes D_1 to D_4, 5 Ω, 10 W resistance and 500 Ω, 1/2 W resistance. When lower plate of the capacitor acquires the positive voltage required at the gate of SCR to turn on the device the SCR is fired. The bulb connected in series with SCR glows. The current flows through DC source, bulb, SCR and secondary of the transformer.

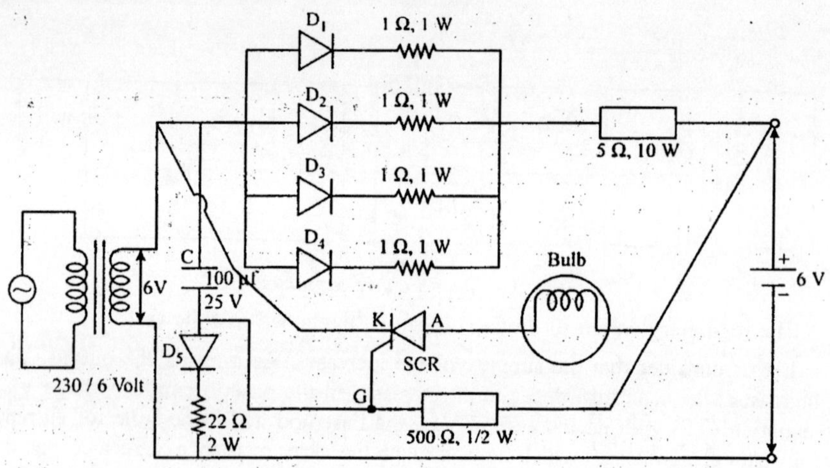

Fig. 3.10. Emergncy light.

When AC power is again available, the capacitor starts discharging with upper plate positive. As a result the thyristor is turned off, thereby disconnecting the bulb from the circuit.

Similar calculations can be made for bulb and DC battery of any rating.

3.5. ALARM CIRCUIT

The alarm circuit is shown in Fig. 3.11. The door switch and the resistor are connected in series with the gate terminal of the thyristor T. The door switch is closed when either the door is open or the swicth is pressed. We assume that the reset switch S is closed initially. With the door closed the signal at the gate of SCR T is zero, hence it is off as no return path is available for the current. However when the door is open or

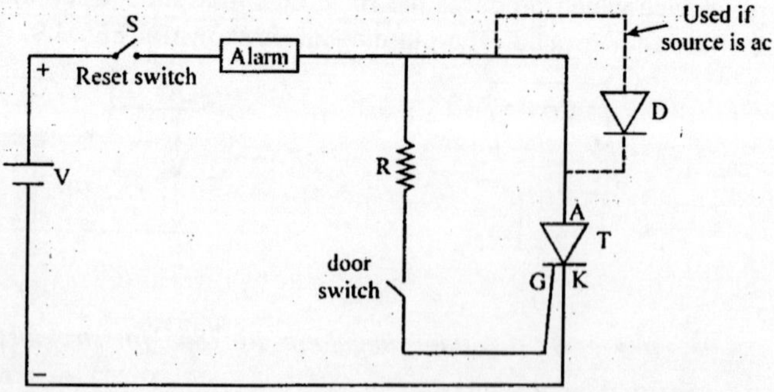

Fig. 3.11. Alarm circuit.

the door switch is closed the signal appears at the gate terminal. The SCR is tuned on. The alarm connected in the circuit is energized which is deenergized by opening the reset switch S.

If the source voltage is AC, a diode D is connected in series with the SCR which protects the device from voltage between anode and cathode during negative half cycle.

Alarm switch can be designed by using phototransistor also as switch. The circuit of such alarm is shown in Fig. 3.12. The operation of the circuit is as follows :

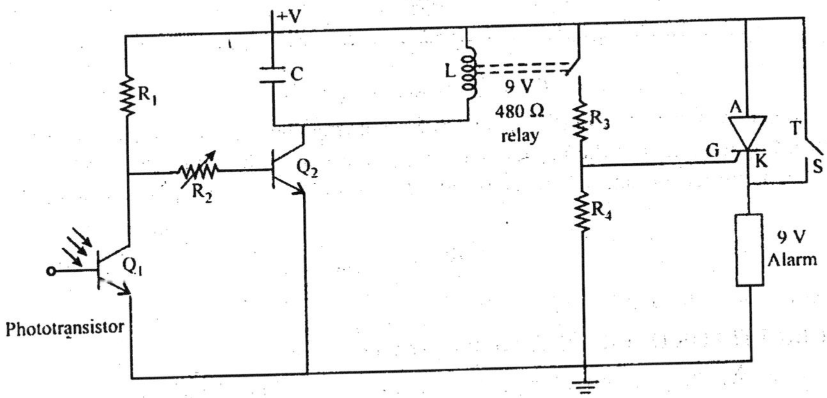

Fig. 3.12. Alarm circuit.

When the light falls on the base of the phototransistor, it starts conducting. As a result the current to the base of transistor Q_2 is reduced to zero. Hence Q_2 is turned off.

Under this condition as the current through the coil of relay is zero the contact of R_3, R_4 network to source battery through relay is broken. Hence gate voltage of the thyristor is zero which keeps it in OFF state. No power is supplied to the alarm.

When shadow falls on the base of phototransistor, it is turned off. As a result, finite current is supplied to transistor Q_2 and it is turned on. Under this condition the current through relay closes it. Hence the current flows through resistive network formed by R_3 and R_4 and a voltage is applied at the gate of SCR. As a result the thyristor is turned ON and the alarm connected in series with SCR starts sounding. We turn OFF the SCR and hence alarm by pressing the switch S across SCR.

3.6. TIME DELAY CIRCUIT

This circuit is used to change the time after which a relay connected in the circuit is closed. The circuit is shown in Fig. 3.13. In the circuit UJT is used to give pulse to the gate of SCR. The capacitor C is charged through resistances R_1, R_2 and source V. The duration of charging of C is dependent on product $(R_1 + R_2)$ C. By increasing R_1 the time required to charge C is increased. When capacitor is charged to a value to turn on UJT, the current flows through, primary of the pulse transformer which induces an emf in secondary L_2 and voltage is applied to gate of SCR. Thus SCR is turned ON and current flows through the relay and closes it.

Thus we can change the delay in the relay operation by changing the value $(R_1 + R_2)$ C . $(R_1 + R_2)$ C can be changed by changing the value of potentiometer R_1 as shown in the figure.

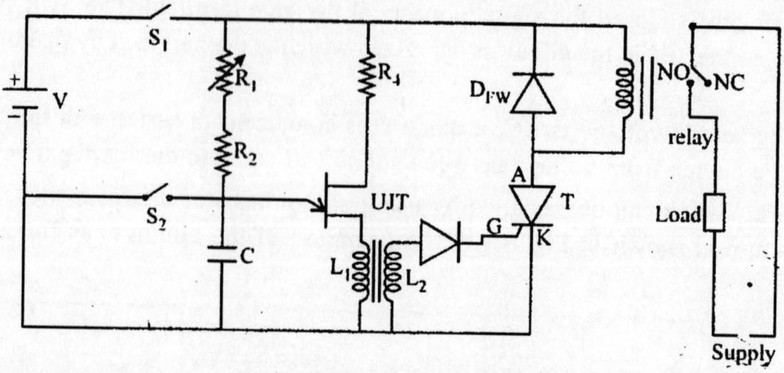

Fig. 3.13. Time delay circuit.

When switch S_1 is closed, the switch S_2 is open. The capacitor starts charging and when the voltage across the capacitor is ηV where η is called intrinsic stand off ratio, the SCR is turned on and the relay is closed. The delay in relay operation is given by

$$T \approx RC \ln\left(\frac{1}{1-\eta}\right)$$

where $R = R_1 + R_2$ and $R_1 + R_2 >> R_3$.

3.7. CIRCUIT FOR OVERVOLTAGE PROTECTION

Fast switching action of thyristor makes it a suitable device for protection of electrical device against overvoltage. The SCR is connected in parallel with the electrical device and the firing circuit is designed such that the pulse at the gate of SCR is applied when the voltage across electrical device exceeds the rated voltage. The circuit is shown in Fig. 3.14.

The operation of the circuit is as follows :

When the voltage across the electrical device is less than or equal to rated voltage, the SCR T will not be fired (turned ON). For this we set the value of potentiometer R so high that no pulse is applied to the gate of SCR when the supply voltage is less than or equal to rated voltage of the electrical device.

When the voltage across the device tries to exceed the rated voltage the SCR is triggered because pulse is applied to its gate. The resistance R_1 has small value hence almost whole current flows through the SCR. Very small current which flows through the electrical device can not damage it. Thus the device is protected against overvoltage.

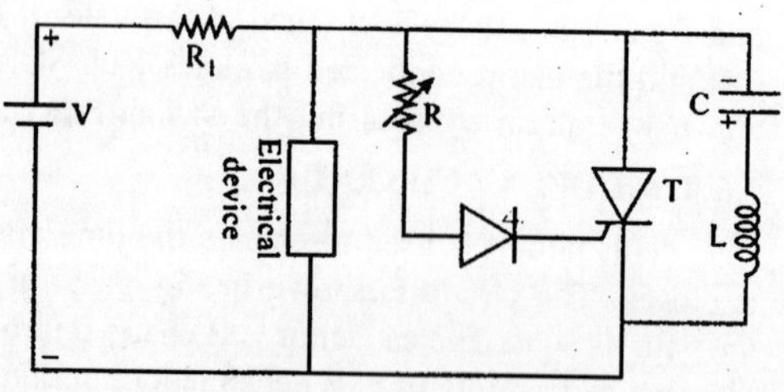

Fig. 3.14. Over voltage protection.

For turning off the thyristor self commutation process is used. For this purpose a combination of capacitor C and inductor L has been used. When thyristor is conducting the capacitor is charged to voltage V with lower plate positive. As the transient (over voltage) is over the capacitor voltage will make the cathode positive and it is turned off. The thyristor used should have high voltage and high current ratings.

3.8. CIRCUIT FOR OVERCURRENT PROTECTION

When the current through the electric device exceeds the rated current, it is protected by changing the path of the current. For this purpose SCR is used. The circuit for overcurrent or overload protection is shown in Fig. 3.11. The working of the circuit is as given below :

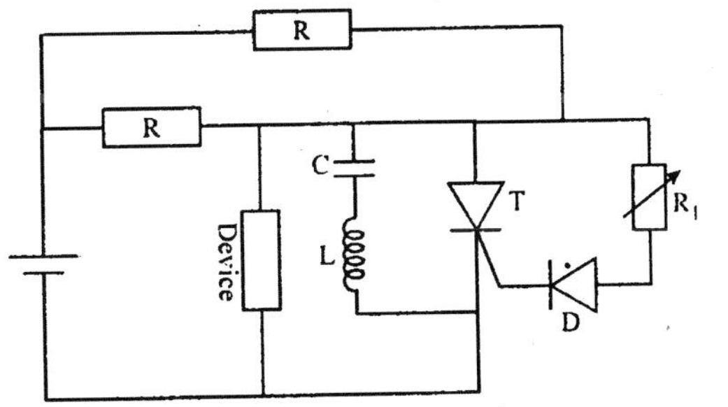

Fig. 3.15. Overcurrent protection.

When the current through the electrical device is less than or equal to the rated value, the SCR T remains off. However when it exceeds the rated value, the current through the gate circuit turns on the SCR. As a result whole current is diverted to the SCR and the device is protected.

The thyristor is turned off again by self commutation technique as explained in Sec. 3.10.

AC circuit breakers: The article 2.9.5 in "Power Electronic" by Sunil Kumar can be taken here.

SUMMARY

Thyristor has various applications. Few of the applications such as alarm circuit, battery charger, over voltage protection, time delay circuit etc. have been discussed in this chapter.

EXERCISE

1. Explain the circuit of automatic battery charger.
2. What is voltage regulator ? Explain the circuit of voltage regulator usign R-load.
3. Explain the output waveforms of a voltage regulator using R-L load.
4. Explain the circuit of emergency light using SCR.
5. Draw an alarm circuit using photo transistor. Explain its working.
6. What is the use of time delay in a circuit? Explain with the help of circuit diagram.
7. Draw overvoltage protection circuit. Explain the working of commutation circuit used.
8. Explain overcurrent protection circuit.
9. Explain the working of dc series regulator.
10. What is meant by dc shunt regulator? Explain.
11. What is the use of AC circuit breaker? Explain.

NUMERICAL PROBLEMS

1. In full wave AC voltage regulator with R-L load input voltage of 230 volts at 50 Hz is used. The load resistance is 10 Ω and inductance 20 mH. Calculate phase angle between load voltage and load current when firing angle is 0°.
2. In an emergency light a 6 volt battery is used as shown in Fig. 3.6. If the gate current required for firing is 3 mA and bulb resistance is 5 Ω, calculate *(a)* heat dissipated by the bulb, *(b)* total resistance in gate circuit.
3. In an alarm circuit shown in Fig. 3.7 a.c. power source of 15 V, 50 Hz is used. What is the average current through the circuit if alarm resistance is 3 ohms ?

4. Refer to Fig. 3.9. If $R_1 = 5\ \Omega$, $R_2 = 10\ \Omega$, $C = 0.47\ \mu F$, what is the time delay under the condition that $R_1 + R_2 >> R_3$? Instrinsic stand off ratio is 0.6.

OBJECTIVE TYPE QUESTIONS

1. In full wave ac regulator
   (a) two thyristors are used (b) one thyristor and one diode is used
   (c) two diodes are used (d) none of the above
2. In half wave ac regulator
   (a) one diode is used (b) one thyristor is used
   (c) one diodes and one thyristor is used
   (d) none of the above
3. In ac voltage regulator with R-L load
   (a) current leads the voltage (b) voltage leads the current
   (c) depends upon values of R and L (d) none of the above
4. At a extinction angle
   (a) load current is zero (b) load voltage is zero
   (c) both (a) and (b) (d) none of the above
5. In time delay circuit delay time is equal to
   (a) RL ln $(1 - \eta)$ (b) RC
   (c) RC ln $\left(\frac{1}{1-\eta}\right)$ (d) none of the above
6. In question 5, η is
   (a) efficiency (b) intrinsic stand off ratio
   (c) a constant depending upon UJT (d) both (b) and (c)
7. In case of overvoltage protection
   (a) thyristor is turned ON (b) thyristor is turned OFF
   (c) its state remains unchanged (d) both (b) and (c)

ANSWERS

1. (a) 2. (c) 3. (b) 4. (a) 5. (c) 6. (d)
7. (a)

❑❑❑

4

Controlled Rectifiers

4.0. INTRODUCTION

Rectifier is used to convert a.c. into d.c.* Rectification can be done by using a number of devices. Simplest way for rectification is by use of simple diodes. However, by using simple diode the desired portions of the input cycle can not be rectified. The desired portion of the input cycle can be rectified by using SCRs. For this reason, the circuit using SCR for rectification is called controlled rectifier.

4.1. SINGLE PHASE RECTIFIERS

As is clear from the name, in single phase rectifier only one phase of a.c. input is rectified. This rectification can be done in the following three ways.

1. Single Phase half wave controlled rectifiers : The circuit diagram is shown in Fig. 4.1 *(a)*. We consider that positive half cycle appears at point A at $\omega t = 0$. Let the pulse is applied at the gate of SCR T at angle α. The input at the point will appear across the load from α to π radians. At $\omega t = \pi$, the SCR is turned off and it will remain off for negative half cycle. The wave forms are shown in Fig. 4.1 *(b)*. We observe that the thyristor starts conducting at angle α, which is called firing angle. The source voltage appears across the load from α to π and from π to $2\pi + \alpha$ the thyristor is off. Since only half cycle of the input is rectified at different firing angles, it is called half wave controlled rectifier.

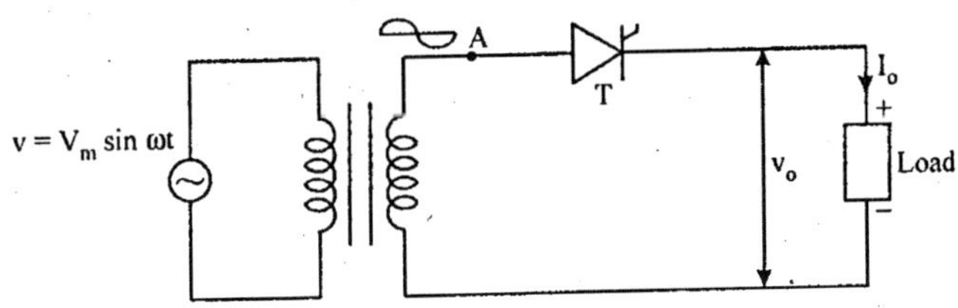

(a) Circuit.

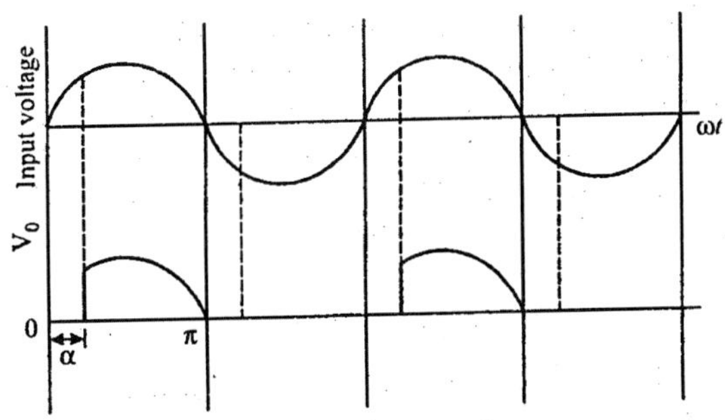

(b) Waveforms.

Fig. 4.1. Half wave controlled rectifier.

*dc produced by rectifier is pulsating in nature which has ripple contents. For getting constant dc ripples are removed by using filter circuits.

The waveform of load current depends upon the type of the load used. We shall consider two case — one when the load is purely resistive and two when the load is resistive as well as inductive. These cases are discussed below :

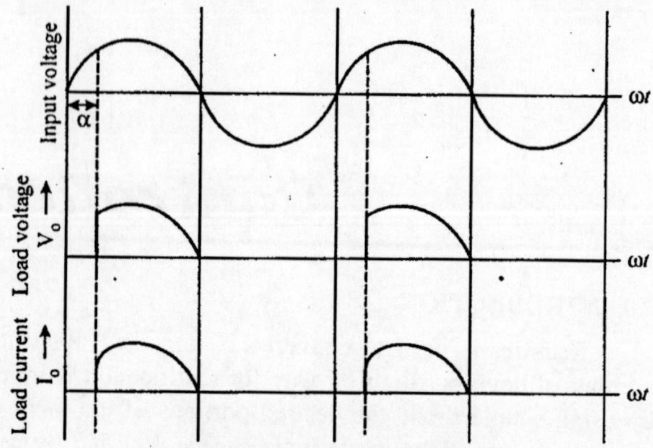

Fig. 4.2. Waveform for R-load.

Half wave rectifier with R load : For pure resistor the current and voltage are in the same phase. Hence the current waveform and voltage waveform will be as shown in Fig. 4.2.

Average output voltage is given by

$$V_{oav} = \frac{1}{2\pi}\int_{\alpha}^{2\pi} V_m \sin\omega t \, d(\omega t)$$

$$= \frac{1}{2\pi}\int_{\alpha}^{\pi} V_m \sin\omega t \, d(\omega t)$$

$$= \frac{V_m}{2\pi}(1+\cos\alpha)$$

Average current $$I_{oav} = \frac{V_m}{2\pi}\frac{(1+\cos\alpha)}{R} \quad ...(1)$$

Half wave controlled rectifier with R-L load : As we all know that the current and voltage are in same phase across a pure resistor whereas across an inductor the current lags the voltage by 90°. If we take the combination of a resistor and an inductor, the voltage and current have some phase difference between them. If the frequency of input is ω radians/sec., L is self inductance of inductor and R is resistance of resistor, the phase angle ϕ between voltage and current is given by

$$\tan\phi = \frac{\omega L}{R} \quad ...(2)$$

The circuit is shown in Fig. 4.3 *(a)*. The working of the circuit is similar to the purely resistive case with the difference in load voltage and load current waveforms. The wave forms are shown in Fig. 4.3 *(b)*. The analysis of the waveforms is as follows :

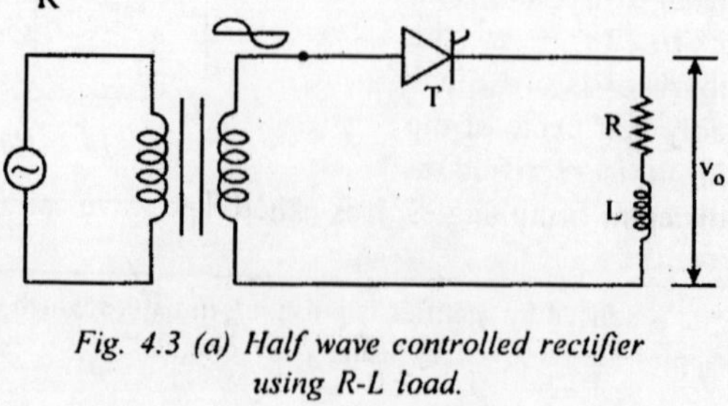

Fig. 4.3 (a) Half wave controlled rectifier using R-L load.

Analysis:

The thyristor is fired at angle α. The load voltage is the same as the source voltage between angle α and π. From π to 2π the SCR is reverse biased hence the load voltage is zero. At firing angle the current starts rising. At $\omega t = \pi$ although the voltage is zero, the load current is not zero as the energy stored in inductor is released through secondary of transformer, thyristor, resistor and inductor. Once the energy is released the current is reduced to zero at angle β. The angle at which the current is zero is called extinction angle. The current maintains zero value from $\omega t = \pi$ to $\omega t = 2\pi + \alpha$. It $\beta = \alpha$, this mode is called continuous conduction. For $\beta < \alpha$ the mode is called discontinuous mode. Fig. 4.3 *(b)* is an example of discontinuous mode of conduction.

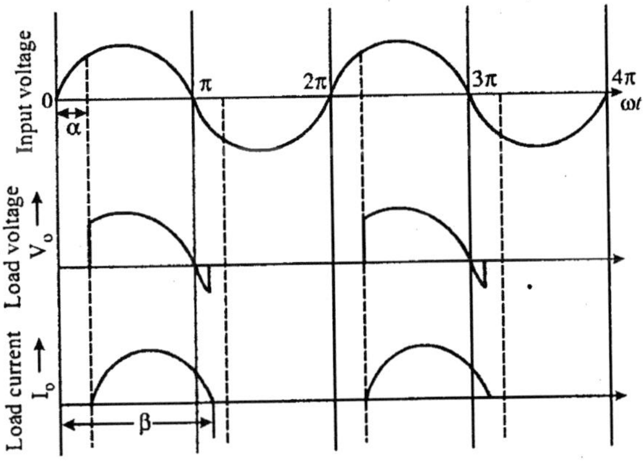

(b) Waveforms.

Fig. 4.3.

Average value of the output voltage is given by

$$V_{oav} = \frac{1}{2\pi}\int_{\alpha}^{\beta} V_m \sin\omega t\, d(\omega t)$$

$$= \frac{1}{2\pi}\int_{\alpha}^{\beta} V_m \sin\omega t\, d(\omega t) = \frac{V_m}{2\pi}\left[-\cos\omega t\right]_{\alpha}^{\beta}$$

$$= \frac{V_m}{2\pi}\left[\cos\alpha - \cos\beta\right]$$

$$V_{oav} = \frac{V_m}{2\pi}(\cos\alpha - \cos\beta) \qquad ...(3)$$

And the average load current is

$$I_{oav} = \frac{V_{oav}}{Z} = \frac{V_m}{2\pi\sqrt{R^2 + \omega^2 L^2}}[\cos(\alpha) - \cos(\beta)] \qquad ...(4)$$

4.2. FULL WAVE CONTROLLED RECTIFIER WITH RESISTIVE LOAD

In full wave controlled rectifier for both of positive and negative half cycle in the input we get output across the load. Desired portion of both the half cycles can be allowed to appear across the load. This is the reason it is called full wave controlled rectifier. The circuit of a mid point full wave controlled rectifier with resistive load is shown in Fig. 4.4 *(a)*.

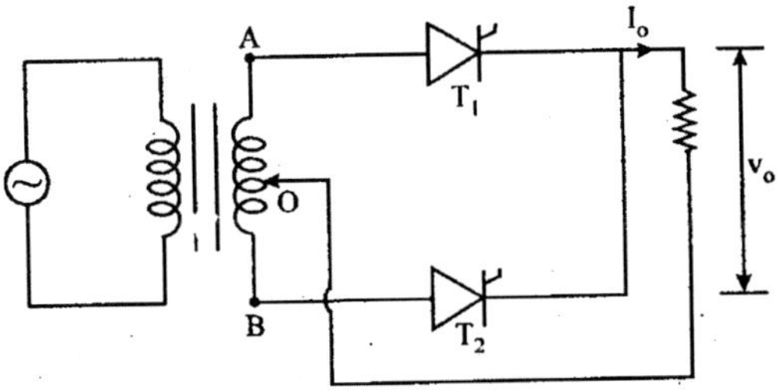

Fig. 4.4 (a) Full wave controlled rectifier using R-load.

In the circuit during positive half cycle point A is positive w.r.t. mid point, point B will be negative. In this case thyristor T_1 will conduct when it is fired. During negative half cycle point B is positive. The waveforms are shown in Fig. 4.4 *(b)*.

It is clear from the waveform that during positive input cycle thyristor T_1 is conducting while during negative input cycle thyristor T_2 does. In each case the current and voltage across the load are positive and in same phase. This is first quardrant rectifier, which means voltage and current both are positive. The average output voltage is given by

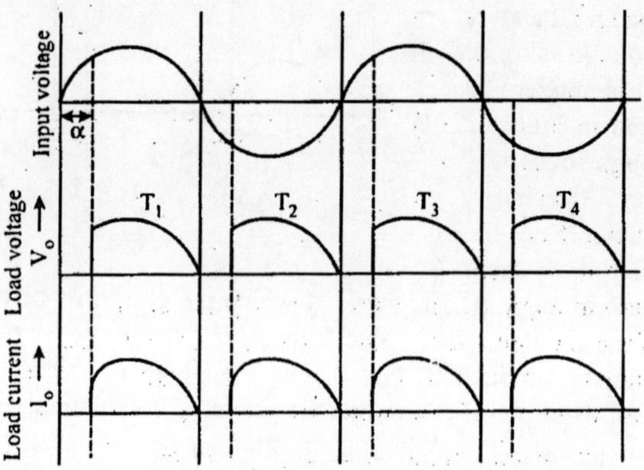

Fig. 4.4. (b) Waveforms.

$$V_{oav} = \frac{1}{2\pi}\int_{\alpha}^{2\pi} V_m \sin\omega t\, d(\omega t)$$

or

$$V_{oav} = \frac{V_m}{\pi}(1+\cos\alpha) \qquad ...(5)$$

and average load current

$$I_{oav} = \frac{V_m}{\pi}\frac{(1+\cos\alpha)}{R} \qquad ...(6)$$

If one of the thyristor is replaced by a simple diode it is called half controlled full wave rectifiers or semi converter. The circuit and waveform of semiconverter are shown in Fig. 4.5 *(a)* and *(b)* respectively.

Analysis:

The average load voltage is given by

$$V_{oav} = \frac{1}{2\pi}\int_{\alpha}^{2\pi} V_m \sin\omega t\, d(\omega t)$$

$$= \frac{1}{2\pi}\left[\int_{\alpha}^{\pi} V_m \sin\omega t\, d(\omega t) + \int_{\pi}^{2\pi} -V_m \sin\omega t\, d(\omega t)\right]$$

$$= \frac{V_m}{2\pi}(-\cos\omega t)_{\alpha}^{\pi} + \frac{-V_m}{2\pi}(-\cos\omega t)_{\pi}^{2\pi}$$

$$= \frac{-V_m}{2\pi}[-1 - \cos\alpha] + \frac{V_m}{2\pi}[2]$$

$$= \frac{V_m}{\pi}\left[\frac{1+\cos\alpha}{2} + 1\right]$$

$$= \frac{V_m}{\pi}\left[\frac{3+\cos\alpha}{2}\right] \quad \text{...(6A)}$$

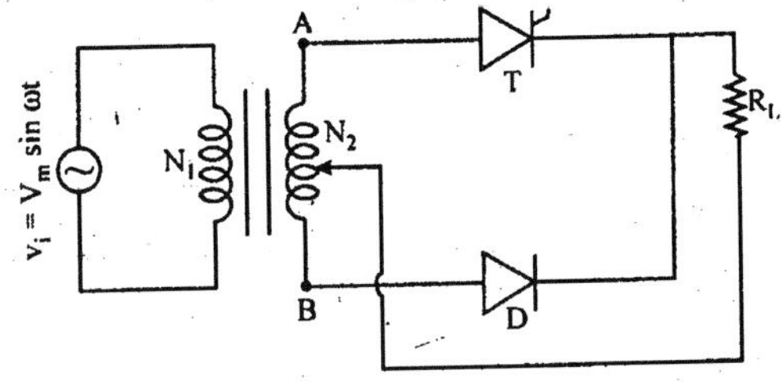

(a)

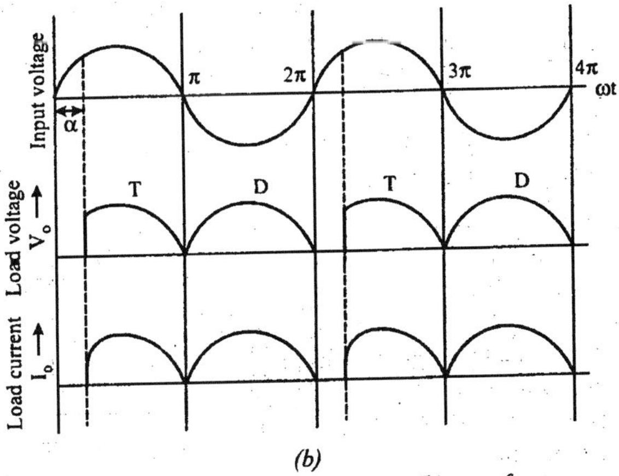

(b)

Fig. 4.5. Semiconverter (a) circuit, (b) waveforms.

4.3. FULL WAVE CONTROLLED RECTIFIER WITH R-L LOAD

As has already been discussed in half wave controlled rectifier that with R-L load the load current and load voltage are not in phase. The circuit of full wave controlled rectifier is shown in Fig. 4.6 *(a)*. Thyristor T_1 conducts for positive half cycle while SCR T_2 conducts for negative half cycle. The wave forms are shown in Fig. 4.6 *(b)*.

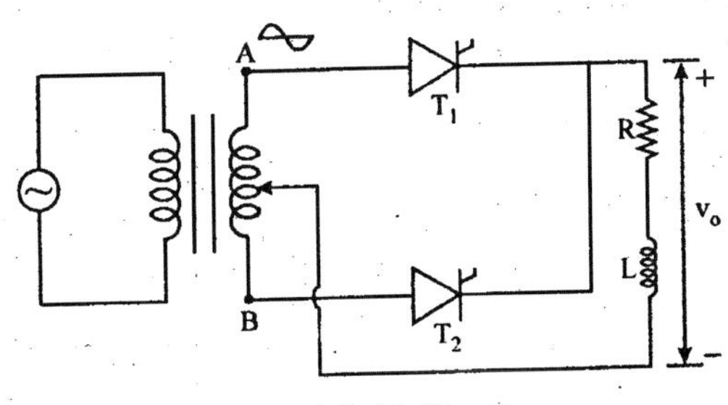

Fig. 4.6. (a) Circuit.

From the waveforms it is observed that the current and voltage are not in phase. When load voltage falls to zero the load current is non zero. The load current falls to zero at angle β, called extinction angle. In this case we have assumed that α = β.

Average load voltage is given by

$$V_{oav} = \frac{1}{2\pi}\int_{\alpha}^{\pi+\beta} V_m \sin\omega t\, d(\omega t)$$

$$= \frac{V_m}{2\pi}\int_{\alpha}^{\pi+\beta} \sin\omega t\, d(\omega t)$$

$$= \frac{V_m}{\pi}[-\cos\omega t]_{\alpha}^{\beta}$$

$$= \frac{V_m}{\pi}[\cos\alpha - \cos\beta]$$

But $\pi + \alpha = \beta$*, giving

$$V_{oav} = \frac{2V_m}{\pi}\cos\alpha \quad ...(7)$$

Average load current will then be

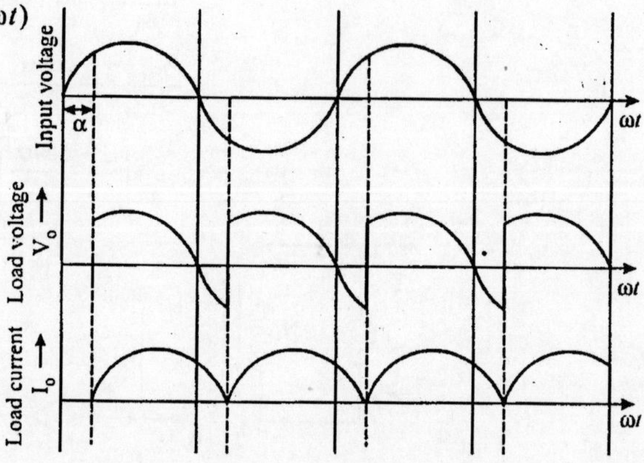

(b) Waveforms.

Fig. 4.6. Full wave controlled rectifier with R-L load.

$$I_{oav} = \frac{V_{oav}}{Z} = \frac{2V_m}{\pi Z}\cos\alpha \quad ...(8)$$

where $Z = \sqrt{R^2 + \omega^2 L^2}$ is the load impedance.

4.4. CONTROLLED BRIDGE RECTIFIER

We have two types of bridge rectifier-half controlled or semiconverter and fully controlled rectifier. In fully controlled converter all the four devices used are SCRs while in semiconverter two of the SCR are replaced by simple diodes. The circuits of half controlled bridge rectifier and fully controlled rectifier are shown in Fig. 4.7 *(b)* and *(a)* respectively.

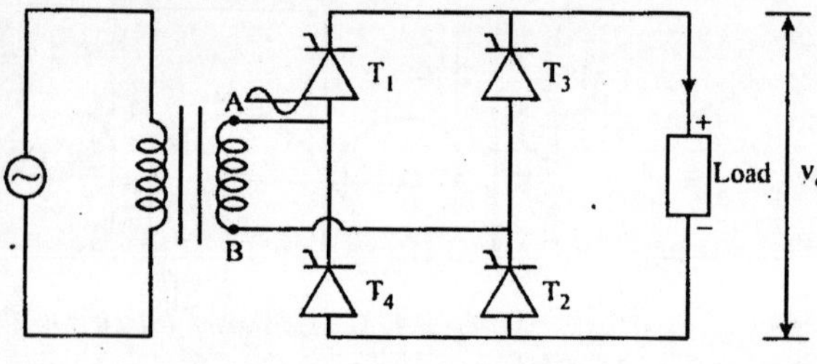

Fig. 4.7 (a) Fully controlled bridge rectifier.

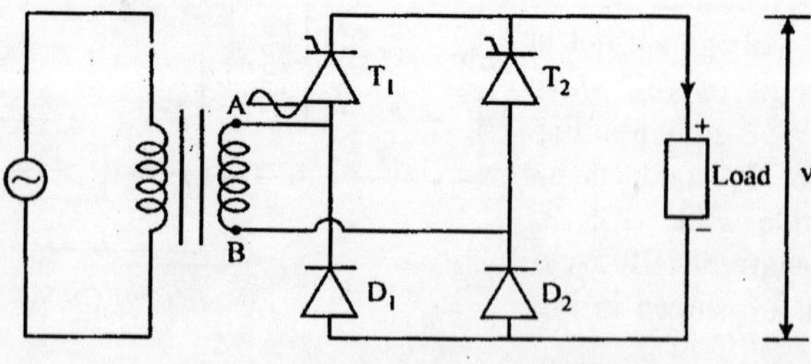

(b)

Fig. 4.7. Half controlled bridge rectifier.

*For continuous mode of conduction $\pi + \alpha = \beta$

In Fig. 4.7 *(a)*, during positive half cycle SCR T_1 and T_2 conduct while during negative half cycle SCRs T_3 and T_4 conduct. In both the cases the direction of current and voltage are as shown in the circuit. The waveforms for the circuit are shown in Fig. 4.8 *(a)*. We are considering purely resistive load.

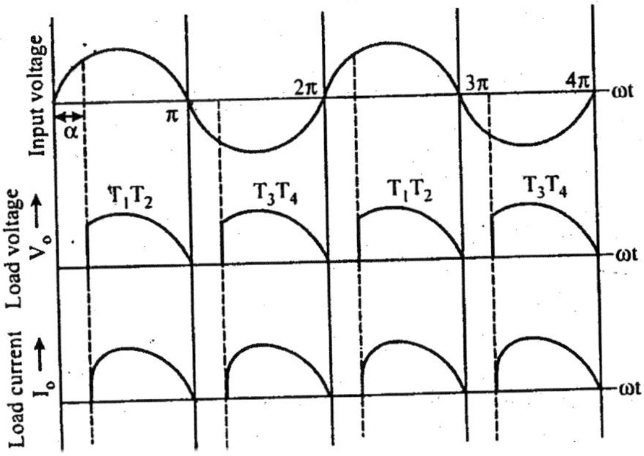

(a) Waveforms are for pure resistive load.

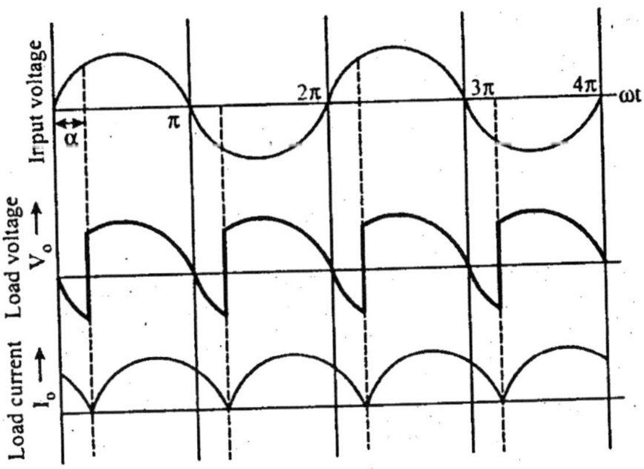

(b) Waveform for R-L load.

Fig. 4.8.

For R-L load the waveforms are shown in Fig. 4.8 *(b)*. We are considering α = β. The average load voltage for a complete cycle is given by

$$V_{oav} = \frac{1}{2\pi}\int_{\alpha}^{\pi+\beta} V_m \sin\omega t \qquad ...(9)$$

$$= \frac{1}{2\pi}\int_{\alpha}^{\beta} V_m \sin\omega t \, d(\omega t) + \frac{1}{2\pi}\int_{\pi+\alpha}^{\pi+\beta} V_m \sin\omega t \, d(\omega t)$$

Since both the half cycles are same. Hence we can write eqn. (9) as

$$V_{oav} = 2\times\frac{1}{2\pi}\int_{\alpha}^{\beta} V_m \sin\omega t \, d(\omega t)$$

$$= \frac{V_m}{\pi}\int_{\alpha}^{\beta} \sin\omega t \; d(\omega t)$$

$$= \frac{V_m}{\pi}(-\cos\omega t)\Big|_{\alpha}^{\beta}$$

$$= -\frac{V_m}{\pi}[\cos\beta - \cos\alpha]$$

$$= -\frac{V_m}{\pi}[\cos\beta - \cos\alpha]$$

$$= \frac{V_m}{\pi}[\cos\alpha - \cos\beta] \qquad ...(10)$$

If $\pi + \alpha = \beta$, we can write eqn. (10) as

$$V_{oav} = \frac{V_m}{\pi}(2\cos\alpha)$$

$$= \frac{2V_m}{\pi}\cos\alpha \qquad ...(11)$$

Similarly the average current is given by

$$I_{oav} = \frac{1}{2\pi}\times 2\int_{\alpha}^{\beta} I_m \sin\omega t$$

$$= \frac{I_m}{\pi}(\cos\alpha - \cos\beta) \qquad ...(12)$$

RMS value of the load voltage is given by

$$V_{orms} = \sqrt{\frac{1}{2\pi}\int_{\alpha}^{\pi+\beta} V_m^2 \sin^2\omega t \; d(\omega t)}$$

$$= \sqrt{\frac{2}{2\pi}\int_{\alpha}^{\beta} V_m^2 \sin^2\omega t \; d(\omega t)}$$

$$= \left[\frac{V_m^2}{\pi}\int_{\alpha}^{\beta}\frac{1}{2}\times 2\sin^2\omega t \; d(\omega t)\right]^{1/2}$$

$$= \left\{\frac{V_m^2}{\pi}\int_{\alpha}^{\beta}\left[\frac{1-\cos 2\omega t}{2}\right] d(\omega t)\right\}^{1/2}$$

$$= \frac{V_m}{\sqrt{2\pi}}\sqrt{\left[\omega t - \frac{1}{2}\sin 2\omega t\right]_{\alpha}^{\beta}}$$

$$= \frac{V_m}{\sqrt{2\pi}}\left[\beta-\alpha-\frac{1}{2}\{\sin 2\beta-\sin 2\alpha\}\right]^{1/2}$$

$$= \frac{V_m}{\sqrt{2\pi}}\left[\beta-\alpha-\frac{1}{2}\{\sin 2\beta-\sin 2\alpha\}\right]^{1/2} \quad ...(13)$$

Eqn. (13) gives the rms load voltage. If $\pi + \alpha = \beta$, eqn. (13) becomes

$$V_{orms} = \frac{V_m}{\sqrt{2\pi}}\left[\pi+0-\frac{1}{2}\times 0\right]^{1/2} = \frac{V_m}{\sqrt{2}}$$

The ripple factor is given by

$$\gamma = \frac{(V_{orms}^2 - V_{oav}^2)^{1/2}}{V_{oav}}$$

$$= \frac{\left[\frac{V_m^2}{2\pi}\left[\beta-\alpha-\frac{1}{2}\{\sin 2\beta-\sin 2\alpha\right]-\left[\frac{V_m}{\pi}(\cos\alpha-\cos\beta\right]^2\right]^{1/2}}{\frac{V_m}{\pi}(\cos\alpha-\cos\beta)}$$

$$= \frac{\left[\frac{\beta-\alpha+\frac{1}{2}(\sin 2\alpha-\sin 2\beta)}{2\pi}-\left(\frac{\cos\alpha-\cos\beta}{\pi}\right)^2\right]^{1/2}}{(\cos\alpha-\cos\beta)/\pi}$$

When $\pi + \alpha = \beta$

$$\gamma = \frac{\left(\frac{1}{2}-\frac{4\cos^2\alpha}{\pi}\right)^{0.5}}{2\cos\alpha/\pi} \quad ...(13A)$$

same as for full wave rectifier using p-n junction diodes.

Above analysis is for fully controlled bridge rectifier. For half controlled bridge rectifier shown in Fig. 4.7 *(b)* the wave forms are shown in Fig. 4.9 *(a)* and *(b)*.

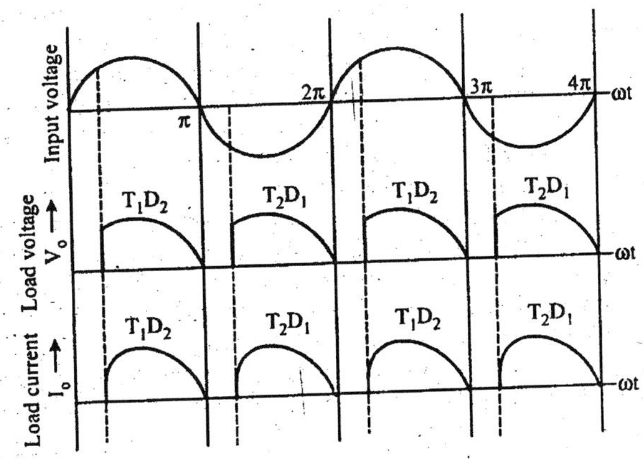

Fig. 4.9. (a) Resistive Load.

Circuit Operation : When point A is positive, pulse is applied at the gate of thyristor T_1. Now the current flows through the path T_1 load D_2 and secondary of the transformer. During negative half cycle the pulse is applied at the gate of thyristor T_2. Now the current flows through the path T_2, load D_1 and secondary of transformer. Thus the cycle is completed.

For a resistive load, the voltage and current across the load are in phase. It means when voltage is zero, current is zero, when voltage is maximum, current is maximum. This is shown in Fig. 4.9 *(a)*.

However, for R-L load the current and voltage do not have same phase. When voltage is maximum, current is not maximum and vice versa. This is shown in Fig. 4.9 *(b)*. At $\omega t = \pi$, the load voltage is zero as it follows the input voltage but the load current is not zero. This is due to energy stored in the inductor which is released to the circuit when input voltage falls to zero, through the path consisting of load $D_1 - T_1$ and back to load. The diode D_1 performs the freewheeling operation. Thus in this case freewheeling operation is carried by diodes D_1 and D_2. Hence the current keeps flowing even when the voltage is zero.

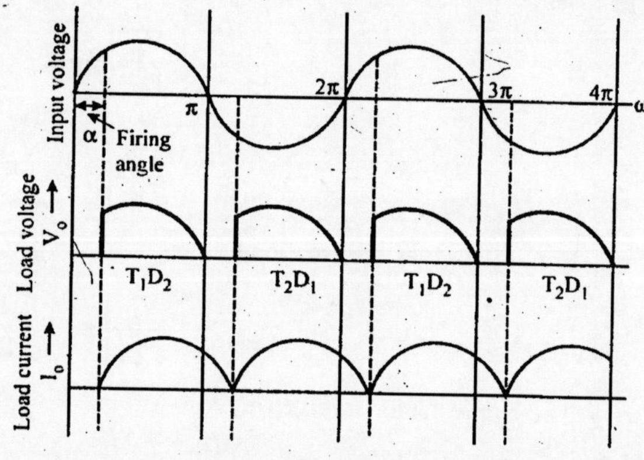

Fig. 4.9. (b) Waveforms for R-L Load.

Example 4.1. For a full wave controlled rectifier shown in Fig. 4.7 *(a)* the firing angle is $\alpha = 60°$, input voltage $v = 100 \sin 200t$. Calculate V_{orms}, V_{oav} and form factor.

Solution : For resistive load $\beta = 0$, hence

$$V_{orms} = \frac{V_m}{\sqrt{2\pi}}\left[\pi - \alpha + \frac{1}{2}\sin 2\alpha\right]^{1/2}$$

Here $V_m = 100$ volt, $\alpha = 60° = \pi/3$ radian, hence

$$V_{orms} = \frac{100}{\sqrt{2\pi}}\left[\pi - \frac{\pi}{3} + \frac{1}{2}\sin 2 \times 60°\right]^{1/2}$$

$$= \frac{100}{\sqrt{2\pi}}\left[\frac{2\pi}{3} + \frac{1}{2}\cos 30\right]^{1/2}$$

$$= \frac{100}{\sqrt{2\pi}}\left[\frac{2\pi}{3} + \frac{\sqrt{3}}{4}\right]^{1/2} \text{ volt} = 63.4 \text{ volts } \textbf{Ans.}$$

$$V_{oav} = \frac{V_m}{\pi}(1 + \cos\alpha)$$

$$= \frac{100}{\pi}(1 + \cos 60°) = \frac{100}{\pi}\left(1 + \frac{1}{2}\right) = \frac{150}{\pi} \text{ volt}$$

$= 47.8$ volts **Ans.**

$$\text{Form factor} = \frac{V_{orms}}{V_{oav}} = \frac{63.4}{47.8} = 1.33 \textbf{ Ans.}$$

Example 4.2. For the circuit of Fig. 4.7 *(b)* input voltage is given by $v = 30 \sin 100\, t$. Calculate for R-L load the average load voltage for firing angle $\alpha = 30°$, extinction angle $\beta = 30°$. What is the time period of input wave.

Solution : Using eqn. (10), we get

$$V_{oav} = \frac{V_m}{\pi}(\cos\alpha - \cos\beta) \quad ; \quad V_m = 30 \text{ volts}$$

$$= \frac{30}{\pi}\left[\cos 30° - \cos(\pi + 30°)\right]$$

$$= \frac{30}{\pi}\left(2 \times \frac{\sqrt{3}}{2}\right) = \frac{3\sqrt{3}}{\pi} \text{ volt} \quad \textbf{Ans.}$$

a.c. voltage is given by

$$v = V_m \sin \omega t$$

The given eqn is

$$v = 30 \sin 100t$$

Comparing the above eqns. we get

$$V_m = 30, \omega = 100$$

Time period is given by $\quad T = \frac{2\pi}{\omega} = \frac{2\pi}{100}$

$$= 2 \times 3.16 \times 10^{-2} \text{ sec.}$$

$$= 62.8 \text{ m sec.} \quad \textbf{Ans.}$$

4.5. SINGLE PHASE SERIES CONTROLLED RECTIFIER

Upto this point we are familiar with single phase half wave and full wave controlled rectifiers. When high output voltage are required, we can connect two or more rectifiers in series. In Fig. 4.10, two rectifiers are connected in series.

In the Fig. one of the rectifiers has SCRs T_1 T_2 T_3 T_4 and the other has SCRs T_5 T_6 T_7 T_8. Let the firing angle for each thyristor in first rectifier be α_1 and for each thyristor in second (lower) rectifier be α_2.

Suppose we turn ON first (upper) rectifier at angle α_1. The current flows through load-diode DF_2-T_2-S_1 and T_1. At an angle α_2 lower rectifier also is turned on. Now the current starts flowing through load, T_7-S_2-T_5 and first rectifier. If output voltage of first rectifier is V_{01} and that of lower rectifier be V_{02} the total voltage across the load

$$V_0 = V_{01} + V_{02} \quad \text{...(14)}$$

Here V_{01} and V_{02} are average values which are given by

$$V_{01} = \frac{V_m}{\pi}(1 + \cos\alpha_1), \text{ and } V_{02} = \frac{V_m}{\pi}(1 + \cos\alpha_2) \quad \text{...(15)}$$

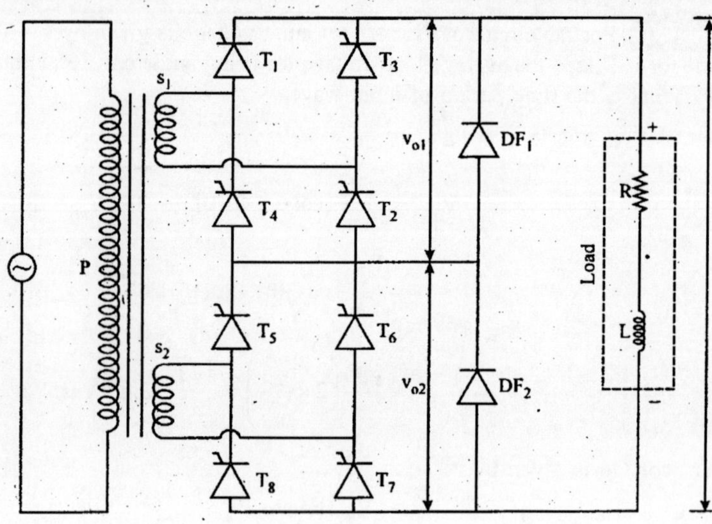

Fig. 4.10. Single phase series converter.

From eqn. (14) and (15) we get

$$V_{oav} = \frac{V_m}{\pi}(1+\cos\alpha_1)+\frac{V_m}{\pi}(1+\cos\alpha_2)$$

$$V_{oav} = \frac{V_m}{\pi}(2+\cos\alpha_1+\cos\alpha_2) \qquad ...(16)$$

Eqn. (16) is the expression for the voltage across the load due to series combination of two rectifiers.

Again from eqn. (16), if $\alpha_1 = \alpha_2 = \alpha$ say; we get

$$V_{oav} = \frac{V_m}{\pi}(2+2\cos\alpha)$$

$$V_{oav} = \frac{2V_m}{\pi}(1+\cos\alpha) \qquad ...(17)$$

twice the value of output voltage due to single rectifier.

In similar manner we can connect a number of rectifiers in series. For *n* thyristors eqn. (17) is modified to

$$V_{oav} = \frac{nV_m}{\pi}(1+\cos\alpha) \qquad ...(18)$$

Example 4.3. Two rectifiers having firing angles 30° and 45° are connected in series. An input supply $v = 100 \sin 100t$ is given to the combination. What is the average value of load current, when R = 100 Ω and L = 20 m H?

Solution : The average load voltage is given by

$$V_{oav} = \frac{V_m}{\pi}(2+\cos\alpha_1+\cos\alpha_2)$$

$$= \frac{V_m}{\pi}(2 + \cos 30° + \cos 45°)$$

$$= \frac{V_m}{\pi}\left(2 + \frac{\sqrt{3}}{2} + \frac{1}{\sqrt{2}}\right) \text{ Volts} = 113.75 \text{ volts } \textbf{Ans.}$$

Now, compare eqn. $v = 100 \sin 100t$ with $v = V_m \sin \omega t$, we get

$$V_m = 100 \text{ V}$$

$$\omega = 100 \text{ radians}$$

$$\text{Load impedance } Z = \sqrt{R^2 + (\omega L)^2}$$

$$= \sqrt{100^2 + (100 \times 20 \times 10^{-3})^2}$$

$$= \sqrt{10000 + 4}$$

The average load current will be

$$I_o = \frac{\frac{V_m}{\pi}\left(2 + \frac{\sqrt{3}}{2} + \frac{1}{\sqrt{2}}\right)}{\sqrt{10004}} \text{ Amp}$$

$$= \frac{100\left(2 + \frac{\sqrt{3}}{2} + \frac{1}{\sqrt{2}}\right)}{\pi\sqrt{10004}} \text{ Amp} = 3.59 \text{ Amp } \textbf{Ans.}$$

4.6. THREE PHASE RECTIFIERS

So far we have studied the rectifier circuit for single phase a.c. input. In three phase the phase difference between any two phase voltage is 120°. We shall discuss few methods to convert three phase a.c. into d.c. as described below :

1. Three phase fully controlled bridge converter : The circuit is shown in Fig. 4.11 *(a)* and the three phases *a, b, c* are shown by waveforms shown in Fig. 4.11 *(b)*. The phase voltages are given by

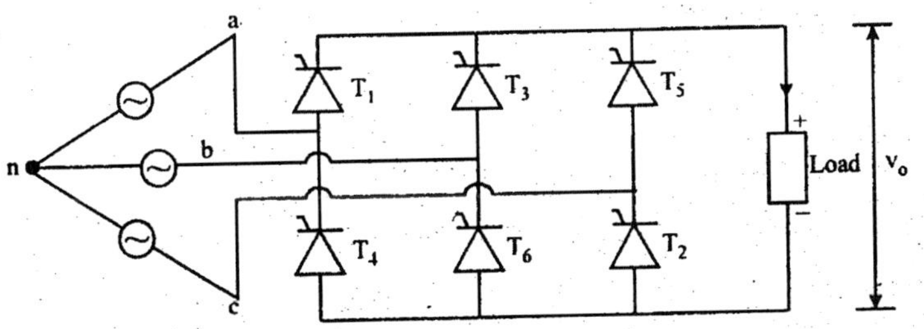

(a) Circuit.

Fig. 4.11. Three phase fully controlled bridge converter.

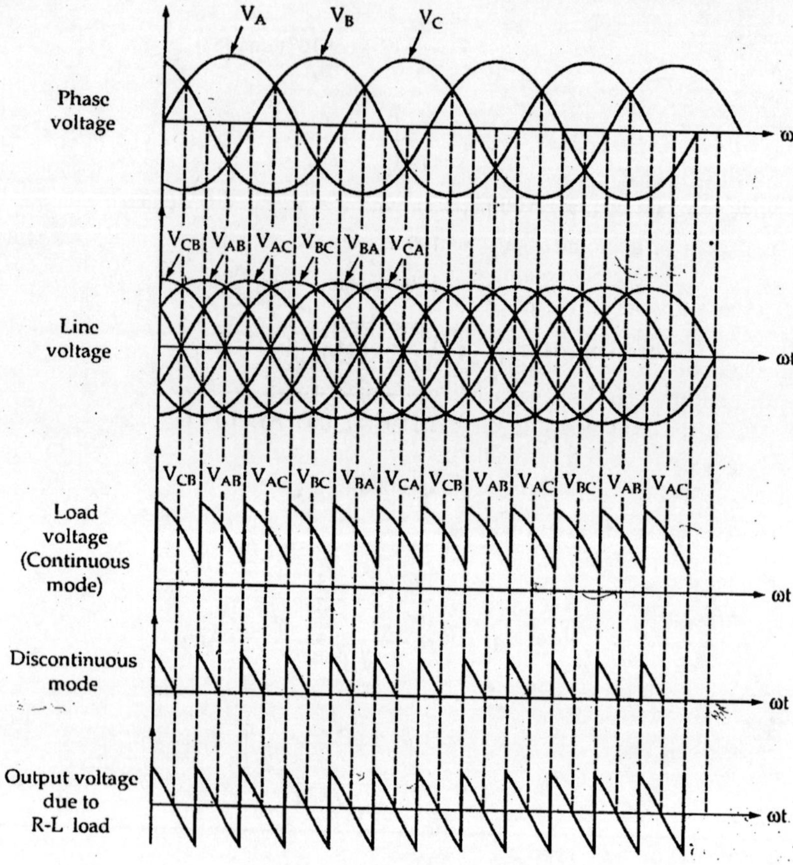

Fig. 4.11. (b) Waveforms.

$$v_{an} = V_m \sin \omega t$$
$$v_{bn} = V_m \sin (\omega t + 120°)^*$$
$$v_{cn} = V_m \sin (\omega t + 240°)$$

and the line voltages are given by

$$v_{ab} = v_{an} - v_{bn}$$
$$= \sqrt{3}V_m \sin\left(\omega t + 30°\right) \quad ...(19)$$
$$v_{bc} = v_{bn} - v_{cn}$$
$$= \sqrt{3}\,V_m \sin(\omega t + 120 + 30°)$$
$$= \sqrt{3}\,V_m \sin(\omega t + 150°) \quad ...(20)$$
$$v_{ca} = v_{cn} - v_{an}$$
$$= \sqrt{3}\,V_m \sin(\omega t + 240 + 30°)$$
$$= \sqrt{3}\,V_m \sin(\omega t + 270°) \quad ...(21)$$

* In clockwise direction $V_{bn} = V_m \sin (\omega t - 120°)$ and $V_{cn} = V_m \sin (\omega t - 240°)$

For positive half cycle thyristors T_1, T_3 and T_5 conduct while for negative half cycle thyristors T_2, T_4 and T_6 conduct. All the six devices in the circuit are thyristors. Hence the desired amount of positive and negative cycle can be switched to the load, *i.e.*, complete cycle can be controlled. Due to above property of the thyristor it is called fully controlled bridge converter.

We observe the waveform of three phase voltages. At $\omega t = 0$, phase a voltage rises from zero value. At $\omega t = \pi/6$, it intersects the wave of voltage of phase c, which was conducting prior to angle $\omega t = \pi/6$. At this angle the thyristor T_1 will start conducting and thyristor T_5 is forcibly commutated. Prior to $\omega_t = \pi/6$, the phase C voltage is more than the phase a voltage. Hence SCR T_1 cannot turn ON even if it remains triggered. Therefore, firing angle is measured from 30° with reference to positive half cycle of voltage of phase a. Similar is the case with the voltage of other two phases *b* and *c*. Each thyristor conducts for 120°. The firing sequence of thyristors is T_1, T_3, T_5 for positive half cycle and T_2, T_4, T_6 for negative half cycle. At $\omega t = \pi/2$ thyristor T_2 is set into conduction and thyristor T_6 is turned off. Thus angle between commutation of T_5 and T_6 is 60°. Hence a complete cycle can be divided into six interval each of 60° duration. Thus commutation occurs every 60°. The pair of thyristors which conduct in different intervals are T_1 T_2, T_2 T_3, T_3 T_4, T_4 T_5 and T_6 T_1 as shown in Fig. 4.11 *(c)*. It is evident from Fig. 4.11 *(b)* that in a complete cycle six pulses are applied.

Average load voltage is given by

$$V_{oav} = 3\times\frac{1}{\pi}\int_{\frac{\pi}{6}+\alpha}^{\frac{\pi}{2}+\alpha} \text{line voltage } d(\omega t) \qquad ...(22)$$

Duration of each pulse is 120°. Hence in a complete cycle we may have three pulses.

$$V_{oav} = 3\times\frac{1}{\pi}\int_{\frac{\pi}{6}+\alpha}^{\frac{\pi}{2}+\alpha} v_{ab}\, d(\omega t)$$

$$= \frac{3}{\pi}\int_{\frac{\pi}{6}+\alpha}^{\frac{\pi}{2}+\alpha} \sqrt{3}\, V_m \sin(\omega t+30°)\, d(\omega t) \text{ from eqn. (19)}$$

$$= -\frac{3\sqrt{3}\, V_m}{\pi}(\cos\omega t)\Big|_{\frac{\pi}{6}+\alpha}^{\frac{\pi}{2}+\alpha}$$

$$= -\frac{3\sqrt{3}\, V_m}{\pi}\left[\cos\left(\frac{\pi}{2}+\frac{\pi}{6}+\alpha\right)-\cos\left(\frac{\pi}{6}+\alpha+\frac{\pi}{6}\right)\right]$$

$$= -\frac{3\sqrt{3}\, V_m}{\pi}\left[-\sin\left(\frac{\pi}{6}+\alpha\right)-\cos\left(\frac{\pi}{3}+\alpha\right)\right]$$

$$= \frac{3\sqrt{3}\, V_m}{\pi}\left[\sin\left(\frac{\pi}{6}+\alpha\right)+\cos\left(\frac{\pi}{3}+\alpha\right)\right]$$

$$= \frac{3\sqrt{3}\, V_m}{\pi}[\sin 30\cos\alpha+\cos 30\sin\alpha$$

$$+ \cos 60 \cos\alpha - \sin 60 \sin\alpha]$$

$$= \frac{3\sqrt{3}\,V_m}{\pi}\left[\frac{\cos\alpha + \cos\alpha}{2}\right]$$

$$V_{oav} = \frac{3\sqrt{3}\,V_m}{\pi}\cos\alpha \qquad \text{...(23)}$$

For $\alpha = 0$, $\quad V_{oav} = \dfrac{3\sqrt{3}\,V_m}{\pi}$

The rms value of load voltage is given by

$$V_{orms} = \sqrt{\frac{3}{\pi}\int_{\frac{\pi}{6}+\alpha}^{\frac{\pi}{2}+\alpha}\left[\sqrt{3}\,V_m \sin(\omega t + 30°)\right]^2 d(\omega t)}$$

$$= \left[\frac{3}{\pi}\int_{\frac{\pi}{6}+\alpha}^{\frac{\pi}{2}+\alpha} 3V_m^2 \sin^2(\omega t + 30°)\, d(\omega t)\right]^{0.5}$$

$$= \frac{3V_m}{\sqrt{\pi}}\left[\int_{\frac{\pi}{6}+\alpha}^{\frac{\pi}{2}+\alpha} \frac{1-\cos 2(\omega t + 30°)}{2} d(\omega t)\right]^{0.5}$$

$$= \frac{3V_m}{\sqrt{\pi}\times\sqrt{2}}\left[\omega t - \frac{1}{2}\sin 2(\omega t + 30)\right]_{\frac{\pi}{6}+\alpha}^{\frac{\pi}{2}+\alpha}$$

$$V_{orms} = \frac{3V_m}{\sqrt{2\pi}}\left[\frac{\pi}{2} + \alpha - \frac{\pi}{6} - \alpha - \frac{1}{2}\sin 2\left\{\frac{\pi}{2} + \alpha + \frac{\pi}{6}\right\} + \frac{1}{2}\sin 2\left\{\frac{\pi}{6} + \alpha + \frac{\pi}{6}\right\}\right]^{0.5}$$

$$= \frac{3V_m}{\sqrt{2\pi}}\left[\frac{\pi}{3} - \frac{1}{2}\sin\left\{\pi + \frac{\pi}{3} + 2\alpha\right\} + \frac{1}{2}\sin\left\{\frac{2\pi}{3} + 2\alpha\right\}\right]^{0.5}$$

$$= \frac{3V_m}{\sqrt{2\pi}}\left[\frac{\pi}{3} + \frac{1}{2}\sin\left\{\frac{\pi}{3} + 2\alpha\right\} + \frac{1}{2}\cos\left\{\frac{\pi}{6} + 2\alpha\right\}\right]^{0.5}$$

$$= \frac{3V_m}{\sqrt{2\pi}}\left[\frac{\pi}{3}+\frac{1}{2}\left\{\sin\frac{\pi}{3}\cos 2\alpha+\cos\frac{\pi}{3}\sin 2\alpha + \cos\frac{\pi}{6}\cos 2\alpha - \sin\frac{\pi}{6}\sin 2\alpha\right\}\right]^{0.5}$$

$$= \frac{3V_m}{\sqrt{2\pi}}\left[\frac{\pi}{3}+\frac{1}{2}\left\{\sin\frac{\pi}{3}\cos 2\alpha+\cos\frac{\pi}{6}\cos 2\alpha\right\}\right]^{0.5}$$

$$= \frac{3V_m}{\sqrt{2\pi}}\left[\frac{\pi}{3}+\frac{\sqrt{3}}{2}\cos 2\alpha\right]^{0.5}$$

$$= \sqrt{3}\,V_m\left[0.5+\frac{3\sqrt{3}}{4\pi}\cos 2\alpha\right]^{0.5}$$

Ripple factor. It is given by

$$\gamma = \frac{(V_{orms}^2 - V_{oav}^2)^{1/2}}{V_{oav}}$$

$$= \frac{\left[3V_m^2\left(0.5+\frac{3\sqrt{3}}{4\pi}\cos 2\alpha\right)-\left(\frac{3\sqrt{3}\,V_m}{\pi}\cos\alpha\right)^2\right]^{1/2}}{\frac{3\sqrt{3}}{\pi}V_m\cos\alpha}$$

$$= \frac{\left[1.5+\frac{9\sqrt{3}}{4\pi}(2\cos^2\alpha-1)-\frac{27}{\pi^2}\cos^2\alpha\right]^{0.5}}{\frac{3\sqrt{3}}{\pi}\cos\alpha}$$

$$= \frac{\left[1.5+\frac{18\sqrt{3}}{4\pi}\cos^2\alpha-\frac{9\sqrt{3}}{4\pi}-\frac{27}{\pi^2}\cos^2\alpha\right]^{0.5}}{\frac{3\sqrt{3}}{\pi}\cos\alpha}$$

$$= \frac{[0.2589-0.2563\cos^2\alpha]}{1.65\cos\alpha}$$

$$\approx 0.15\sin\alpha\tan\alpha.$$

Example 4.4. A 3 ϕ full converter is fed by 400 V 3 ϕ 50 Hz supply. The average load current is 150 A and the load is highly inductive. The firing angle is 60°. Find *(a)* output power, *(b)* average rms and peak current through thyristor, *(c)* Peak inverse voltage.

Solution : Given

$$\text{Line voltage} = 400\text{ V}$$

$$\text{Phase voltage} = \frac{400}{\sqrt{3}}$$

Peak value of phase voltage

$$V_m = \sqrt{2} \times \frac{400}{\sqrt{3}} = 325.56 \text{ V}$$

$$V_{dc} = \frac{3\sqrt{3}\, V_m}{\pi} \cos(\alpha),\ \alpha = 60^\circ$$

$$= \frac{3\sqrt{3}\, V_m}{\pi} \cos\left(\frac{\pi}{3}\right)$$

$$= 269.23 \text{ V}$$

(*a*) Output power, $P_{av} = V_{dc} \times I_{dc}$ watt

$$= 293.23 \times 150 = 40384.5 \text{ W } \textbf{Ans.}$$

(*b*) Average thyristor current $= \dfrac{150}{3} = 50$ Amp **Ans.**

(The thyristor conducts for $\frac{1}{3}$ of the total time period.)

rms value of thyristor current $= \dfrac{150\sqrt{2}}{\sqrt{6}} = 86.6$ A

Peak current through thyristor = average load current = 150 A

(*c*) Peak reverse voltage = Peak value of line to line voltage

$$= 400\sqrt{2} \text{ V} = 565.5 \text{ V } \textbf{Ans.}$$

Example 4.5 A three phase full wave converter is fed by 415 V 3 phase 50 Hz input. The firing angle is 50° for each thyristor. A load containing resistance 50 ohms and inductance 10 m H is connected in the output. Calculate average current through the load.

Solution : Phase voltage $= \dfrac{\text{line voltage}}{\sqrt{3}} = \dfrac{415}{\sqrt{3}}$

Peak voltage $= \sqrt{2} \times$ phase voltage $= \dfrac{415\sqrt{2}}{\sqrt{3}}$

Average load voltage, $V_{oav} = \dfrac{3\sqrt{3}\, V_m}{\pi} \cos\alpha$

$$= \frac{3\sqrt{3} \times 415\sqrt{2}}{\pi} \cos 30$$

$$= \frac{1245\sqrt{2}}{\pi} \times \frac{\sqrt{3}}{2} = 485.6$$

Average load current, $I_{oav} = \dfrac{1245\sqrt{3/2}}{\sqrt{R^2 + \omega^2 L^2}}$

$$= \frac{1245\sqrt{3/2}}{\pi\sqrt{(50)^2 + (10 \times 10^{-3} \times 3)^2}} = 9.69 \text{ Amp } \textbf{Ans.}$$

2. Three phase half controlled bridge converter : In half controlled converter either for positive half or negative half of input cycle the output power is controlled. This is realized by replacing three of the SCRs by diodes as shown in Fig. 4.12 *(a)*. Thyristors T_1, T_2, T_3 conduct for positive cycle while diodes D_1, D_2, D_3 conducts for negative cycles. Each diode and thyristor conduct for 120°. The waveforms are shown in Fig. 4.12 *(b)*.

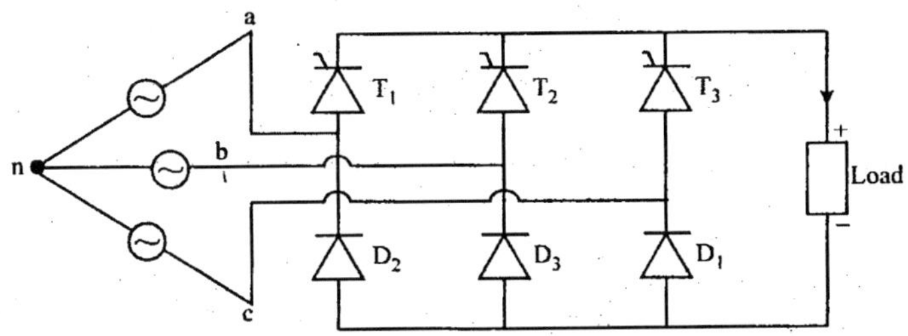

(a) Circuit.

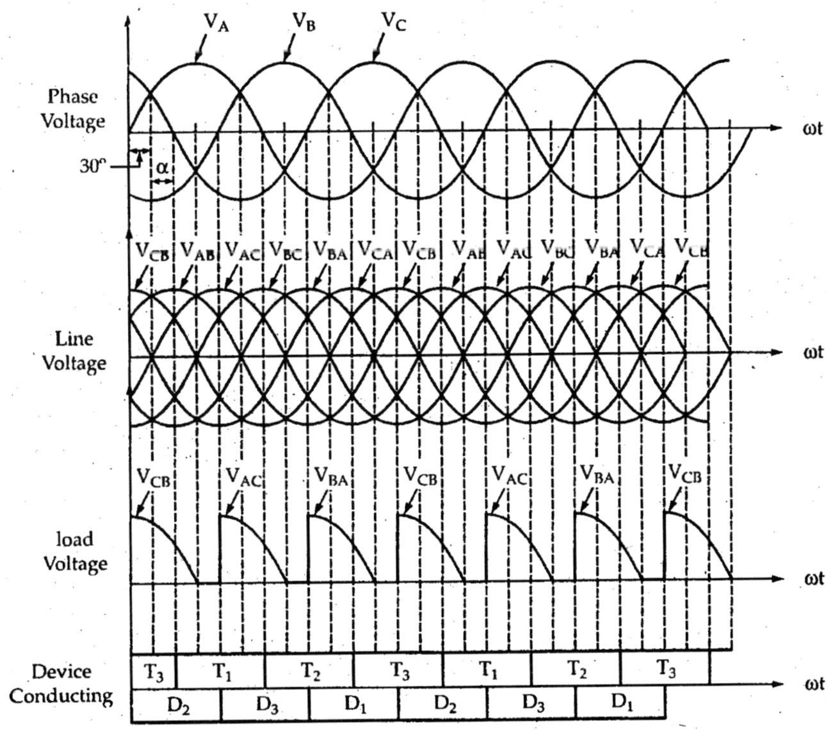

(b) Waveforms.

Fig. 4.12. Three phase half controlled bridge converter.

3. Three phase half wave rectifiers : The circuit of a three phase half wave rectifier is shown in Fig. 4.13 *(a)*.

At time $t = 0$ positive cycle appears at diode D_1 while diode D_3 is already in the conduction state as is obvious from the waveforms shown in Fig. 4.13 *(b)*. At an angle $\omega t = 30°$ the phase '*a*' and phase '*c*' voltages are equal and hence the diode D_3 is reverse

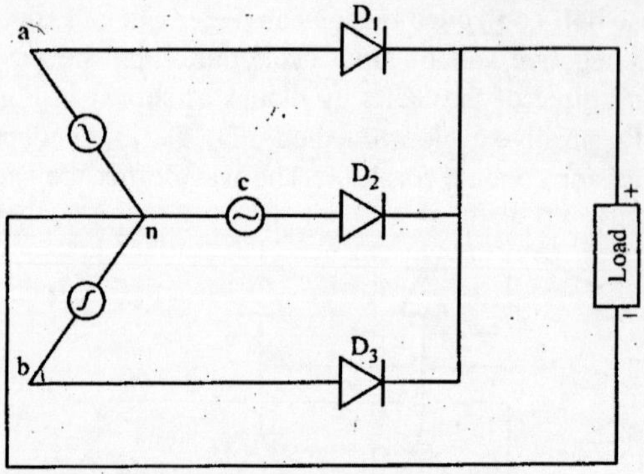

Fig. 4.13 (a) Three phase half wave rectifier.

biased at this angle. From angle $\omega t = 30°$ to $\omega t = 150°$ only diode D_1 conducts. At $\omega t = 150°$ this diode is reverse biased and diode D_2 takes part in conduction from this angle to $\omega t = 270°$. It is clear, therefore, that every diode conducts for 120° angle. The load voltage waveforms are shown in the figure.

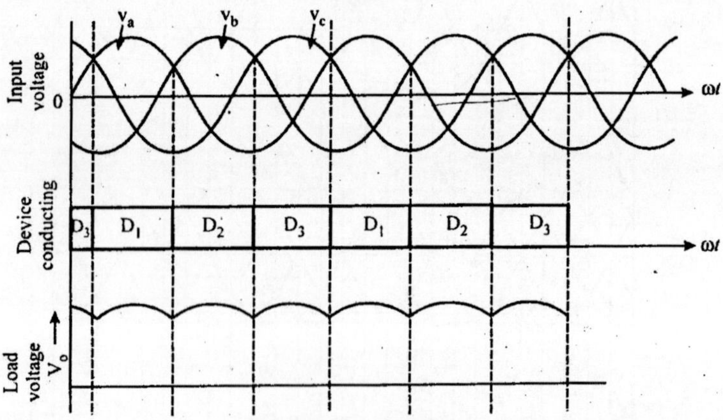

Fig. 4.13 (b) Waveforms.

4. Three phase full wave rectifier : In full wave rectifier, as the name implies, rectified output is obtained for both positive and negative half of the input cycle. The circuit of a three phase full wave rectifier is shown in Fig. 4.14 *(a)*. Diodes D_1, D_3 and D_5 conduct for positive half cycles while D_2, D_4, D_6 conduct for negative half cycles.

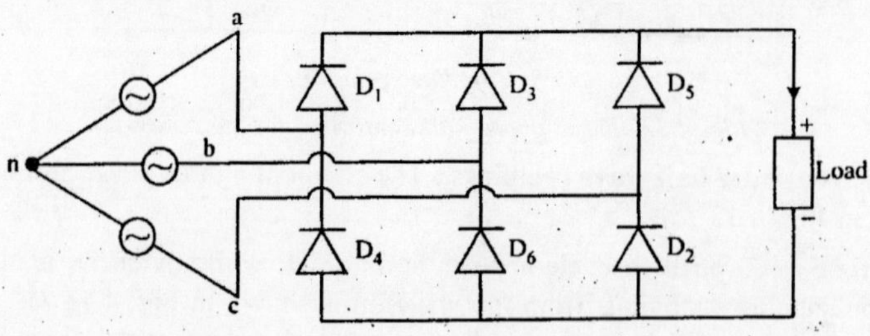

Fig. 4.14. (a) Circuit.

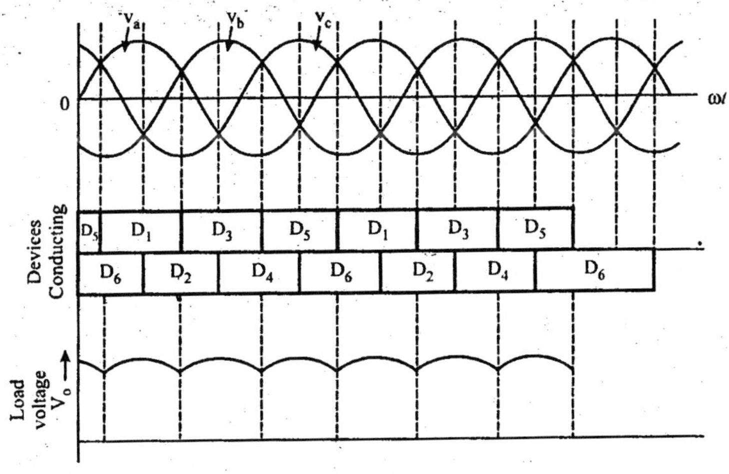

(b) Waveforms.

Fig. 4.14. Three phase full wave rectifier.

The conducting sequence of the diodes is D_1, D_2, D_3, D_4, D_5, D_6. We begin with time $t = 0$ or $\omega t = 0$. The waveforms have been drawn in Fig. 4.14 *(b)*. We observe from the waveforms that at $\omega t = 0$ only phase voltage V_c is positive which means that diode D_5 is conducting. At $\omega t = 30°$, voltage V_a and V_c are equal with V_a rising while V_c falling. Beyond this angle the diode D_5 is reverse biased due to higher positive voltage developed across load due to voltage V_a. Hence diode D_5 is turned OFF while D_1 is set into conduction. Similarly in negative half cycle at $\omega t = 0$ voltage V_b is negative which means diode D_6 is in conduction state. At $\omega t = 90°$ phase voltage V_c and V_b have same value with V_c increasing in negative direction while V_b decreasing. As a result D_6 is turned OFF while D_2 is set into conduction. Same reason can be applied to explain as to which diode takes part in conduction. Every diode conducts for 120° angle. The output voltages are same as in case of full wave rectifier using SCR except the fact that in present case $\alpha = 0°$.

SUMMARY

In some cases machines are required to be driven by dc inputs. For conversion of ac into dc rectifier circuits are used. Various rectifier circuits using simple diodes and SCRs have been discussed in this chapter. Also three phase rectifiers have been studied in the chapter.

EXERCISE

1. Explain the circuit of single phase half wave rectifier using resistive load.
2. Compare the waveforms of single phase half wave rectifier using resistive and resistive inductive (R-L) load.
3. Explain the working of full wave semiconverter.
4. Draw the output waveforms of fully controlled full wave bridge rectifier using R-L load in discontinuous conduction mode. Explain it.
5. Explain the circuit of half wave controlled rectifier using resistive load.
6. Draw the output waveforms of a half wave cotrolled rectifier using R-L load.

7. What is the need of series combination of rectifier ? Derive expression for output voltage when two rectifiers are connected in series.
8. Derive expression for average load voltage for a three phase full wave controlled rectifier.
9. Explain the working of three phase half controlled full wave rectifier.
10. Draw the circuit of three phase half wave rectifier. Explain in detail the waveforms.
11. Explain in detail the working of three phase full wave rectifier.

NUMERICAL PROBLEMS

1. 220 volts is fed to a single phase controlled rectifier at 50 Hz. If the firing angle of the thyristor is 20°, calculate *(a)* average load voltage, *(b)* average power across load. Assume load resistance to be equal to 50 Ω.
2. In the above problem what will be the values of above quantities if load resistance is 40 Ω and load inductance is 20 mH. Assume continuous conduction mode.
3. A full wave controlled rectifier is fed by a supply of 220 volts at 50 Hz. The rectifier drives a load of R = 50 Ω and L = 10 mH. Calculate averagee load voltage, average load current, average output power. Assume $\beta = 20°$, $\alpha = 30°$.
4. A voltage given by $v = 200 \sin 100\,t$ is applied as an input to a controlled full wave rectifier. The output of the rectifier is fed to a load having resistance of 10 Ω and an inductance of L henry. If firing angle is 35° calculate the value of L. Given the average current through the load is 9 amp. Assume continuous conduction mode.
5. For a controlled full wave rectifier the firing angle is 45°, input voltage $V = 200 \sin 150t$ calculate V_{orms}, V_{oav} and form factor for resistive load.
6. Two controlled rectifiers whose firing angles are 30° and 45° are connected in series. The combination is excited by an input voltage $v = 300 \sin 200t$ calculate *(a)* average load voltage, *(b)* average load current, *(c)* output average power. Assume that load has a resistance of 40 Ω.
7. In the above problem make the calculations for R = 50 Ω and L = 25 mH $\beta = 10°$.
8. For a three phase full wave rectifier the firing angle is 34.6°. Calculate average voltage for $v = 230 \sin 100t$.
9. In the above problem if R = 100 Ω and L = 10 mH, calculate average load current and output power. Assume continuous conduction mode.
10. A three phase full converter is fed by 415 V 3ϕ, 50 Hz supply. The average load current is 100 amp and load is highly inductive, calculate *(a)* output average power, *(b)* rms and average current through thyristor, *(c)* peak reverse voltage. The firing angle is 35°.
11. Repeat the above problem for 3 ϕ rectifier.
12. An input voltage $v = 100 \sin \omega t$ is fed to a full wave rectifier whose output feeds a load having resistance R = 10 Ω and inductance 25 mH. If the average current through the load is 6 amp, calculate the time period of input wave.

OBJECTIVE TYPE QUESTIONS

1. A rectifier
   (a) converts ac into fixed dc *(b)* converts ac into pulsating dc
   (c) converts variable ac into variable dc
   (d) either *(b)* or *(c)*.
2. In the output of controlled rectifier

(a) harmonics are present (b) most of the part is dc

(c) fundamental component is present (d) none of the above.

3. In half wave controlled rectifier

(a) one diode and one thyristor is required

(b) only one thyristor is required

(c) either a diode or a thyristor may be used

(d) diode is an essential part.

4. In full wave controlled bridge rectifier

(a) four diodes are used (b) two diodes and two SCRs are used

(c) four thyristors are required (d) none of the above.

5. In full wave controlled rectifier with resistive load.

(a) $V_{oav} = \frac{V_m}{\pi}(1+\cos\alpha)$ (b) $V_{oav} = \frac{V_m}{2\pi}(1+\cos\alpha)$

(c) $V = \frac{V_m}{\pi}\left(1+\frac{\cos 2\alpha}{2}\right)$ (d) none of the above

6. In full wave semiconverter

(a) two diodes are used (b) two thyristors are used

(c) two diodes and two SCRs are used (d) one diode is used.

7. In half wave controlled rectifier

(a) $V_{oav} = \frac{V_m}{2\pi}(1+\cos\alpha)$ (b) $\frac{V_m}{2\sqrt{\pi}}(\pi+\cos\alpha)$

(c) $\frac{V_m}{\sqrt{2\pi}}\left(\pi+\alpha+\frac{\cos 2\alpha}{2}\right)$ (d) (b) and (c)

8. In three phase fully controlled full wave rectifier

(a) each thyristor conducts for 60° (b) each thyristor conducts for 120°

(c) conduction angle of thyristor depends upon firing angle

(d) some of thyristors conduct while others not.

9. Average load voltage in case of three phase fully controlled full wave rectifier is

(a) $\frac{3\sqrt{3}V_m}{\pi}$ (b) $\frac{3\sqrt{3}V_m}{\pi}\cos\alpha$

(c) $\frac{3\sqrt{3}V_m}{2\sqrt{\pi}}\cos\alpha$ (d) $\frac{3\sqrt{3}V_m}{\sqrt{2\pi}}\cos\alpha$

10. In a three phase full wave rectifier

(a) each diodes conducts for 120° (b) each diode conducts for 60°

(c) each thyristor conducts for 120° (d) none of the above is correct.

11. If two converters whose firing angles are a_1 and a_2 are connected in series the average value of load voltage is

(a) $V_{oav} = \frac{V_m}{\pi}(1+\cos\alpha_1+\cos\alpha_2)$ (b) $V_{oav} = \frac{V_m}{\pi}(2+\cos\alpha_1+\cos\alpha_2)$

(c) $V_{oav} = \frac{V_m}{2\pi}(2+\cos\alpha_1+\cos\alpha_2)$ (d) $V_{oav} = \frac{V_m}{\sqrt{2\pi}}(2\pi+\cos\alpha_1+\cos\alpha_2)$

12. **If n converters whose firing angles are same are connected in series the average load voltage is given by**

(a) $V_{oav} = \frac{V_m}{\pi}(n+\cos\alpha)$ (b) $V_{oav} = \frac{V_m}{2\pi}(n+\cos\alpha)$

(c) $V_{oav} = \frac{nV_m}{\pi}(1+\cos\alpha)$ (d) $V_{oav} = \frac{nV_m}{\pi}(2+\cos\alpha)$

ANSWERS

1. (b)	2. (a)	3. (b)	4. (c)	5. (a)	6. (c)
7. (a)	8. (b)	9. (b)	10. (a)	11. (b)	12. (c)

❑❑❑

5

AC Phase Control

5.0. INTRODUCTION

As we all know that for turning on a thyristor or a triac a minimum value of gate current and gate voltage should be supplied to the gate circuit of the device. When a.c. input is applied across the SCR the gate circuit also is excited by a.c. The gate current ($I_{g\,min}$) can be obtained for different values of a.c. inputs by changing the circuit parameter such as resistance etc. For example if we increase the value of input voltage so that minimum gate current ($I_{g\,min}$) can be obtained to turn on the device, we may increase the potentiometer resistance connected in gate circuit. The value of this voltage corresponds to an angle as shown in Fig. 5.1. From the Fig. it is clear that for small resistance voltage v_1 is sufficient to turn on the device whereas voltage v_2 is needed for higher resistance. v_1 and v_2 correspond, respectively, to phase angles α_1 and α_2 α_2. It means we have to change phase angle to turn on the thyristor by changing the resistance in gate circuit. This method is called phase angle control.

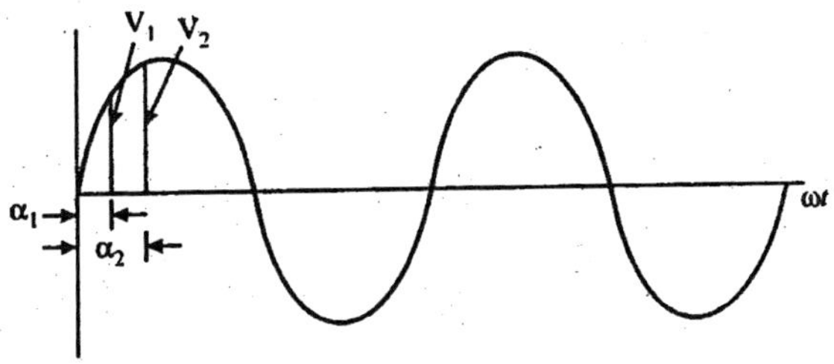

Fig. 5.1. Illustration of phase control.

Example 5.1. In the circuit shown in Fig. 5.2 an input voltage given by $v = 30 \sin 100 t$ is applied. What is the firing angle ? Given R = 20 Ω and C = 47 μF.

Solution : Firing will occur when $V_c = 1.4$ volt. Hence we write

$$V_c = V_0\left(e^{-\frac{t}{RC}}\right)$$

$$1.4 = 30\left(e^{-\frac{t}{20\times47\times10^{-6}}}\right)$$

$$1.4 = 30\left(e^{-\frac{t}{9.4\times10^{-4}}}\right)$$

or $$\frac{t}{9.4\times10^{-4}} = \ln\frac{30}{1.4}$$

or $$t = 9.4\times10^{-4}\ln\frac{30}{1.4} = 2.88 \text{ m sec}$$

$$\text{Firing angle} = \omega t = \frac{100\times180}{\pi}t = 16.51° \quad \textbf{Ans.}$$

5.1. BASIC PRINCIPLE OF AC PHASE CONTROL

Let us consider Fig. 5.2 *(a)*. During negative half cycle P is –ve. The capacitor is charged to peak value of the input through diode D_2 with its upper plate negative. During positive half cycle point P is positive and the thyristor is forward biased. After the capacitor is charged to peak negative value in quarter cycle, it will start discharging as the negative voltage starts falling from peak value to zero [See Fig. 5.2 *(a2)*]. The rate of discharge of the capacitor is governed by the time constant given by $\tau = RC$. The capacitor voltage starts rising from zero to positive value in the positive input cycle. When $v_c = 1.4$ V, the thyristor is triggered and the bulb starts glowing. The waveforms are shown in Fig. 5.2. *(b)*.

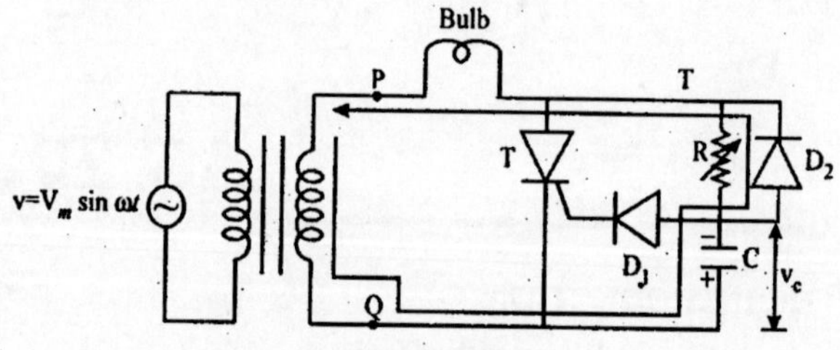

Fig. 5.2 (a1) Circuit.

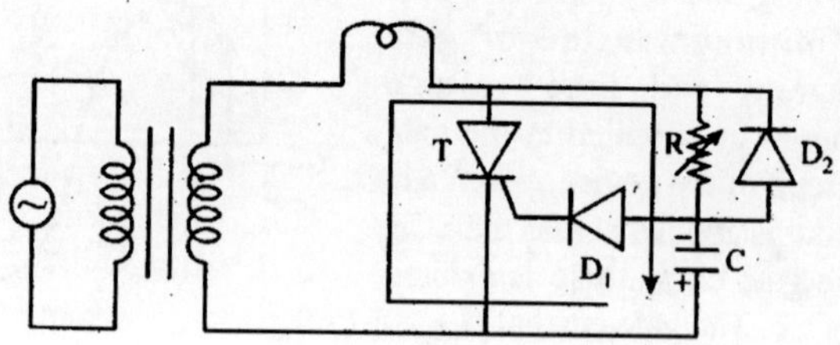

Fig. 5.2 (a2) Circuit.

From the Fig. it is clear that the capacitor voltage falls to zero at an angle greater than π radians. This angle can be changed by changing the product RC. In other words the phase angle at which the thyristor is turned on can be changed by changing the product RC. This is the principle of AC phase control.

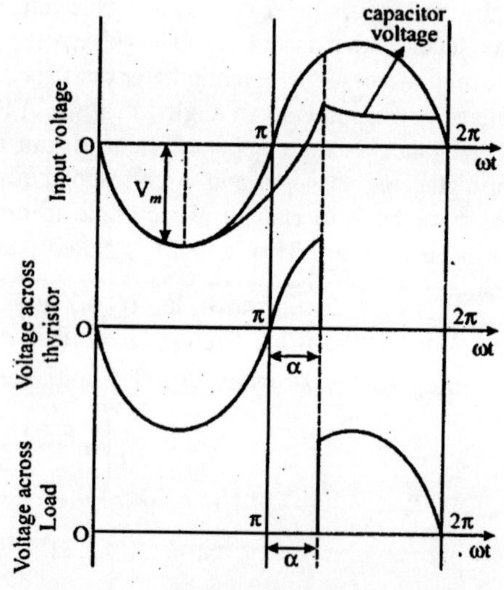

(b) Waveforms.

Fig. 5.2. AC phase control.

These days phase angle control method is used in most of the cases to deliver controlled power to the load. Like thyristor other devices in this family such as diac, triac, IGBT etc. also can be used. In the following sections we shall have discussion on AC phase control methods by using Triacs.

5.2. PHASE ANGLE CONTROL USING TRIAC

The circuit is shown in Fig. 5.3 *(a)*. RC network is used for changing firing angle as explained in sec. 5.1. The triac conducts for both the half of the input wave. For this reason it is called full wave phase angle control method. When the voltage across the capacitor is quite enough to turn ON the diac, the triac also is turned ON as the pulse is sent to its gate by the diac. By changing the value of potentiometer resistance R, firing angle of triac is changed. The waveforms are shown in Fig. 5.3 *(b)*.

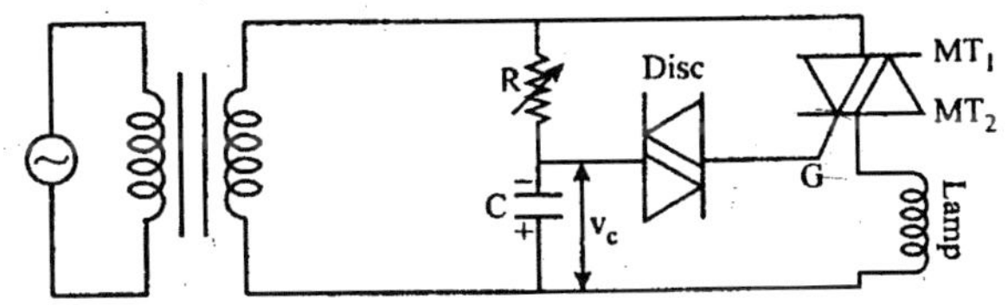

Fig. 5.3. (a) Phase control using Diac-triac pair.

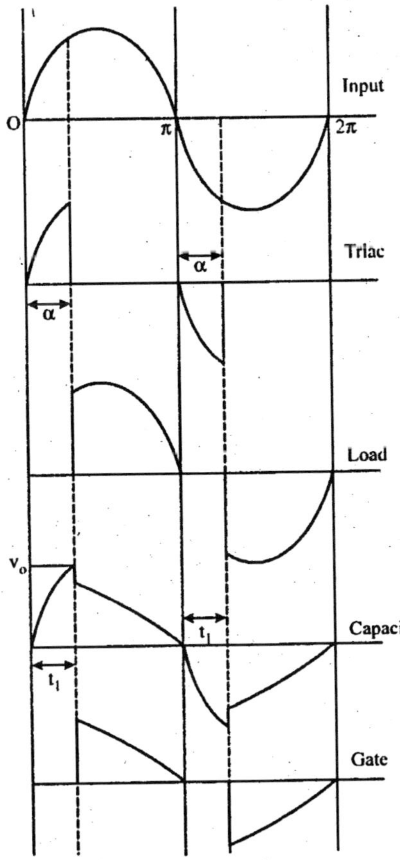

Fig. 5.3. (b) Waveforms.

Circuit Operation : In the positive half cycle the capacitor is charged to voltage V_0 which is the voltage required to turn on the diac. The time taken by the capacitor to attain value V_0 depends upon the product RC. Once the diac is turned ON the triac is also turned ON. The capacitor discharges through the diac and gate circuit of the triac, to zero voltage. In the negative half cycle the capacitor is charged first to $-V_0$ value in the time t_1. The diac is again turned ON and hence the triac. This is how the triac conducts during negative cycle. It is shown in Fig. 5.3. *(b)*.

Example 5.2. In Fig. 5.3 *(a)*, input wave has peak voltage of 100 volts and frequency 50 Hz. If R = 100 Ω and C = 10 μF calculate the average load voltage.

Solution : The time constant RC = 1000×10^{-6} sec = 1 m sec.

1 m sec is required by the capacitor to charge to the voltage required to turn on diac. In other words the phase at which the thyristor is turned on can be changed by changing RC. After positive half cycle the thyristor is turned off.

$$\text{Angle will be} = \omega t = 50 \times 2\pi \times 1 \times 10^{-3} \text{ radian} = 18°$$

$$\text{Average load voltage} = V_{oav} = \frac{V_m}{\pi}(1+\cos\alpha)$$

$$= \frac{100}{\pi}(1+\cos 18°) = 62.14 \text{ volts}$$

5.3. FULL WAVE PHASE CONTROL USING SCR AND UJT

The AC phase control circuit employing UJT is shown in Fig. 5.4 *(a)*. AC voltage is rectified by bridge rectifier formed by diodes D_1, D_2, D_3 and D_4. The rectified output is passed through the thyristor as it is fired by the pulse generated by UJT. The waveforms are shown in Fig. 5.4 *(b)*.

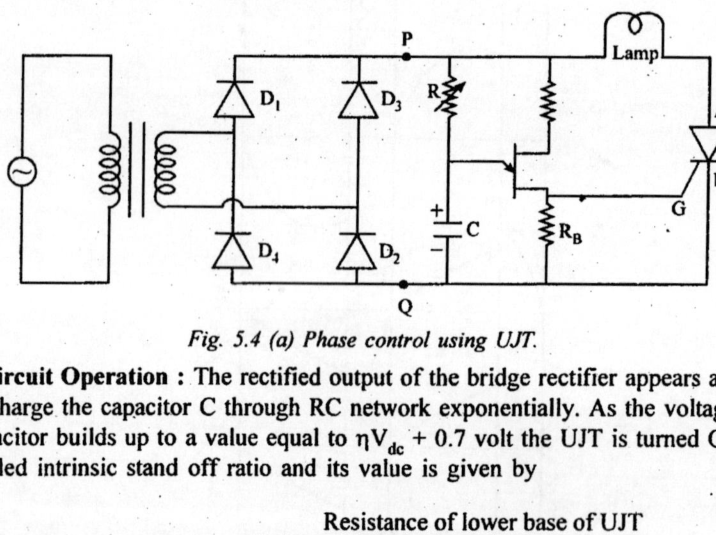

Fig. 5.4 (a) Phase control using UJT.

Circuit Operation : The rectified output of the bridge rectifier appears at point P which charge the capacitor C through RC network exponentially. As the voltage across the capacitor builds up to a value equal to ηV_{dc} + 0.7 volt the UJT is turned ON. Here η is called intrinsic stand off ratio and its value is given by

$$\eta = \frac{\text{Resistance of lower base of UJT}}{\text{Total base resistance}}$$

$$= \frac{R_{B1}}{R_{B1}+R_{B2}}$$

where R_{B1} = Resistance of lower base B_1, and

R_{B2} = Resistance of upper base B_2.

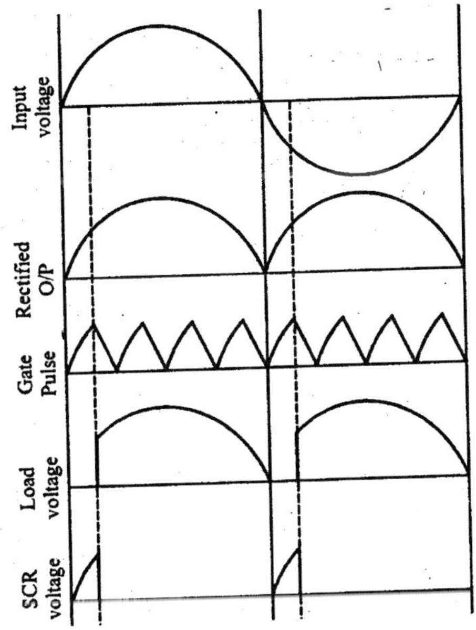

Fig. 5.4. (b) Waveforms.

At this value of capacitor voltage UJT is turned on and the current flows through resistance R_B. The voltage across R_B turns on the thyristor. Let us examine the waveforms. We observe that during half cycle the thyristor is turned on by first pulse at the gate. The remaining pulses do not affect the SCR working in half cycle. These are redundant.

Circuit Analysis : Let the voltage at point P be V_{dc}, where $V_{dc} = \frac{2V_m}{\pi}$, capacitor C is charged exponentially, the voltage across capacitor at any time t is given by

$$v_c = V_{dc}\left(1 - e^{-\frac{t}{RC}}\right)$$

If time taken by capacitor to charge to a voltage $\eta V_{dc} + 0.7$ V be T_1, then

$$\eta Vd_c + 0.7 = V_{dc}\left(1 - e^{-\frac{T_1}{RC}}\right) \quad ...(1)$$

As the capacitor is discharged through resistance R_B, the minimum voltage across UJT will be called valley voltage designated V_v. Let the time taken by capacitor to discharge to V_v is T_2, we can write

$$V_v = (\eta Vd_c + 0.7)\, e^{-\frac{T_2}{R_B C}} \quad ...(2)$$

We assume that $T_1 >> T_2$ and $\eta Vd_c >> 0.7$, eqn. (1) is modified to

$$\eta Vd_c = Vd_c\left(1 - e^{-\frac{T_1}{RC}}\right)$$

or $$1 - \eta = - e^{-\frac{T_1}{RC}}$$

or $$+ (1 - \eta) = e^{-\frac{T_1}{RC}}$$

or $$1 - \eta = \frac{1}{e^{\frac{T_1}{RC}}}$$

$$e^{\frac{T_1}{RC}} = \frac{1}{1-\eta}$$

Taking natural log on both sides, we get

$$\frac{T_1}{RC} = \ln\left(\frac{1}{1-\eta}\right)$$

or $$T_1 = RC \ln\left(\frac{1}{1-\eta}\right) \quad ...(3)$$

Duration of the pulse will be given by

$$T = T_1 + T_2$$
$$\approx T_1$$

$$T = RC \ln\left(\frac{1}{1-\eta}\right) \quad ...(4)$$

5.4. FULL WAVE PHASE CONTROL CIRCUIT USING UJT AND PULSE TRANSFORMER

The above circuits suffer from one drawback — there is no isolation between gate and anode circuit. For isolation between anode and gate circuit pulse transformer is used. The circuit is shown in Fig. 5.5 *(a)*.

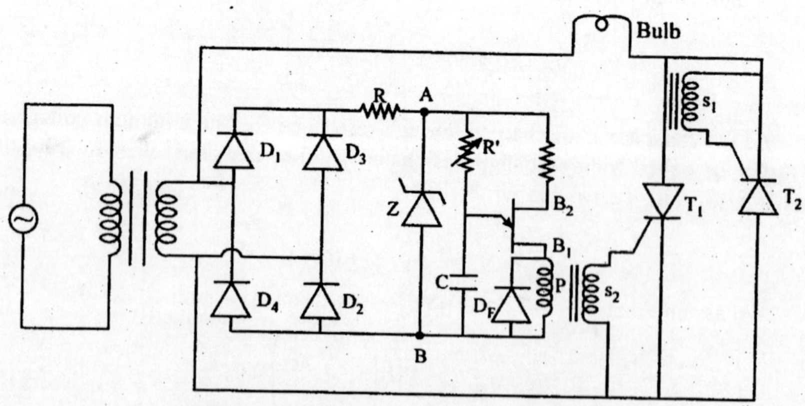

Fig. 5.5. (a) Phase control pulse transformer.

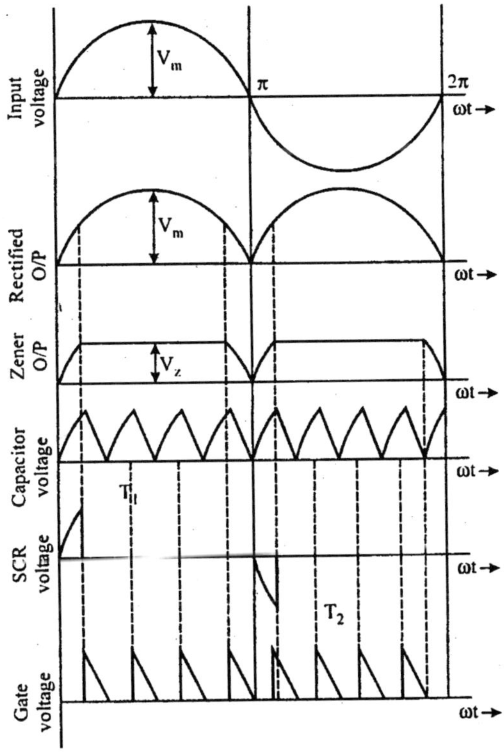

Fig. 5.5 (b) Waveforms.

In the Fig. gate circuit and the anode circuit are isolated from each other. Zener diode is used to convert pulsating d.c. into constant d.c except at zero crossings. The function of UJT is the same as has been discussed in previous section. As the current flows through the primary of pulse transformer, e.m.f. is induced in secondaries which turns on the SCR. Diode DF is freewheeling diode. The energy of coil P is released through this diode when UJT goes from on to off state.

Input ac voltage appears across the thyristors. During positive half cycle SCR T_1 conducts as the pulse is applied at its gate. Thyristor T_2 conducts for negative half cycle. The waveforms are shown in Fig. 5.5 *(b)*.

From eqn. (4) we observe that the duration of the pulse generated by relaxation oscillator or the frequency can be changed by changing the product RC and hence the controlled power can be obtained across the load for positive as well as negative half cycle. This method of ac phase control is sometimes referred as open loop control or normal control.

We can replace the two SCRs by a single TRIAC for the same purpose. The circuit employing a triac for full wave phase control is shown in Fig. 5.6 *(a)*. The operation of the circuit is same as that of the circuit shown in Fig. 5.5 *(a)* and 5.5 *(b)*.

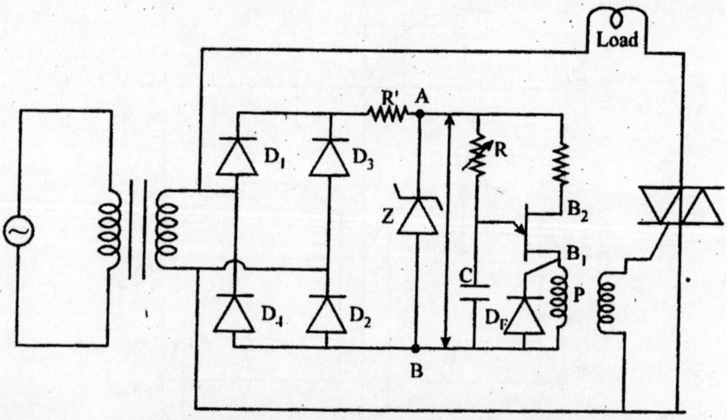

Fig. 5.6. (a) Phase control using UJT and triac.

Example 5.3. A UJT relaxation oscillator has potentiometer resistance 100 Ω and the capacitance of 4.7 μF as shown in Fig. 5.6 *(a)*. If total resistance of the UJT is 300 Ω and resistance of base is 150 Ω, calculate the firing angle. Input wave has frequency of 50 Hz.

Solution : Duration of the pulse is given by eqn. (4).

$$T = RC \ln\left(\frac{1}{1-\eta}\right)$$

Here
$$\eta = \frac{R_{B1}}{R_{BB}} = \frac{150}{300} = 0.5$$

$$T = 100 \times 4.7 \times 10^{-6} \ln\left(\frac{1}{0.5}\right)$$
$$= 0.47 \times 10^{-3} \ln 2$$
$$= 0.47 \times 10^{-3} \times 2.3 \times 0.3010$$
$$= 0.33 \text{ m sec.}$$

$$\text{Firing angle} = \omega t$$
$$= 50 \times 2\pi \times 0.33 \times 10^{-3} \text{ radian}$$
$$= 16.50 \times 2\pi \times 10^{-3} \text{ radian}$$
$$= .33\pi \times 10^{-3} \text{ radian} = 5.94°$$

Example 5.4. In the above example if peak input voltage is 200 V, calculate the average load voltage and average power dissipated by the bulb. Resistance of filament of bulb is 40 Ω.

Solution : Average load voltage,

$$V_{oav} = \frac{V_m}{\pi}(1+\cos\alpha) = \frac{200}{\pi}(1+\cos 5.94°) = 127.05 \text{ volts}$$

$$\text{Average power dissipated} = \frac{V_{oav}^2}{R} \frac{(127.5)^2}{40} = 403.5 \text{ watt } \textbf{Ans.}$$

5.5. APPLICATIONS

AC phase control methods find various applications e.g. AC power control, AC motor speed control, AC to DC converters, power intverter and uninterruptible power supply (UPS) etc. Some of the applications will be discussed in the following section.

1. AC voltage regulator* : AC voltage regulator converts constant AC into variable AC. Two methods are used for AC regulation — integral cycle control or on off control and phase control. We shall describe phase control method using SCR and triac.

1 ϕ half wave regulator : The circuit of a half wave regulator is shown in Fig. 5.7 *(a)*. As the name suggests, only half cycle can be controlled by using this circuit. During positive half cycle SCR T conducts when it is turned on by applying pulse at its gate. After the completion of positive half cycle it is naturally turned off and diode D conducts in negative half cycle. The waveforms are shown in Fig. 5.7 *(b)*. By changing firing angle or phase the power switched to the load can be controlled. We can write

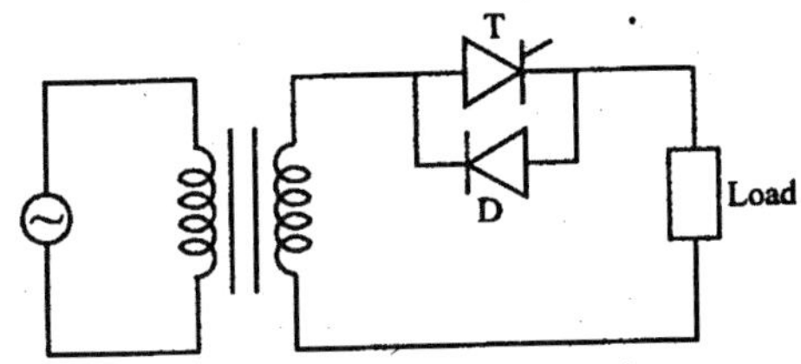

Fig. 5.7 (a) Half wave ac voltage regulator.

$$V_{oav} = \frac{1}{2\pi}\int_{\alpha}^{2\pi} V_m \sin\omega t = \frac{V_m}{2\pi}(\cos\alpha - 1)$$

rms load voltage is given as

$$V_{orms} = \left[\frac{1}{2\pi}\int_{\alpha}^{2\pi} V_m^2 \sin^2 \omega t\, d(\omega t)\right]^{0.5} = \frac{V_m}{2\sqrt{\pi}}\left[2\pi - \alpha + \frac{\sin\alpha}{2}\right]^{0.5}$$

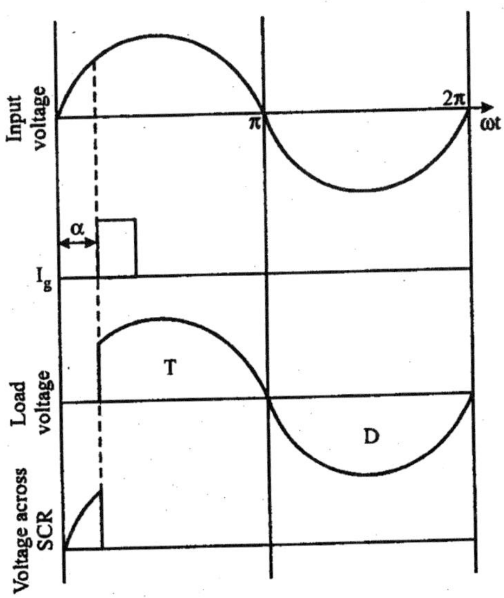

Fig. 5.7. (b) Waveforms.

---

*AC voltage regulator is sometimes called ac voltage controller as it gives controlled power to the load.

Single phase full wave regulator : For full wave regulation two SCRs or a triac can be used. The circuit of single phase full wave regulator using a triac is shown in Fig. 5.8 *(a)*. The waveforms are shown in Fig. 5.8 *(b)*.

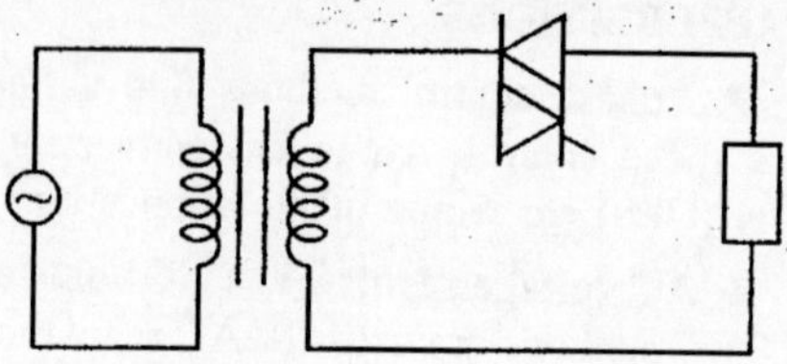

Fig. 5.8 (a) Full wave ac regulator.

Average load voltage is given by

$$V_{oav} = \frac{1}{2\pi}\int_{\alpha}^{2\pi} V_m \sin \omega t \; d(\omega t) = 0$$

rms load voltage. It is given by (for resistive load)

$$V_{orms} = \left[\frac{1}{2\pi}\int_{\alpha}^{2\pi} V_m^2 \sin^2 \omega t \; d(\omega t)\right]^{0.5}$$

$$= \left[\frac{1}{2\pi}\int_{\alpha}^{\pi} V_m^2 \sin^2 \omega t \; d(\omega t) + \int_{\pi+\alpha}^{2\pi} V_m^2 \sin^2 \omega t \; d(\omega t)\right]^{0.5}$$

$$= \frac{V_m}{\sqrt{2\pi}}\left[2\int_{\alpha}^{\pi} \frac{1-\cos 2\omega t}{2} d(\omega t)\right]^{0.5}$$

$$= \frac{V_m}{\sqrt{2\pi}}\left[\omega t\Big|_{\alpha}^{\pi} - \frac{1}{2}\sin 2\omega t\Big|_{\alpha}^{\pi}\right]^{0.5} = \frac{V_m}{\sqrt{2\pi}}\left[\pi - \alpha + \frac{1}{2}\sin 2\alpha\right]^{0.5}$$

$$= V_m\sqrt{\frac{\pi-\alpha}{2\pi} + \frac{\sin 2\alpha}{4\pi}} \qquad ...(5)$$

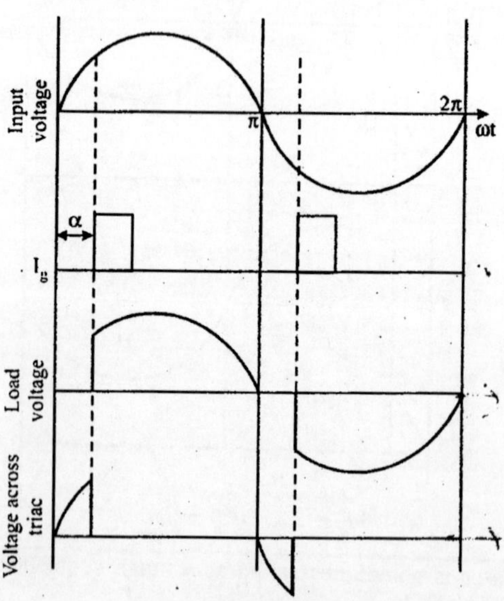

Fig. 5.8. (b) Waveforms.

Three phase full wave regulator : For AC regulation in three phase three triacs are used — one in each phase. The circuit is shown in Fig. 5.9.

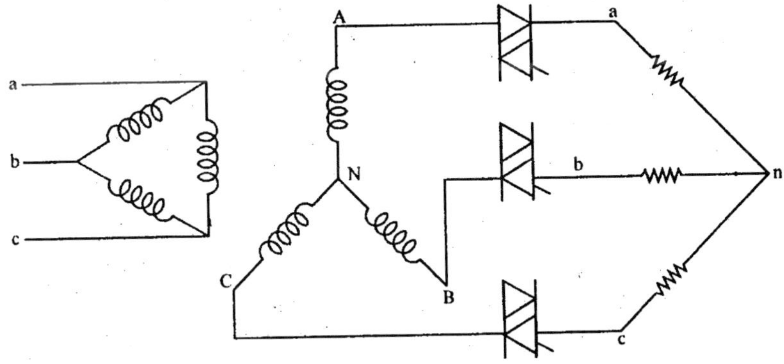

Fig. 5.9. Three phase ac regulator using triac.

2. AC phase control in AC to DC converter : AC to DC converter using SCR has been dealt with in chapter 4.

3. Illumination Control : Fig. 5.10 shows the circuit to control the illumination of incandescent lamp. The circuit has been described in Fig. 5.10. By changing the potentiometer resistance R, firing angle of the triac can be changed. Corresponding change in the conduction angle and hence in average power takes place. The rms power is given by

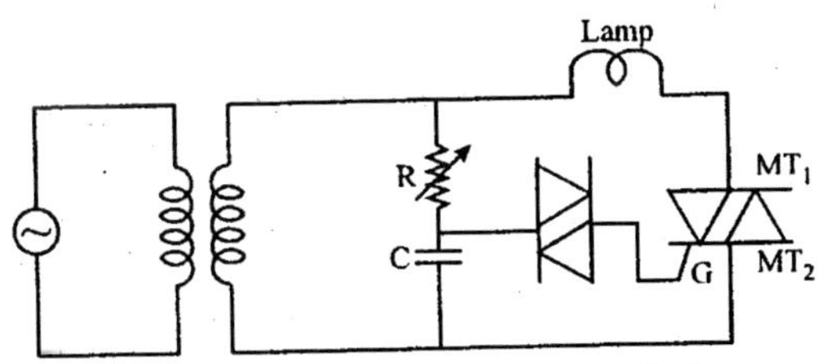

Fig. 5.10. Illumination control method.

$$P_{orms} = \text{rms voltage across load} \times \text{rms load current}$$

$$= V_m\sqrt{\frac{\pi-\alpha}{2\pi}+\frac{\sin 2\alpha}{4\pi}} \times \frac{V_m}{R}\sqrt{\frac{\pi-\alpha}{2\pi}+\frac{\sin 2\alpha}{4\pi}}$$

$$= \frac{V_m^2}{R}\left(\frac{\pi-\alpha}{2\pi}+\frac{\sin 2\alpha}{4\pi}\right) \quad ...(6)$$

Max rms power across the load is given by solving $\frac{dP_{rms}}{d\alpha} = 0$

$$\frac{dP_{orms}}{d\alpha} = \frac{V_m^2}{R}\left(-\frac{1}{2\pi}+\frac{2\cos 2\alpha}{4\pi}\right) = 0$$

or $$\frac{2\cos 2\alpha}{8\pi} = \frac{1}{4\pi}$$

or $$\cos 2\alpha = 1$$

or $$\alpha = 0.$$

Putting $\alpha = 0$ in eqn. (6) one obtains

$$P_{orms}/\max = \frac{V_m^2}{2R} \quad ...(7)$$

From above eqn. it is clear that the rms power delivered to the load can be varied by a variation in α and R.

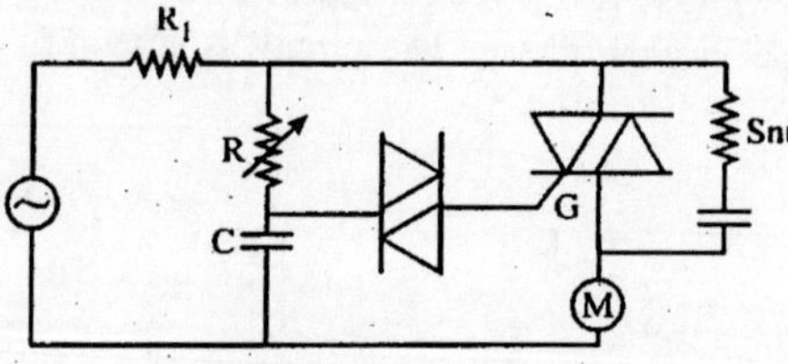

Fig. 5.11. Speed control schematic.

4. Speed control of single phase induction motor using triac (regulator) : Conventional speed regulator for a fan uses a resistance as regulator. The regulator resistance is in series with the motor. By increasing the resistance the voltage drop across it is increased so that the voltage developed and hence the speed of the fan is reduced. In this method the power loss across the regulator is higher at lower fan speed as voltage developed across the regulator is higher. This drawback is overcome by AC phase control method using DIAC-TIAC pair. The circuit is shown in Fig. 5.11. RC network forms the firing network. As the voltage across the capacitor triggers the diac it sends triggering pulse to the triac. When triac is turned ON the motor starts running. By changing the value of resistance, firing angle for triac is changed to control the power to the motor. The speed of the motor designated N r.p.m. is given by relation

$$N \propto V^2_{orms}$$

V_{orms} depends upon the firing angle.

Thus the speed of the motor can be controlled by changing potentiometer resistance R. Snubber circuit is used in parallel with the triac to protect it against unintended $\frac{dv}{dt}$ turn on.

5. AC phase control in UPS : UPS is the acronym for uninterruptible power supply, *i.e.,* a power supply which continues to feed the load even in event of failure of mains supply.

A number of equipments and processes such as computer systems, communication system, continuous process industries require constant voltage supply. Failure of mains supply can cause loss of data in computers, wastage of material in continuous process industries etc. Such loads are called controlled loads. These loads are fed from UPS. Main components of a UPS are rectifier, battery, inverter, static switches, filter, automatic voltage regulator.

The rectifier converts ac into dc. The dc supply charges the battery as well as feeds the inverter. The filter removes ripple from dc supply. Inverter feeds the controlled loads. The static switches connect / disconnect the required components. The automatic voltage regulator keeps the output ac voltage constant. It uses negative feedback to maintain constant input voltage.

The battery works as standby source of supply. It is generally nickel cadmium battery.

UPS can be classified as off line and on line.

Off line UPS : The block diagram of off line UPS is shown in Fig. 5.12.

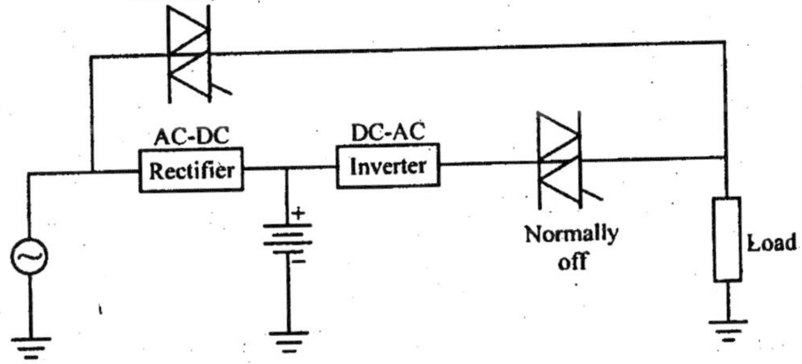

Fig. 5.12. OFF line UPS.

When mains ac supply is available the inverter remains off. In the event of failure of mains supply the inverter is automatically turned on by a static switch. Triac switches are used to control the power to the load. The drawback of this UPS is that there is some time lag between failure of mains supply and turning on of the inverter which may result in loss of data in the computer (load).

On line UPS : In this UPS the supply is fed always through the inverter. Triac acts like a controlled switch. In case the inverter becomes faulty, mains supply is directly connected to the load through the switch which remains normally off. The speed of on line UPS is higher.

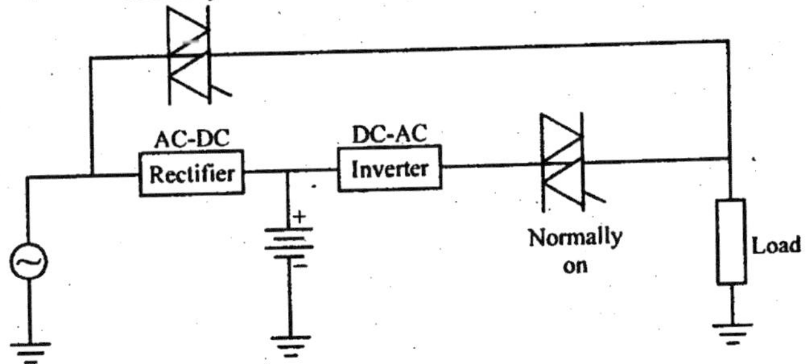

Fig. 5.13. ON line UPS.

Example 5.5. In an experiment on illumination control using triac an input voltage of 200 sin 100*t* is used, resistance of the filament of bulb used is 30 Ω. Calculate rms power dissipated when phase angle is 30°. Also plot the graph between rms power and firing angle.

Solution : rms power dissipated is

$$P_{orms} = \frac{V_{orms}^2}{R} = \frac{V_m^2}{R}\left[\frac{\pi-\alpha}{2\pi}+\frac{\sin 2\alpha}{4\pi}\right]$$

$$= \frac{(200)^2}{30}\left[\frac{5\pi}{12\pi}+\frac{\sin 60^\circ}{4\pi}\right]$$

$$\approx 640 \text{ watts.}$$

The graph between P_{orms} and α is shown in Fig. 5.14.

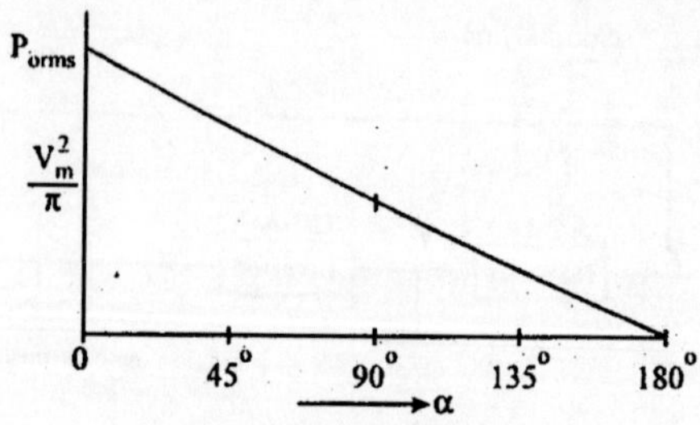

Fig. 5.14.

SUMMARY

Thyristor can be tuned ON in a number of ways. In this chapter methods employing ac inputs to turn ON the SCR have been discussed. These are called ac phase control methods. Details of the methods is given in the chapter.

EXERCISES

1. Write a short note on AC phase angle control.
2. Explain basic principle of phase angle control.
3. Draw the circuit of full wave phase angle control using UJT, what is the drawback of this circuit ? Explain.
4. Explain the full wave phase angle control using UJT, pulse transformer and triac.
5. What is the application of phase angle control in AC voltage regulation ? Explain.
6. Explain how illumination of incandescent lamp is controlled by using phase angle control method ?
7. What is the limitation of speed control of single phase induction motor using variable resistance ? Explain.
8. Draw the circuit of fan regulator using triac. Describe the circuit operation.
9. What is the use of AC phase control in UPS ? Describe off line UPS.
10. What are limitation of off line UPS ? How is it overcome by using on line UPS ?

NUMERICAL PROBLEMS

1 In a half wave AC phase control method, input voltage has the value 100 V at 50 Hz. What is the average voltage across the load if R = 100 Ω, C = 15 μF ?

2. In Fig. 5.5 *(a)*, the value of R and C are 30 Ω and 4.7 μF respectively. If input voltage is 70 V at 50 Hz. calculate rms load voltage. Intrinsic stand off ratio = 0.5.

3. In a single phase half wave regulator firing angle of thyristor is 30°. Calculate the average power at the load. Assume load resistance 10 Ω, input voltage 100 V at 50 Hz.

4. A single phase induction motor has winding resistance of 20 Ω. Its rotor speed at firing angle α_1 is N_1 and at angle α_2 is N_2. What is the relationship between N_1 and N_2. Draw graph between N and α.

OBJECTIVE TYPE QUESTIONS

1. A triac is a

(a) four layer three terminal device

(b) four layer, three terminal, three junction device

(c) three layer, three terminal, three junction device

(d) four layer, two terminal, three junction device

2. In phase control method

(a) output is controlled by changing firing angle

(b) phase difference always occurs between input and output voltage

(c) the output is controlled by phase difference between load voltage and load current

(d) output depends upon input power factor.

3. Phase angle control is used for

(a) ac power (b) dc power

(c) ac and dc both (d) none of the above

4. In illumination control R-C network is used

(a) to control impedance of gate circuit

(b) to change phase angle between load voltage and load current

(c) to control power to the load

(d) none of the above

5. Pulse transformer is used

(a) to produce pulses (b) to isolate gate circuit from anode circuit

(c) to step up or step down voltage (d) to reduce wastage of power in the gate circuit

6. Diac is a

(a) bidirectional switch (b) unidirectional switch

(c) three terminal, three layer device

(d) three terminal four layer, three junction device

7. AC regulator

(a) converts ac into ac (b) keeps the output constant

(c) converts constant ac into variable ac

(d) keeps ac constant.

8. In fan speed control method

(a) RC network change the inpedence of induction motor

(b) RC network controls the conduction angle

(c) Triac resistance controls the current through motor

(d) RC network changes the impedance of triac as it is connected in parallel with triac.

9. The frequency of firing pulses generated by UJT is

(a) Approximately $\dfrac{1}{RC \ln\left(\dfrac{1}{1-\eta}\right)}$ (b) Exactly $\dfrac{1}{RC \ln\left(\dfrac{1}{1-\eta}\right)}$

(c) $RC \ln (1 - \eta)$ (d) none of the above.

10. η has the values

(a) between 0.5 and 0.8 (b) 0.5

(c) not defined (d) any value.

11. The drawback of OFF line UPS is

(a) The time lag between the failure of main supply and turn on of the UPS is large

(b) It leads to the loss of data in the computer

(c) both (a) and (b)

(d) only (b).

ANSWERS

1. (b)	2. (a)	3. (a)	4. (c)	5. (b)	6. (a)
7. (c)	8. (b)	9. (a)	10. (a)	11. (c)	

❑❑❑

6

Applications of Phase Controlled Rectification And AC Phase Control Circuits

6.0. INTRODUCTION

In chapters 4 and 5 we have studied in detail the rectifiers and ac phase control methods. These are few of the very useful tools in our hands to control the power to the load. In this chapter some of the applications of controlled rectification and ac phase control will be discussed.

A phase controlled rectifier gives controlled dc output for ac input. This output can be made use of for driving any load such as dc motors, lamp etc. In what follows we shall discuss usefulness of this method to drive the load.

6.1. ILLUMINATION CONTROL BY USING THE PHASE CONTROLLED RECTIFICATION

We shall discuss this application under the following heads.

1. Half wave controlled rectifier or single pulse rectifier to control illumination: The circuit is shown in Fig. 6.1 *(a)*. The thyristor T conducts for positive half cycle. The firing angle can be changed by changing the potentiometer resistance R. When the SCR is turned ON the current will flow through the bulb and it starts glowing. At zero crossing the SCR is turned OFF for negative half cycle. Average load voltage is given by

(a) Circuit. *(b) Waveform.*

Fig. 6.1. Illumination control.

$$V_{oav} = \frac{V_m}{2\pi}(1+\cos\alpha)$$

and the average power

$$p_{av} = V_{oav} \times I_{oav}.$$

$$= \frac{V_{oav} \times V_{oav}}{R_b}, R_b \text{ is bulb resistance}$$

$$P_{av} = \frac{\left[\frac{V_m}{2\pi}(1+\cos\alpha)\right]^2}{R_b} \quad ...(1)$$

The graph between average power and firing angle is shown Fig. 6.1 *(b)* which shows that the power to the load can be controlled by changing phase angle or firing angle.

2. Illumination control by mid point full wave fully controlled rectifier : The circuit is shown in Fig 6.2 *(a)*. This circuit switches controlled power to the load in positive as

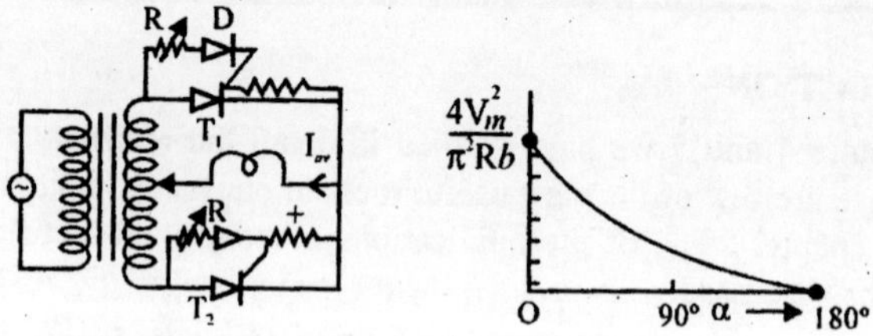

(a) Circuit. *(b) Waveform.*

Fig. 6.2. Illumination control.

well as negative half cycle SCR T_1 conducts for positive half cycle while SCR T_2 conducts for negative half cycle. Firing angle for T_1 and T_2 may be different or same. In the circuit the load is connected to midpoint of the secondary of the transformer and two pulses are obtained in one cycle. It is, therefore called M-2 converter also. The direction of the current through the lamp is shown in the Fig. The average load voltage is given by

$$V_{oav} = \frac{V_m}{\pi}(1+\cos\alpha).$$

and average power $$P_{av} = \frac{\left[\frac{V_m}{\pi}(1+\cos\alpha)\right]^2}{R_b} \quad ...(2)$$

The graph between average load power and firing angle is given in Fig 6.2 *(b)*.

3. Illumination control by using half controlled bridge converter : The circuit is shown in Fig. 6.3 *(a)*. The operation is the same as that of mid point full wave rectifier. The average output power is plotted with α in Fig. 6.3 *(b)*.

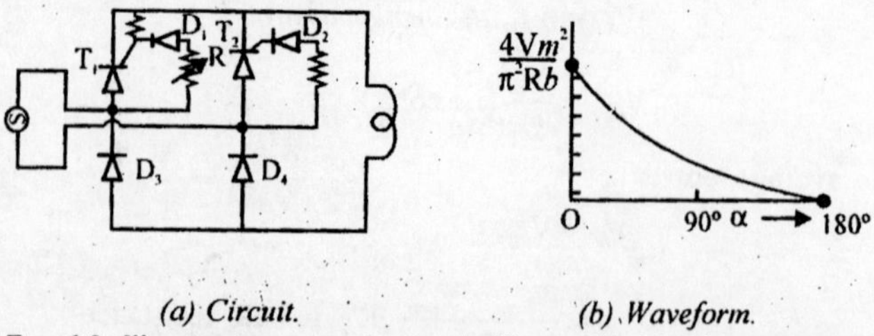

(a) Circuit. *(b) Waveform.*

Fig. 6.3. Illumination control using half controlled bridge converter.

6.2. ILLUMINATION CONTROL BY PHASE ANGLE CONTROL METHODS

In chapter 5 we have discussed illumination control by phase angle control using triac. Few other methods will be discussed is this section, to control illumination of a lamp. The methods are given as under:

1. Full wave a.c. regulator for illumination control : The circuit is shown in Fig 6.4 *(a)*. SCR, T_1 conducts for positive while SCR T_2 conducts for negative half cycle. The

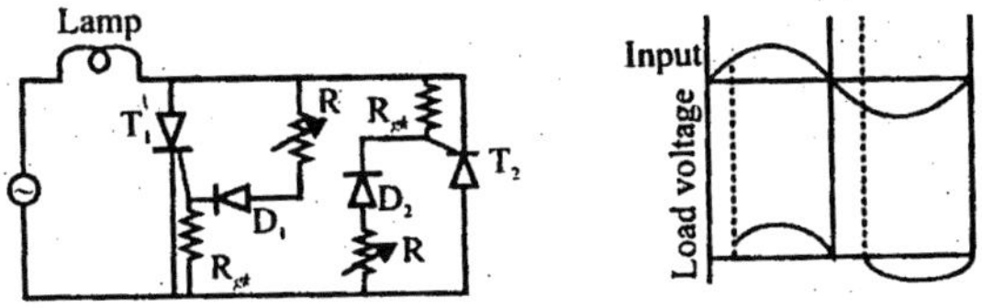

(a) Circuit. *(b) waveforms.*

Fig. 6.4. Illumination control by ac regulator.

direction of current through lamp is opposite in both half cycles. The waveforms are shown in Fig 6.4 *(b)*. The rms value of load voltage is given by

$$V_{orms} = \left[\frac{1}{\pi}\int_{\alpha}^{\pi} V_m{}^2 \sin^2 \omega t (d\omega t)\right]^{1/2}$$

$$= \frac{V_m}{\sqrt{2\pi}}\left[\pi - \alpha + \frac{1}{2}\sin 2\alpha\right]^{0.5}$$

The power dissipated by the load is

$$P = \frac{\left(\frac{V_m}{\sqrt{2\pi}}\right)^2 \left[\pi - \alpha + \frac{1}{2}\sin 2\alpha\right]}{R_h}$$

where R_h = lamp resistance $= \frac{V_m^2}{2\pi R_h}\left(\pi - \alpha + \frac{1}{2}\sin 2\alpha\right)$ ..(3)

A change in potentiometer resistance R causes a change in the instantaneous value of input voltage at which the thyristor is turned ON. It means the firing angle changes with a change in R. At low value of R the minimum gate current required to trigger the SCR is obtained at low voltage or low firing angle As a result most of the input power appears across the load and it glows brighter. At high resistance value the firing angle is higher. Hence a small fraction of input power appears across the load. Consequently it glows dimmer. Thus variation in phase angle causes a corresponding change in the illumination level.

Example 6.1. The resistance of filament of a lamp is 40 Ω. It is connected in the output of an ac regulator having input voltage 200 sin 100 t. What is the power dissipated by the lamp at firing angle of *(a)* 30° *(b)* 60° ?

Solution : *(a)* $V_m = 200$ volt

Power dissipated $= \frac{V_m^2}{2\pi R_h}\left(\pi - \alpha + \frac{1}{2}\sin 2\alpha\right)$

$$= \frac{(200)^2}{2\pi \times 40}\left(\pi - \pi/6 + \frac{1}{2}\sin 60^\circ\right)$$

$$p_1 = \frac{500}{\pi}\left(\frac{5\pi}{6} + \frac{1}{2} \times \frac{\sqrt{3}}{2}\right) \text{ watt} = 485.54 \text{ watt}$$

$$p_2 = \frac{(200)^2}{2\pi \times 40}\left(\pi - \pi/3 + \frac{1}{2}\sin 120^\circ\right)$$

$$= \frac{500}{\pi}\left(\frac{2\pi}{3} + \frac{1}{2} \times \frac{\sqrt{3}}{2}\right)$$

$$= 402.20 \text{ watt}$$

2. Illumination control by three phase a.c. regulator : The circuit of a three phase full wave regulator to control the illumination of three lamps is shown in Fig. 6.5.

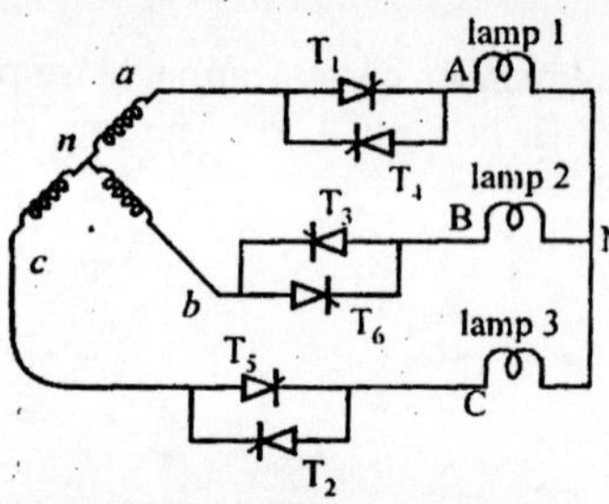

Fig. 6.5. Illumination control using three phase ac regulator.

Thyristors T_1, T_3, T_5 conduct for positive half cycle while thyristors T_2, T_4, T_6 conduct for negative half cycles. By changing firing angle the power to the lamps can be controlled as in case of single phase regulator.

3. Illumination control by using chopper : A chopper circuit used to control illumination level of any lamp is shown in Fig 6.6. By changing turn ON time of the transistor switch the duration of current and hence the average power dissipated by the lamp may be changed. Let T_{ON} be the duration of conduction of transistor and T_{OFF} be duration of OFF state then output voltage is given by

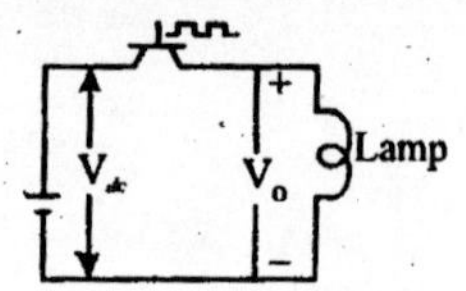

Fig. 6.6. Use of chopper in illumination control.

$$V_0 = \frac{1}{T}\int_0^{T_{ON}} V_{dc}\,dt$$

$$= \frac{V_{dc}\,T_{ON}}{T} \quad \text{...(4)}$$

where $T = T_{ON} + T_{OFF}$ and $\frac{T_{ON}}{T} = \alpha$, where α is called duty cycle.

Average power dissipated will be

$$P_{av} = \frac{V_0^2}{R_h} = \frac{(\alpha V_{dc})^2}{R_h}$$

$$P_{av} = \frac{\alpha^2 V_{dc}^{\,2}}{R_h} \quad \text{...(5)}$$

It is clear from eqn. (5) that the illumination level can be controlled by changing the duty cycle α.

6.3. FAN SPEED CONTROL

The working principle of fan speed control is the same as that of illumination control. Various methods used for controlling the fan speed are described below:

1. Speed control of fan, using dc motor, by half wave controlled rectifier : The circuit is shown in Fig 6.7 *(a)*. RC network is used to control the firing angle of the SCR T. At different values of product RC the firing angle and hence the voltage across the fan are different. The speed of the fan which depends upon the voltage is thus controlled. The load voltage is given by

$$V_{oav} = \frac{v_m}{2\pi}(1+\cos\alpha)$$

$$= \frac{v_m}{2\pi} + \frac{v_m}{2\pi}\cos\alpha$$

It is clear that speed of the fan

$N \propto \cos\alpha$

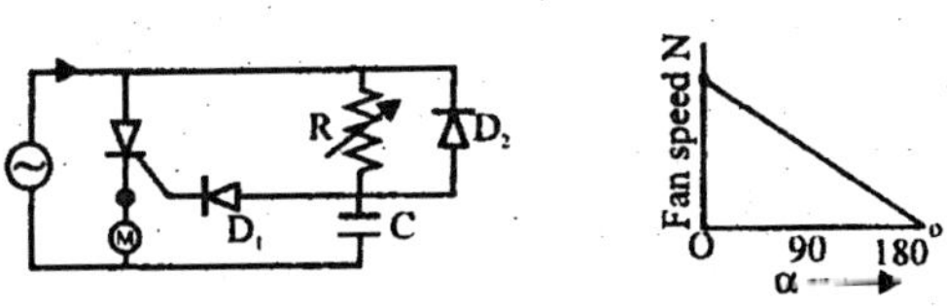

(a) (b)

Fig. 6.7. Speed control of fan.

The graph representing the variation of speed with firing angle is shown in Fig. 6.7 *(b)*. One drawback of this method is that the presence of harmonics in the output generates the heat in the fan. The harmonics should be reduced to minimum value so that wastage of power can be avoided. Maximum power to the motor should be contributed by fundamental component which is the useful power.

2. Fan speed control by using DIAC-TRIAC pair : The circuit is given in Fig 5.10 under section 5.5.

3. Speed control by using half wave AC regulators : The circuit of ac regulator to control fan spped is shown in Fig 6.8. In the Fig. half wave AC regulator has been shown where SCR T conducts for positive half while diode D_1 conducts for negative half of input cycle. By changing the potentiometer resistance R, voltage across fan motor changes which controls the speed of the fan.

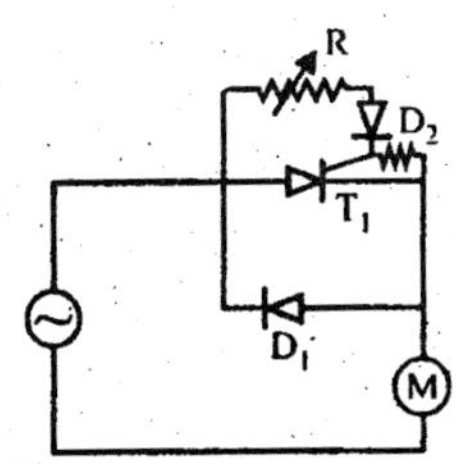

Fig. 6.8. Half wave ac regulator to control speed.

4. Fan speed control by full wave ac regulator : By using this method large variation in fan speed can be attained. The circuit is shown in Fig. 6.9. Thyristor T_1 controls the power to fan motor in positive half cycle while thyristor T_2 works in negative half of input cycle.

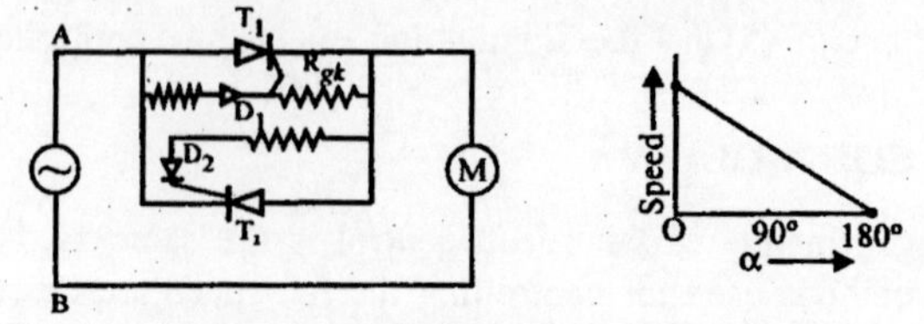

(a) (b)

Fig. 6.9. Speed control full wave ac regulator.

Operation : When positive cycle appears at point A thyristor T_1 is forward biased. It can be turned ON at desired angle by changing potentiometer resistance R. As this SCR is turned ON the fan starts running. At zero crossing this SCR is turned OFF and thyristor T_2 is turned ON at the disired angle in the negative half cycle. The speed of the motor depends upon the voltage across it which can be adjusted by changing the conduction angle. The variation of speed with firing angle is expressed as,

$N \propto \sin \alpha$. Here α is firing angle.

The graph between α and speed is shown in Fig. 6.9 *(b)*. Various other methods can be used to control the speed of different motor. In the above analysis we have taken into consideration single phase induction motor only. Other methods to control speed of DC motors and AC motors will be discussed in the later sections of this chapter.

6.4. TEMPERATURE CONTROL

The temperature of any object (load) can be controlled by controlled electric heating. Various methods can be used to control the temp of any object. Some of the methods used to control the temperature manually are described as under.

1. Temp control by using controlled rectifier : The circuits to control the temp of any object by using half wave and full wave controlled rectification are shown in Fig. 6.10 *(a), (b)* and *(c)* respectively. Half wave controlled rectifier is not used much as most of the power remains unused. Fig 6.10 *(b)* shows centre tapped full wave rectifier. In the circuit SCR T_1 conducts for positive cycle when it is turned ON by applying pulse to its gate whereas thyristor T_2 is fired to conduct in negative cycle. Fig. 6.10 *(c)* shows full wave controlled bridge rectifier. Thyristors T_1 and T_2 supply power to the heater in positive half cycle while thyristor pair T_3, T_4 in the negative half cycle.

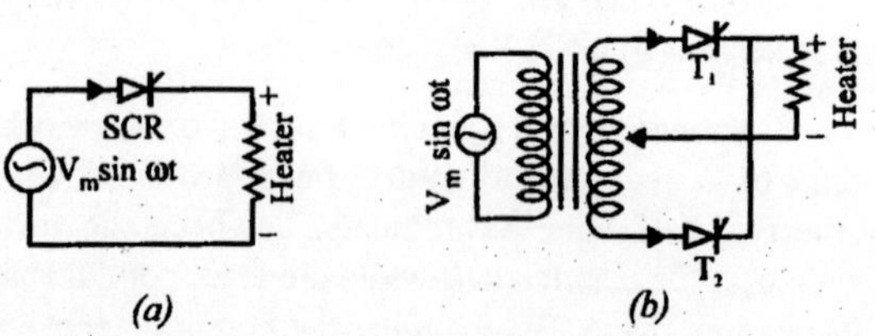

(a) (b)

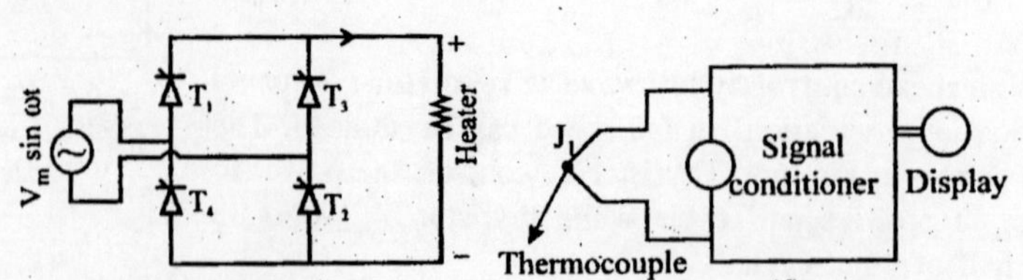

(c) (d)

Fig. 6.10. Temperature control using controlled rectifier.

Average load voltage for half wave rectifier is given by

$$V_{oav} = \frac{V_m}{2\pi}(1+\cos\alpha)$$

And for full wave controlled rectifier

$$V_{oav} = \frac{V_m}{\pi}(1+\cos\alpha).$$

The power dissipated by the heater will then be,

$$P_{oav} = \frac{V_{oav}^2}{R_h} \quad \text{...(6)}$$

where R_h is the resistance of heater element. If energy radiated per unit area per unit time by the heater is E, then by using Stephen Boltzman's law, we can write

$$E = \sigma\left(T^4 - T_o^4\right) \quad \text{...(7)}$$

Here T is the temp of heater element and T_o is that of surroundings Energy radiated for area A.

$$E_A = \sigma\left(T^4 - T_o^4\right)A$$

Using the definition of power, we get from eqn. (6) and (7).

$$P_{oav} = \sigma\left(T^4 - T_o^4\right)A \quad \text{...(8)}$$

or
$$\frac{V_{oav}^2}{R_h} = \sigma\left(T^4 - T_o^4\right)A$$

or
$$\frac{V_m^2}{4\pi^2 R_h}(1+\cos\alpha)^2 = \sigma\left(T^4 - T_o^4\right)A \quad \text{...(9)}$$

Eqn. (9) is for half wave rectifier.

For full wave rectifier eqn. (8) is changed to

$$\frac{V_m^2}{\pi^2 R_h}(1+\cos\alpha)^2 = \sigma\left(T^4 - T_o^4\right)A \quad \text{...(10)}$$

From eqn. (8) and (9) we find that

$$T^4 \alpha (1+\cos\alpha)^2$$

or
$$T \alpha (1+\cos\alpha)^{1/2}$$

Thus by changing the firing angle the temp of the body can be controlled. It can be measured by placing a suitable temp meter either in contact or in viscinity of the specimen as shown in Fig. 6.10 *(d)*.

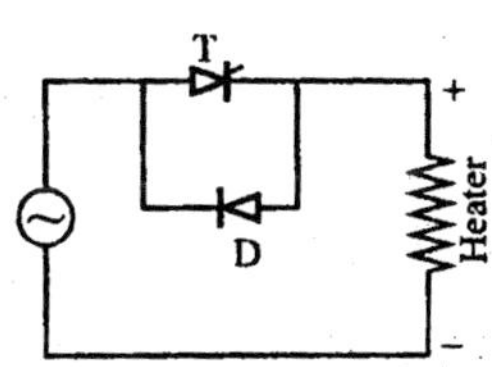

Fig. 6.11. Temperature control by half wave ac regulator.

2. Temperature control by half wave AC regualtor : The circuit to control temperature by using, half wave AC regualator is shown in Fig 6.11. Diode conducts during negative cycle which is not controlled. The thyristor conducts for positive cycle which is controllable. The rms value of load voltage is given by

$$V_{orms} = \left[\frac{1}{2\pi}\int_{\alpha}^{2\pi} V^2 \omega t\ d(\omega t)\right]^{0.5}$$

$$= \frac{V_m}{\sqrt{2\pi}}\left[\int_{\alpha}^{\pi}\sin^2\omega t + \int_{\pi}^{2\pi}\sin^2\omega t\, d\omega t\right]^{0.5}$$

$$= \frac{V_m}{\sqrt{2\pi}}\left[\int_{\alpha}^{\pi}\frac{1-\cos 2\omega t}{2}d(\omega t) + \int_{\pi}^{2\pi}\frac{1-\cos 2\omega t}{2}d(\omega t)\right]^{0.5}$$

$$= \frac{V_m}{\sqrt{2\pi \times 2}}\left[\left(\omega t - \frac{1}{2}\sin 2\omega t\right)_{\alpha}^{\pi} + \left(\omega t - \frac{1}{2}\sin 2\omega t\right)_{\pi}^{2\pi}\right]^{0.5}$$

$$= \frac{V_m}{2\sqrt{\pi}}\left[\pi - \alpha - \frac{1}{2}\{\sin\pi - \sin\alpha\} + (2\pi - \pi) - \frac{1}{2}(\sin 2\pi - \sin\pi)\right]^{0.5}$$

$$= \frac{V_m}{2\sqrt{\pi}}\left[\pi - \alpha + \frac{1}{2}\sin 2\alpha + \pi - 0\right]^{0.5}$$

$$\boxed{V_{orms} = \frac{V_m}{2\sqrt{\pi}}\left[2\pi - \alpha + \frac{\sin 2\alpha}{2}\right]^{0.5}} \quad ...(11)$$

Eqn. (11) gives the rms value of load voltage. The power dissipated by the heater will be

$$P_{oav} = \frac{V_{orms}^2}{R_h}$$

The temp of the heater element is agian given by using eqn. (8) which gives.

$$\frac{V_{orms}^2}{R_h} = \sigma\left(T^4 - T_o^4\right)A$$

From eqn. (11) putting the value of V_{orms} we get

$$\frac{V_m^2}{4\pi R_h}\left[2\pi - \alpha + \frac{\sin 2\alpha}{2}\right] = \sigma\left(T^4 - T_o^4\right)A \quad ...(12)$$

Eqn. (12) gives the relationship between temp of the object and the firing angle α. By changing the firing angle the temp of the object can be controlled.

3. Temp control by full wave regulator : The circuit is shown is Fig 6.12 *(a)* and *(b)*. In Fig. 6.12 *(a)* two thyristors are used which conduct for half cycle of the input. The two thyristors can be replaced by a single triac as shown in Fig. 6.12 *(b)*. The waveforms are shown in Fig 6.12 *(c)*.

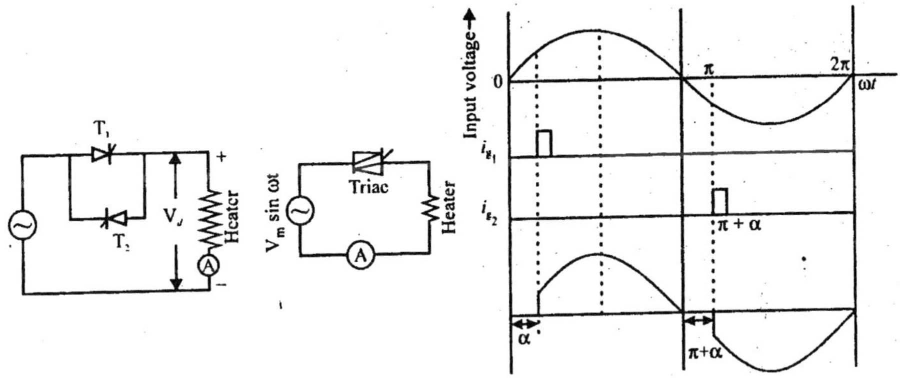

(a) Circuit (b) Circuit. (c) Waveforms.

Fig. 6.12. Temperature control by full wave ac regulator.

The rms value of load voltage is given by

$$V_{orms} = \frac{V_m}{\sqrt{2\pi}}\left[\pi - \alpha + \frac{\sin 2\alpha}{2}\right]^{0.5}$$

By carrying the same analysis as in case of half wave a.c. regulator, we get

$$\frac{V_m^2}{2\pi R_h}\left[\pi - \alpha + \frac{\sin\alpha}{2}\right] = \sigma\left(T^4 - T_o^4\right)A \qquad ...(13)$$

Example 6.2. A Chopper is used to control the illumination of a lamp. The dc source has e.m.f. of 200 V, duty cycle of pulse is 0.7. What is the average power dissipated if the lamp has a resistance of 30 Ω

Solution : The average output voltage of a chopper is given by

$$V_{oav} = \alpha\, Vdc = 0.7 \times 200 = 140 \text{ V}$$

Average power $P_{av} = \frac{V_{0av}^2}{R_h}$, R_h the resistance of lamp.

$$= \frac{(140)^2}{30} \text{ watt}$$

$$= \frac{19600}{30} = 653.3 \text{ watt} \quad \textbf{Ans.}$$

Example 6.3. An electric bulb is illuminated by a single phase full wave regulator. The intensity of light at a distance of 10 cm from the bulb is I units for firing angle of 30°. What will be the intensity if the firing angle is 60°.

Solution : Intensity at any point α power dissipated.

Hence from eqn. (3), we get

$$I \alpha P_{oav} \alpha \frac{V_m^2}{2\pi R_b}\left(\pi - \alpha + \frac{\sin 2\alpha}{2}\right)$$

or
$$I \alpha \left(\pi - \alpha + \frac{sin^2 \alpha}{2}\right)$$

using first condition

$$I\alpha \left(\pi - \pi/6 + \frac{\sin 60^0}{2}\right) \quad ...(14)$$

Using second condition, one obtains.

$$I^1 \alpha \left(\pi - \pi/3 + \frac{\sin 120^0}{2}\right)$$

$$\alpha \left(2\pi/3 + \frac{\cos 30^0}{2}\right) \quad ...(15)$$

Divide eqn. (14) by eqn. (15) we obtain.

$$\frac{I}{I^1} = \frac{(5\pi/6 + \sqrt{3}/4)}{(2\pi/3 + \sqrt{3}/4)}$$

or
$$I^1 = \left(\frac{2\pi/3 + \sqrt{3}/4}{5\pi/6 + \sqrt{3}/4}\right) I = 0.45\ I$$

Example 6.4. Speed of a table fan controlled by half wave rectifier at a firing angle α is 600 r.p.m. At 60° the speed is 470 r.p.m. what is the value of α ?

Solution : Speed
$$N \propto (1 + \cos \alpha)$$

$$600 \propto (1 + \cos \alpha) \quad ...(16)$$

Under second condition

$$470 \propto (1 + \cos 60°) \quad ...(17)$$

Divide eqn. (16) by eqn. (17), we obtain

$$\frac{600}{470} = \frac{\cos\alpha + 1}{\cos 60^0 + 1}$$

or
$$\cos \alpha = \frac{600 \times \cos 60^0 + 600 - 470}{470}$$

or
$$\cos \alpha = \frac{300 + 130}{470} = \frac{43}{47} \text{ or } \alpha = 23.8° \textbf{ Ans.}$$

Example 6.5. A voltage 100 sin 120 t is fed as input to a full wave regulator which is used to control the speed of a fan. The triac is fired at an angle of 45°. What is the ratio of power to the motor at this angle and 0°?

Solution : rms value of load voltage when firing angle 45° is

$$V_{orms_{45^\circ}} = \frac{v_m}{\sqrt{2\pi}}\left[\pi - \alpha + \frac{\sin 2\alpha}{2}\right]^{0.5}$$

Power fed to the motor

$$P_{0_{45^\circ}} = \frac{V_m^2}{2\pi R_f}\left(\pi - \alpha + \frac{\sin 2\alpha}{2}\right) \quad ...(18)$$

Here, R_f is resistance of fan.

when firing takes place at 0°

$$P_{0_{0^\circ}} = \frac{v_m^2}{2 R_f} \quad ...(19)$$

Dividing eqn. (18) by eqn. (19), one gets.

$$\frac{P_{0_{45^\circ}}}{P_{0_{0^\circ}}} = \frac{\dfrac{V_m^2}{2\pi R_f}\left(\pi - \alpha + \dfrac{\sin 2\alpha}{2}\right)}{\dfrac{V_m^2}{2R_f}}$$

$$= \frac{\pi - 2 + \dfrac{\sin 2\alpha}{2}}{\pi} = \frac{\pi - \pi/4 + \dfrac{\sin 90}{2}}{\pi}$$

$$= \frac{3\pi/4 + \dfrac{1}{2}}{\pi} = \frac{2.84}{\pi} = 0.91$$

Above example is taken to show how the power is transferred and hence the temperature of load changes with change in the firing angle.

6.5. AUTOMATIC TEMPERATURE CONTROL

The methods discussed under section 6.4. to control temp are manual. The temperature of the heater element is measured by any temperature measuring device and corresponding change in the firing angle is made. Such methods are not very accurate. To ensure reliability of the method automatic control system is used. The block diagram of automatic control system is given in Fig 6.13.

Temp sensor converts the temp into equivalent electrical quantity which is compared with the electrical signal corresponding to Ref temperature. If measured temperature is more than the desired temperature cooler is turned ON to reduce the temperature of the system (furnace, oven etc). In case the temp of the system under test is less than desired

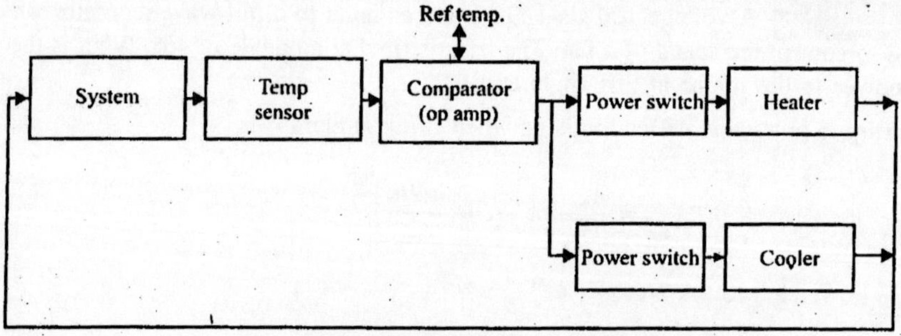

Fig. 6.13. Automatic temperature control.

one, the heater is turned ON to increase the temp of the system. If both the temp are same, neither heater nor cooler is turned ON and the system remains under the desired temperature. In what follows we shall discuss the phase control method for regulation of temperature.

1. Phase control method to control temp : The circuit is shown in Fig. 6.14. Heater and thermistor having resistance R_{Th} are placed in an oven or other system whose temperature is to be controlled. Resistance R_1 is used to set the desired temperature marked on the panel of the oven. We use a zener diode Z_1 of rating decided by the breakover voltage of diac for the purpose of making the voltage free of fluctuations. The voltage fixed by zener diode appears across thermistor R_{Th} and charges capacitor C. When capacitor is charged to a voltage equal to the breakover voltage of diac it sends triggering pulse to the triac, which connects the heater in the circuit. As the temperature of the oven is more than the set temperature, the thermistor resistance decreases. As a result the voltage across capacitor goes low. The voltage across the capacitor is maintained fixed by higher input voltage. Thus instantaneous value of the voltage in the input should be higher which is equivalent to saying that the phase angle is increased. With increased phase angle, the voltage across heater is decreased thereby lowering the heat generated and hence the temperature. In case the temp of the oven decreases, the resistance of the thermistor R_{Th} increases and as such the voltage across the capacitor C inceases. The firing angle decreases to maintain the desired voltage across capacitor. As a result the voltage across heater increases thereby increasing the heat generated and hence temperature. Thus the temperature of the system is maintained constant.

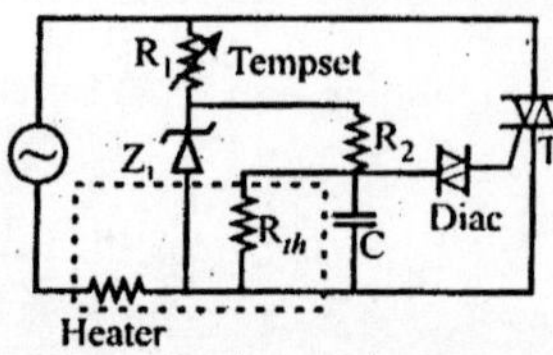

Fig. 6.14. Temp control by phase control method.

Example 6.6. Temperature inside an electric oven is maintained at 80°C by using diac-triac pair as shown in Fig. 6.14. At this temp the resistance of thermistor is 20 Ω. If the temp is set to 60°C what will be change in thermistor resistance. Assume constant '*a*' in thermistor eqn. equal to 0.91 Ω.

Solution : The equation giving the relationship between thermister resistance with temp is given by

$$R_{TH} = ae^{b/T}$$

where $a = 0.91\ \Omega$.

In first case

$$\therefore R_{TH} = R_{TH_1} \text{ say } 20\ \Omega$$

$$T = 80 + 273 = 353 \text{ K}$$

Putting the values $20 = 0.91 e \frac{b}{353}$

Solving, we get

$$b = 1090.78\,\Omega.$$

under 2nd condition,

Temperature = 60°C = 333 K

Hence resistance

$$R_{TH_2} = 0.91 e \frac{1090.78}{333} = 24\,\Omega$$

Change in resistance $R_{Th_2} - R_{Th_1} = 24 - 20 = 4\Omega$.

Example 6.7. A full wave ac regulator is fed by a voltage of 220 V at 50Hz. What is the power dissipated by a lamp having resistance 100 Ω when the firing angle for positive and negative cycle are 30° and 60° respectively?

Solution : The rms load voltage in this case will be different for each half cycle. Total r.m.s. voltage is given by

$$V_{orms} = \frac{V_m}{2\sqrt{\pi}}\left[2\pi - \alpha_1 - \alpha_2 + \frac{\sin 2\alpha_1}{2} + \frac{\sin 2\alpha_2}{2}\right]^{0.5}$$

Putting the values of variables, we get

$$V_{orms} = \frac{220\sqrt{2}}{2\sqrt{\pi}}\left[2\pi - \frac{\pi}{6} - \frac{\pi}{3} + \frac{\sin 60}{2} + \frac{\sin 120}{2}\right]^{0.5}$$

$$= \frac{220}{\sqrt{2\pi}}\left[\frac{3\pi}{2} + \frac{\sqrt{3}}{4} + \frac{\sqrt{3}}{4}\right]^{0.5} = 610.7 \text{ volts}$$

6.6. SPEED CONTROL OF DC MOTORS

We are familiar with a number of ways to control the speed of dc and ac motors. For example armature control method, flux control method etc. These methods to control the speed are not very accurate and reliable and are now being replaced by semiconductor devices such as thyristor, triac etc. In this section we shall have discussion on some methods employing semiconductor devices to control the speed of the motors

D.C. motor fundamentals : The torque developed by dc motor when current flows through it is given by

$$T = \frac{Z\phi I_a P}{2\pi A} \qquad ...(20)$$

where Z = No of armature conductors.

p = no of poles

A = Number of parallel paths.

ϕ = flux per pole, wb.

I_a = Armature current. Amp.

The induced e.m.f. in the conductor as they rotate in magnetic field, called back e.m.f., is given by

$$E_B = \frac{N\phi PZ}{60\,A} \quad ...(21)$$

where E_B = back emf,

N = speed in r.p.m

Z = number of conductors

P = number of poles

A = no of parallel paths which are

= 2 for wave winding.

= p for lap winding.

For a separately excited dc shunt motor, the KVL eqn. is given as

$$V = E_B + I_a R_a + \text{brush contact voltage drop.}$$

$$= \frac{N\phi PZ}{60\,A} + I_a R_a + \text{ brush contact voltage drop} \quad ...(22)$$

where I_a = armature current

and R_a = armature resistance.

In series motor field coil and armature are in series, the KVL for this motor is written as.

$$V = E_b + I_a(R_a + R_{se}) + \text{ brush contact voltage drop} \quad ...(23)$$

We observe from eq (20) that the torque varies a I_a^2 in series motor. For this reason it is used for heavy loads such as electric traction.

Various methods will be discussed to drive dc motors. The methods which are generally used for speed control of DC motor are described below:

2. Speed control of dc motor by single phase half wave converter : The circuit to control the speed of separately excitied dc motor is shown in Fig. 6.15. AC-DC controlled converter is used to excite field coil as well as armature of the motor. We generally avoid

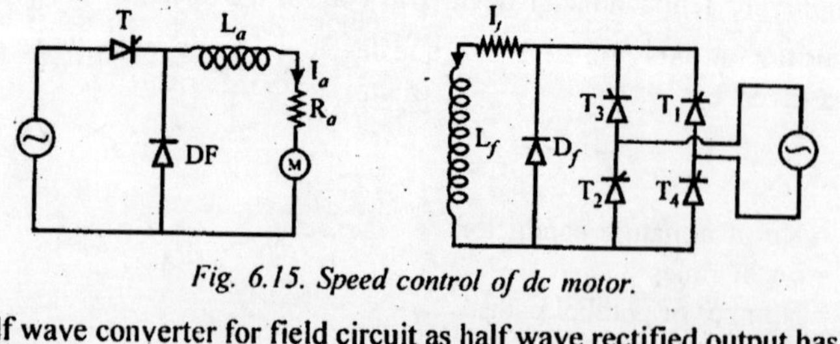

Fig. 6.15. Speed control of dc motor.

the use of half wave converter for field circuit as half wave rectified output has high ripple content which gives rise to large iron loss in the coil. We have used half wave converter for armature circuit. By changing the firing angles voltage across field coil and armature can be changed. As a result the speed of the motor changes. If speed is controlled by changing armature voltage, the method is known as armature control whereas the method to control the speed by changing field voltage is called flux control. The expression for average voltage across armature and field coil is given by

$$V_{oa} = \frac{v_m}{2\pi}(1+\cos\alpha_a);\ 0 \le \alpha_a \le \pi$$

$$V_{of} = \frac{V_m}{\pi}\left(1+\cos\alpha_f\right);\ 0 \le \alpha_f \le \pi.$$

Using eqn. (22), we obtain

$$V_{oa} = \frac{\phi NPZ}{60A} + I_a R_a \quad \text{...(24)}$$

By keeping field current constant, we conclude from eqn. (24) that

$$N \propto (V_{oa} - I_a R_a)$$

$$\propto \left[\frac{v_m}{2\pi}(1+\cos\alpha_a) - I_a R_a\right] \quad \text{...(25)}$$

As armature coil has inductance, the thyristor keeps conducting during some part of negative cycle. This is avoided by using free wheeling diode across armature coil as well as field coil.

3. A semiconverter dc drive : The system is shown in Fig. 6.16. Semi converter have been used for armature as well as field circuits. The operation of the circuit is described as under.

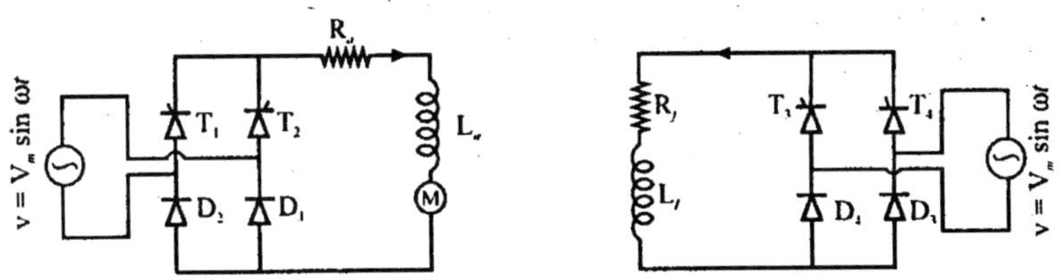

Fig. 6.16. Semiconverter dc drive.

Operation : The power is supplied to the coil at any anlge α_f and to the armature at angle α_a. The load voltage across field coil and armature is given respectively as

$$V_{of} = \frac{v_m}{\pi}\left(1+\cos\alpha_f\right); 0 \le \alpha_f \le \pi$$

and

$$V_{oa} = \frac{v_m}{\pi}(1+\cos\alpha_a); 0 \le \alpha_a \le \pi .$$

The freewheeling operation is performed by diodes D_1, D_2, D_3 and D_4

This drive is used to drive loads up to 15 kw ratings.

4. Single phase full converter drive : The circuit is shown in Fig 6.17. Full converters are used for both armature and field supply. The operation of the circuit can be explained in the same way as for semi converter drive. This circuit has two quadrant operations. When V_a is positive it has first quadrant operation, when V_a is –ve it has fourth quadrant operation. Let α_a and α_f be the firing angles of armature and field circuit then we can write

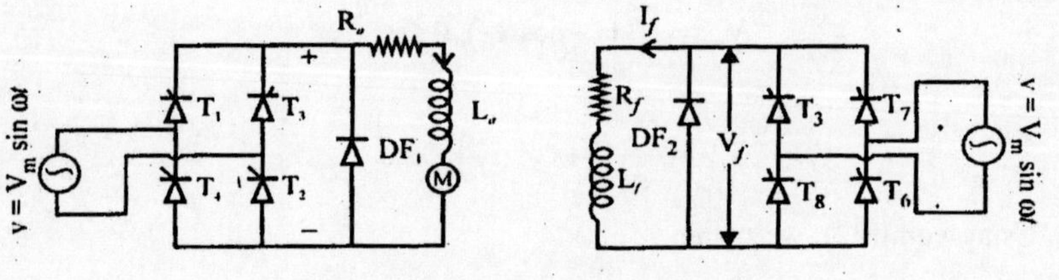

Fig. 6.17. Full converter drive.

$$V_{oa} = \frac{V_m}{\pi}(1+\cos\alpha_a) \quad 0 \le \alpha_a \le \pi$$

and
$$V_{of} = \frac{V_m}{\pi}(1+\cos\alpha_f) \quad 0 \le \alpha_f \le \pi$$

Since armature and field coil both have inductance, the current flows through the thyristors during a part of negative cycle also which is avoided by using free wheeling diodes DF_1 and DF_2. This drive also is used for motors upto 15 kw.

5. Single phase dual converter drives : We can use two converters working opposite to each other. The circuit is called dual converter circuit. It is shown in Fig 6.18. From the circuit it is clear that the output voltages of both the converters v_{o1} and v_{o2} are opposite and equal to each other, *i.e.*,

$$v_{01} = -v_{02} \qquad ...(26)$$

where $v_{01} = \frac{2v_m}{\pi}\cos\alpha_1$ and $v_{02} = \frac{2v_m}{\pi}\cos\alpha_2$

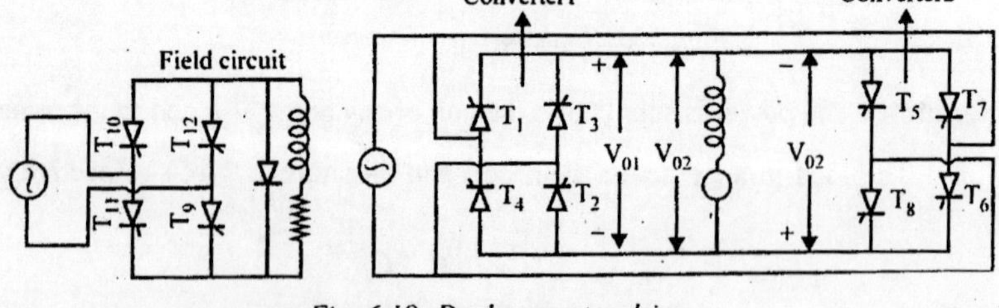

Fig. 6.18. Dual converter drive.

From eqn. (26), we get

$$\frac{2v_m}{\pi}\cos\alpha_1 = -\frac{2v_m}{\pi}\cos\alpha_2$$

or
$$\cos\alpha_1 = \cos(\pi - \alpha_2)$$

i.e., $\alpha_l = \pi - \alpha_2$

or $\alpha_1 + \alpha_2 = \pi$ ...(27)

Eqn. (27) is the condition on the firing of two converters. This circuit gives four quardrant operation. In first quadrant power is supplied from converter 1 to the motor. It is called forward motoring mode. In second quadrant current is –ve which indicates that power is supplied from load (motor) to converter 2. The motor acts like generator in forward mode. It is forward regenerative braking mode. In third quadrant current and voltage both are negative. It indicates that the current flows through the load in a direction opposite to one indicated in the Fig. This is called reverse motoring mode. In fourth quadrant current is positive while voltage is negative. The current flows through converter 1 due to voltage across load. This is reverse regenerative braking mode. Thus it is a four quadrant operation as shown in Fig 6.19. This drive also is used to drive load upto 15 kW.

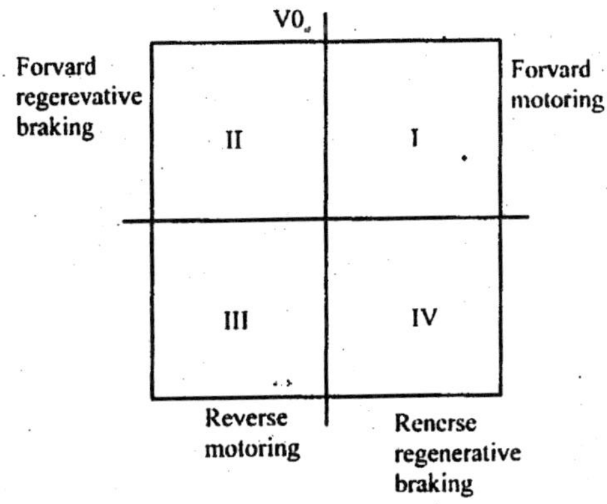

Fig. 6.19. Operations of dual converter drives.

6. Three phase drives : For motors having size more than 15 kW, three phase converters are used. Three phase converters are classified as

(a) Three phase semiconverter drives.

(b) Three phase full converter drives.

(c) Three phase dual converter drives.

Out of the three types of converters described above we shall discuss here only three phase semiconverter drives.

The circuit diagram of three phase semiconnecter drive is shown in Fig 6.20. The armature and field voltage are given by

$$V_{oa} = \frac{3\sqrt{3}}{2\pi} V_m (1 + \cos\alpha_a);\ 0 \le \alpha_a \le \pi$$

$$V_{of} = \frac{3\sqrt{3}V_m}{2\pi} (1 + \cos\alpha_f);\ 0 \le \alpha_f \le \pi.$$

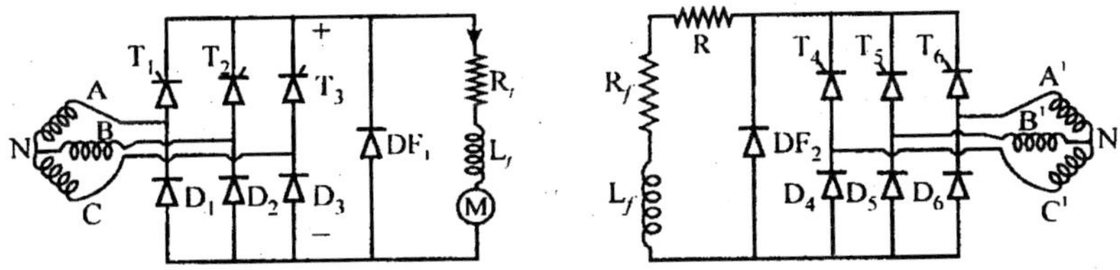

Fig. 6.20. Three phase semiconductor drive.

In line with the single phase full wave converter, we can write expression for armature and field voltage fed by three phase full converter. The expressions are

$$V_{oa} = \frac{3\sqrt{3}V_m}{\pi}\cos\alpha_a; \quad 0 \le \alpha_a \le \pi$$

$$V_{of} = \frac{3\sqrt{3}V_m}{\pi}\cos\alpha_f; \quad 0 \le \alpha_f \le \pi.$$

Thus the speed of the motors as is clear from above eqns., can be controlled by a corresponding changes in α_a or α_f or both.

7. Chopper drives : By using chopper drive, variable supply can be fed to the armature of motor. Chopper fed drives can be operated in the following modes.

(*a*) Motoring mode.

(*b*) Regenerative braking mode.

(*c*) Rheostatic braking mode.

We shall discuss here motoring mode.

The motoring mode is shown in Fig. 6.21. (a) The o/p voltage across the motor is given by $V_0 = \alpha V$, where α is called duty cycle of the chopper, defined as $\alpha = \frac{T_{ON}}{T}$.

Here T_{ON} is the time for which switch T is closed and T is the duration of a cycle. The power supplied to the motor is

$$P = V_o I_a = \alpha V I_a$$

The current drawn from the supply is $I_a = \frac{I_s}{\alpha}$. By changing α the speed of the motor can be controlled. The torque speed curve in shown in Fig. 6.21 (b).

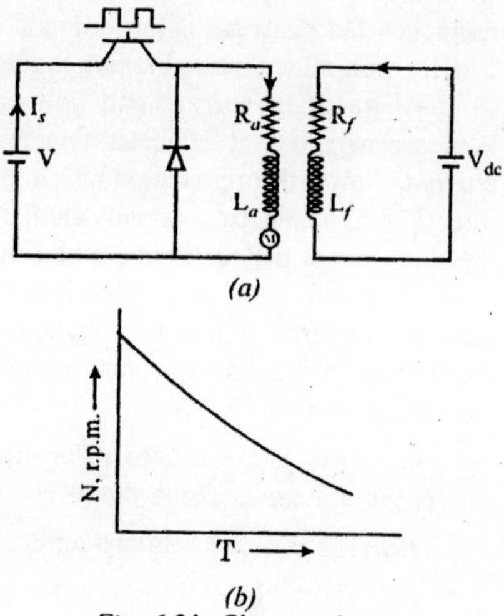

Fig. 6.21. Chopper drive.

Example 6.8. A single phase semi converter is fed by 230 V, 50 Hz power supply. The o/p of the converter is fed to a dc motor having armature ressistance 0.3 Ω. Other semi converter output is fed to field coil having coil resistance of 200 Ω. This converter is fixed at 0°. The load torque is 50 N-m at 900 r.p.m. The voltage constant is 0.8 V/A-rad/sec and the torque cosntant is 0.8 N-m/A^2. Assume armature and field current continuous and neglect losses. Find *(a)* field current *(b)* firing angle of converter in firing circuit *(c)* power factor of converter in the armature circuit.

Solution : *(a)*

$$V_f = \frac{2V_m}{\pi}\cos\alpha_f = \frac{2V_m}{\pi}, \alpha_f = 0:$$

$$= \frac{2 \times 230\sqrt{2}}{\pi} = 207.04 \text{ volt}$$

$$\text{Field current M} = \frac{207.04}{200} = 1.0352 \text{ Amp}$$

(b) Torque $T = K_f I_f I_a$

$$\therefore \quad I_a = \frac{T}{K_f I_f} = \frac{50}{0.8 \times 1.03252} = 60.375 \text{ Amp}$$

$$\omega = \frac{900 \times 2\pi}{60} = 94.248 \text{ rad/sec.}$$

Back e.m.f. $E_b = K_s \omega I_f = 0.8 \times 94.248 \times 1.0352 = 78.05$ V

For separately excited dc motor

$$V_a = E_b + I_a R_a$$
$$= 78.05 + 60 \times .03 = 96.05 \text{ volt}$$

Armature voltage is given as

$$V_{oa} = \frac{V_m}{\pi}(1 + \cos\alpha_a)$$

$$= \frac{230\sqrt{2}}{\pi}(1 + \cos\alpha_a)$$

or $$96.162 = \frac{230\sqrt{2}}{\pi}(1 + \cos\alpha_a)$$

or $$\frac{96.162 \times \pi}{230\sqrt{2}} - 1 = \cos\alpha_a$$

or $$\alpha_a = 94.076^\circ$$

(c) Power output of armature $= V_a I_a = 96.162 \times 60.375 = 58050.78$ watt

$$\text{Input current} = \left[\frac{2}{2\pi}\int_{\alpha_a}^{\pi} I_a^2 d\omega t\right]^{1/2}$$

$$= I_a\left(\frac{\pi - \alpha_a}{\pi}\right)^{0.5} = 41.71 \text{ Amp}$$

Input volt ampere $= 230 \times 4.71$ VA $= 9593.3$ VA

$$\text{Input power factor} = \frac{\text{output power}}{\text{Total input power}} = \frac{5805.78}{9595.3} = 0.605$$

Example 6.9. Armature and field coils of a separately excited dc motor are fed from two full converters. Input voltage to converters is 230 V at 50 Hz. The armature and field coil have resistance 0.25 and 200 Ω. The firing angle for field coil is zero while for armature it is 45°. The torque and voltage constants are both 1each and armature current is 50 A. If brush contact voltage is 0 volt, find torque developed and motor speed.

Solution : $$V_{of} = \frac{2V_m}{\pi}\cos\alpha_f = \frac{2 \times 230\sqrt{2}}{3.14} \times 1 = 20$$

$$V_{oa} = \frac{2V_m}{\pi}\cos\alpha_a = \frac{2\times 230\sqrt{2}}{3.14}\cos 45^\circ = 206.56 \text{ volts}$$

$$V_{oa} = E_b + I_a R_a$$

$$\text{or} \quad 206.56 = E_b + 50 \times 0.025$$

$$\text{or} \quad E_b = 206.56 - 12.5$$

$$= 194.06 \text{ Volts}$$

$$I_f = \frac{V_{of}}{R_f}$$

$$= \frac{20}{200} = 0.1 \text{ amp}$$

$$T = K_f I_f I_a$$

$$= 1 \times 0.1 \times 50 = 5 \text{ N-m}$$

motor speed $N \alpha E_b$

$$= K_b E_b = 1 \times 194.06 \text{ r.p.m.}$$

$$\approx 194 \text{ r.p.m} \quad \textbf{Ans.}$$

Example 6.10. A Separately excited dc motor is fed from a 500 V dc source through a one quadrant chopper. Armature resistance is 0.1 Ω and armature current 200 A. The voltage and torque constants are 1.4 V/A-rad/sec and 1.4 N-M/A^2. The field current is 2 Amp. The duty cycle of chopper is 0.5. Find *(a)* input power *(b)* speed *(c)* torque.

Solution : *(a)* Power input is given by

$$P_{in} = V_{oa} I_a = \alpha_v V I_a = 0.5 \times 500 \times 200 = 50 \text{ kW}$$

(b) The voltage across armature is given by

$$V_{oa} = \alpha V = 0.5 \times 500 = 250 \text{ V}$$

Voltage eqn for separately excited motor is

$$V_{oa} = E_b + I_a R_a$$

$$\text{or} \quad E_b = V_{oa} - I_a R_a = 250 - 200 \times 0.1 = 230 \text{ V}$$

$$E_b = K_f \omega I_f$$

$$\text{or} \quad E_b = 1.4 \times \omega \times 2$$

$$\text{or} \quad \omega = \frac{E_b}{1.4 \times 2} = \frac{230}{1.4 \times 2} = 82.14 \text{ rad/sec.}$$

$$\text{But} \quad \omega = \frac{2\pi N}{60}, \text{ we get}$$

$$N = \frac{60\omega}{2\pi} = \frac{60 \times 82.14}{2 \times 3.14} = 784.38 \text{ rpm}$$

(c) Torque $T = K_t I_f I_a = 1.4 \times 2 \times 200 = 560$ N-m

6.7. SPEED CONTROL OF SMALL AC MOTORS

The speed of ac motors can be contolled by a number of methods. Depending upon the application of the motor under consideration different methods used to control the speed of small ac motors are discussed as under :

1. Fundamentals of a.c. motors : We consider polyphase (two phase in this case) supply to the stator of an induction motor. The magnetic fields produced by the current flowing in motor windings differs in phase by 90° . Hence the magnetic flux cutting the rotor of the motor are also 90° apart in phase. The resultant of these two fluxes produces a rotating flux. As this rotating flux cuts the rotor windings an e.m.f. is induced in the rotor which by Lenz's law tries to catch up the rotating flux. As a result the rotor starts rotating in the direction of the flux rotation. The relative speed between the flux and the rotor is responsible for the rotation of the motor when this speed falls to zero the rotor rotation is stopped. In the following sections we shall describle methods to control speed of a.c. motors.

2. Single phase half wave ac motor drive: The circuit is shown in Fig. 6.22 *(a)* Thyristor T conducts during positive half cycle when it is fired at any desired angle. Thus the desired part of the positive half cycle can be switched to the motor. During negative half cycle the diode D conducts. The rms value of output voltage is given by

$$V_{orms} = \frac{V_m}{2\sqrt{\pi}}\left(2\pi - \alpha + \frac{\sin 2\alpha}{2}\right)^{0.5} \quad ...(28)$$

In the positive half cycle the flow of current is from top to bottom. During this cycle the thyristor conducts. This SCR is turned OFF at zero crossing and when diode D conducts in negative cycle the direction of the current is from bottom to top. A major drawback of this circuit is that due to presence of inductance in the stator windings the thyristor and diode keep conducting even in some portions of negative and positive cycles. SCR blocks the conduction during negative cycle and the stator winding behaves as open circuited. By changing the firing angle voltage across the stator winding and hence the speed of the motor can be controlled. The waveforms are shown in Fig 6.22 *(b)*.

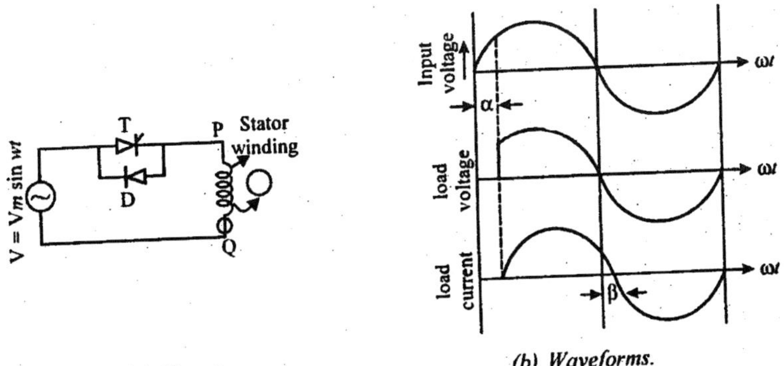

(a) Circuit. *(b) Waveforms.*

Fig. 6.22. Single phase ac motor drive.

3. Single phase full wave a.c. motor speed controller : The circuit is obtained by replacing diode D in Fig. 6.22 *(a)* by an SCR. The circuit is shown in Fig 6.23 *(a)*. Two SCR s T_1 and T_2 are connected in antiparallel which control the power through the stator winding of the motor. When SCR T_1 is turned ON during positive half cycle the current flows from top to bottom in the circuit. During negative cycle SCR T_2 conducts and the current flows from bottom to top through the stator winding. The rms load voltage for full wave ac regulator is given as

$$V_{orms} = \frac{v_m}{\sqrt{2\pi}} \left(\pi - \alpha + \frac{\sin 2\alpha}{2} \right)^{0.5}$$

From above eqn it is clear that the voltage across stator can be controlled by changing firing angle which in trun controls the speed of the motor. The waveforms are shown in Fig 6.23 *(b)*.

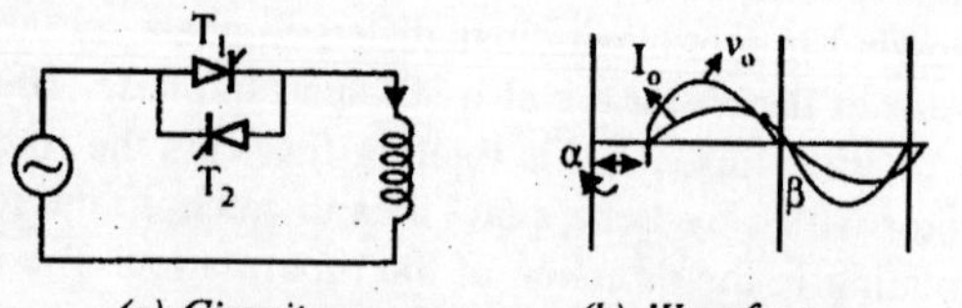

(a) Circuit. *(b) Waveform.*

Fig. 6.23. Full wave ac motor drive.

The limitations of this method again are due to the presence of inductance in the stator windings. From Fig. 6.23 *(b)* we observe that even when the load voltage falls to zero the load current does not fall to zero. During this period the stator windings are open. This is made smoother by using diac triac pair to control the speed of the motor as discussed below.

4. Diac triac pair to control speed of single phase ac motor : The circuit is shown in Fig. 6.24. Matched diac-traic pair is used for the purpose. The diac provides symmetrical pulses on both halves of a.c. input so that triggering pulses can be sent to the gate of triac in both the halves. It is used to control a very wide range of speeds.

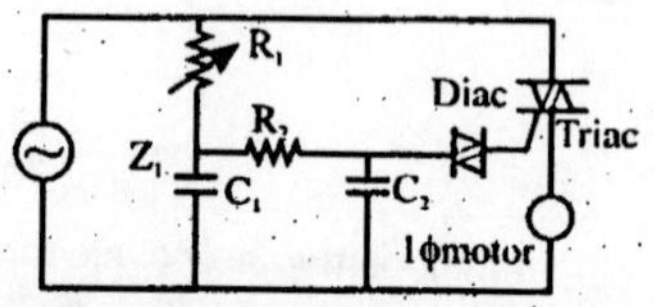

Fig. 6.24. Use of diac-triac pair for speed control.

In the circuit, network $R_1 - C_1$ forms the triggering circuit. The duration of the pulse is determined by the product R_1C_1. The network $C_1 - R_2 - C_2$ forms π filter which bypasses any spike available in the input signal. The resistance. R_2 also limits the current through the diac.

By varying potentiometer R_1 rate of charging of capacitor C_1 and the voltage across C_1 can be varied. When the voltage across capacitor becomes equal to breakover voltage of the diac it is triggered and sends triggering pulse to the gate of triac. As the triac is turned ON the current flows through it and the induction motor. As R_1 changes the firing angle and hence voltage across the motor changes which in turn controls the speed of the motor.

5. Speed control of ac motor by inverter : Inverter converts dc into a.c. The circuit used is shown in Fig. 6.25. $T_1T_2T_3T_4$ form controlled bridge rectifier whose output is d.c. The ripple present in the output of rectifier are removed by using LC filter. Thus almost constant dc is obtained which is now fed to the inverter circuit constituted by SCRs $T_5T_6T_7T_8$. The frequencyof inverter output can be changed by changing duration of pulses at the gates of the thyristors. Variation in frequency of input wave to the motor leads to variation in its speed.

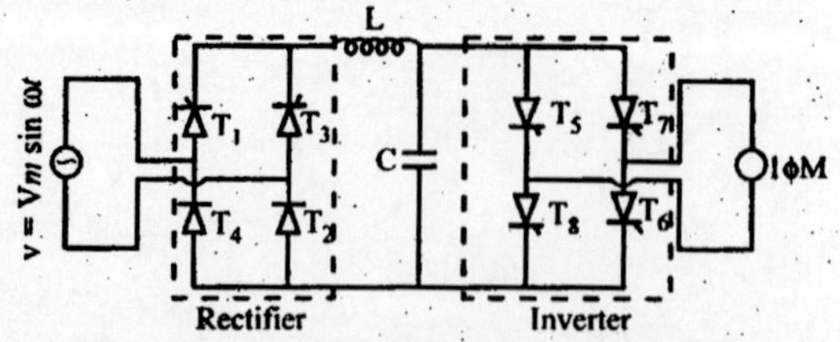

Fig. 6.25. Use of inverter for speed control.

Analysis : The current flowing through the stator windings of induction motor produces a magnetic field which generates flux through rotor circuit. For single phase induction motor the flux is not rotary in nature. As a result the net torque on the rotor in zero. We assume here that the flux has been made rotary by some method. The flux can be made rotary by splitting the given phase into two phase using a capacitor as in ceiling fans. Other methods also can be used. The speed of rotation of the magnetic flux is called synchronous speed. It is denoted by n_s

The synchronous speed is given by $n_s = \dfrac{120f}{p}$ ...(30)

where f = frequency of input supply to the motor

p = no of poles in the motor

The rotating magnetic flux induces an e.m.f. in rotor circuit which makes the rotor to rotate. The speed of the rotor is given in terms of synchronous speed as

$$n = n_s(1-s) \quad ...(31)$$

where s is called slip and is given by

$$s = \frac{n_s - n}{n_s} \quad ...(32)$$

From eqn. (29) and (30) we get

$$n = \frac{120f}{p}(1-s) \quad (33)$$

Clearly the speed of the motor depends upon the frequency of source voltage which can be changed by varying duration of pulses to SCRs used in inverter circuit. This is how inverter circuit is useful to control speed of the the motor.

This method can be used to control the speed at fixed frequency also. For fixed freqeuncy output of inverter the duration of pulses at SCRs used in inverters are kept of fixed duration while the output voltage of rectifier is changed by changing firing angles. Due to variation in voltage to stator windings its speed also varies. This method is called fixed frequency variable voltage control.

Example 6.11. In a single phase induction motor a supply of V = 200 sin 120t is given. The motor has 2 poles. The speed of the motor is 800 r.p.m. What is the slip in the motor?

Solution : The slip in the motor is given by

$$s = \frac{n_s - n}{n_s}$$

n_s is synchronous speed given by $n_s = \dfrac{120f}{p}$, $f = \dfrac{120}{2\pi}$ cycles/sec

or $$n_s = \frac{120 \times f}{2} = \frac{120 \times 120}{2\pi \times 2} \text{ r.p.m.} = \frac{3600}{\pi} \approx 1146.5 \text{ r.p.m.}$$

$$s = \frac{1146.5 - 800}{1146.5} = 0.3 \quad \textbf{Ans.}$$

6.8. SPEED CONTROL OF SYNCHRONOUS MOTORS

In synchronous motor the stator windings are fed by ac supply while rotor windings are excited by dc supply. The speed of the rotor is either equal to the synchronous speed or zero. It means therefore, that the speed of the motor can be changed by changing the synchronous speed.

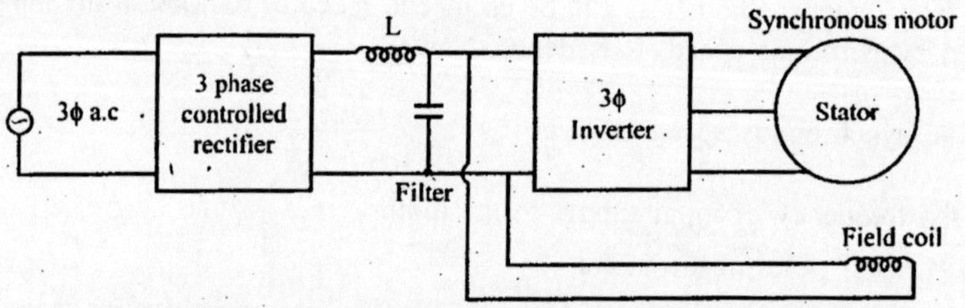

Fig. 6.26. Schematic of speed control of synchronous motor.

The circuit to control the speed of synchronous motor is shown in Fig 6.26.

The working of the circuit has already been dircussed in section 6.7 (4). By changing the frequency of output of inverter the rotor speed can be changed.

SUMMARY

Phase controlled rectification and ac phase control are used in this chapter for various applications. The applications covered in the present chapter include illumination control, temperature control, speed control etc. Various methods for each application have been dealt with in this chapter.

EXERCISE

1. Describe half wave rectifier method to control illumination level of electric lamp.
2. Derive expression for the power dissipated by an electric bulb connected as load for a full wave fully controlled rectifier:
3. Explain how the illumination level of electric bulb is controlled by using DIAC-TRIAC pair.
4. A halfwave a.c. regulator is used to change illumination of a lamp. Derive expression for power dissipated by the bulb.
5. What is the advantage of full wave a.c. regulator over half wave a.c. regulator in illumination control and speed control?
6. Explain briefly the method of illumination control by using chopper. Derive expression for power dissipated by the lamp.
7. Describe half wave and full wave a.c. regulators to control speed of a fan. Compare the two methods.
8. Explain how temperature of a heater can be controlled by using a full wave controlled rectifier.
9. Derive relationship between the temperature of heater and the heat dissipated by using half wave and full wave ac regulators.
10. Draw the circuit for automatic temperature control. Explain its working.
11. What are various methods to control the speed of d.c. and a.c. motors ?
12. Explain semiconverter dc drive.
13. Draw the circuit of single phase dual converter drive. Explain it.

14. Elaborate the need of chopper to control speed of the motor. Explain the related circuit operation.
15. Describe in detail single phase full wave a.c. motor speed controller. How is the drawback of this method overcome by using DIAC-TRIAC pair?
16. Explain how inverter can be used to control the speed of induction motor and synchronous motor.

NUMERICAL PROBLEMS

1. A separately excited dc motor is controlled by step down chopper. The motor specifications are rated current 40 A, armature resistance 0.5 ohm, rated speed 1000 rpm, rated voltage 250 V find *(a)* duty cycle of chopper to reduce the speed to 700 rpm, the torque being constant *(b)* Duty cycle of chopper so that motor may develop half the rated torque at 700 rpm.
2. An electric heating system uses ON-OFF control. The electric supply is ON in 8 cycles and OFF for 16 cycles and the input voltage is 230 V. Find rms value of the output voltage
3. A temperature control system uses half wave ac regulator to control temperature of a heater. The resistance of heater element is 40 Ω and input voltage is 230 V. Find the power dissipated by heater when firing angle is *(a)* 30° *(b)* 60°.
4. In above problem what will be the power dissipated if full wave AC regulator is used?
5. A step down chopper is used to control the temperature of an oven. The chopper output feeds in electric heater placed in the ovens. When duty cycle of chopper is 0.70 the temperature of the oven is T_1 and when the duty cycle is changed to 0.80 the temperature changes to T_2. What is the relation between T_1 and T_2 ?
6. The speed of separately excited dc motor is controlled by a half wave converter. The speed of the motor is 700 rpm at $\alpha = 30°$. What will be the speed at 150° ? $I_a = 10$ amp, $R_a = 0.4\ \Omega$. $V_m = 230$ V, $v_i = 230 \sin \omega t$.
7. A three phase semiconverter drive is used to control the speed of a dc motor. The speed at 30° is 600 rpm, no of conductor 36, no of poles are 4, wave winding has been used. Field flux is 100 mwb. If armature resistance is 5 Ω, find armature current. Neglect losses. $V_1 = 230 \sin 100t$.
8. Speed of an ac motor is controlled by single phase half wave ac regulator. If the motor speed at 30° firing angle is 600 rpm what will be the speed at 75°.

OBJECTIVE TYPE QUESTIONS

1. In mid point full wave fully controlled converter
   (a) two thyristors are used *(b)* one thyristor is used
   (c) one diode is used *(d)* both *(b)* and *(c)*.
2. In mid-point full wave semiconverter.
   (a) one diode is used *(b)* one thyristor is used
   (c) no thyristor is used *(d)* one thyristor one diode is used.
3. Average load voltage of full wave bridge semi converter is
   (a) $\frac{v_m}{\pi}(1+\cos\alpha)$ *(b)* $\frac{v_m}{2\pi}(1+\cos\alpha)$
   (c) $\frac{v_m}{\pi}\left(1+\frac{1+\cos\alpha}{2}\right)$ *(d)* both *(b)* and *(c)*

4. RMS load voaltage of a full wave ac regulator is given by

(a) $\frac{v_m}{\pi}(1+\cos\alpha)^{0.5}$ (b) $\frac{v_m}{\sqrt{2\pi}}\left(\pi-\alpha+\frac{\sin 2\alpha}{2}\right)^{0.5}$

(c) $\frac{v_m}{\sqrt{2\pi}}\left(\pi-\alpha+\frac{\cos 2\alpha}{2}\right)$ (b) none of the above

5. Three phase full wave a.c. regulator uses

(a) three diodes (b) three thyristors

(c) three diodes, three thyristors (d) six thyristors.

6. In star connected loads.

(a) line voltage leads the phase voltage by 30°

(b) line voltage is $\sqrt{3}$ times the phase voltage

(c) line current and phase currant are same

(d) All above (a), (b) and (c) only.

7. Fan speed is proportional to

(a) V_{orms} (b) $(V_{orms})^{1/2}$

(c) $(V_{orms})^{3/2}$ (d) none of the above

8. RMS value of load voltage for half controlled A-C regulator is equal to.

(a) $\frac{v_m}{2\sqrt{\pi}}\left(\pi-\alpha+\frac{\sin 2\alpha}{2}\right)^{0.5}$ (b) $\frac{v_m}{\sqrt{2\pi}}\left(2\pi-\alpha+\frac{\sin 2\alpha}{2}\right)^{0.5}$

(c) $\frac{v_m}{2\sqrt{\pi}}\left(2\pi-\alpha+\frac{\sin 2\alpha}{2}\right)^{0.5}$ (d) $\frac{v_m}{\sqrt{2\pi}}\left(2\pi-\alpha+\frac{\cos 2\alpha}{2}\right)^{0.5}$

9. Which of the following is correct ?

(a) Speed of D.C. motor can be contolled by using choppers

(b) A.C. regulator can be used to control speed of D.C. motor.

(c) In dc motor torque developed is directly proportional to ϕ^2

(d) None of the above

10. Maximum rating of loads that can be driven by semiconductor DC drive is

(a) 12.5 kW (b) 30 kW

(c) 15 kW (d) 17.5 kW.

11. The speed of a dc motor is given by

(a) $E_b = \frac{NPZ}{60A}$ (b) $E_b = \frac{NP\phi Z}{60A}$

(c) $E_b = \frac{NP\phi}{60Z}$ (d) $E_b = \frac{NPA}{60\phi}$

12. For electric traction

(a) DC shunt motor is used
(b) DC series motor is used
(c) none of the above
(d) both of the above.

13. Duty cycle of a chopper is gives as

(a) $\frac{T_{ON}}{T_{ON}+T_{OFF}}$
(b) $\frac{T_{ON}}{T_{OFF}}$
(c) $\frac{T_{OFF}}{T_{ON}}$
(d) $\frac{T_{OFF}}{T_{ON}+T_{OFF}}$

14. Synchronous speed is given by

(a) $\frac{120f}{p}$
(b) $\frac{240f}{p}$
(c) $120\,fp$
(d) $120\frac{p}{f}$

15. An a.c. motor with four poles has speed of 1400 r.p.m. The speed of the motor having 2 poles will be

(a) 3000 rpm
(b) 750 rpm
(c) 2800 rpm
(d) 700 rpm.

16. Slip is given by

(a) 3000 rpm
(b) $1-\frac{n_s}{n}$
(c) $\frac{1-n}{n_s}$
(d) none of the above.

17. In synchronous motors

(a) rotor speed is half of the synchronous speed
(b) rotor speed is 95% of synchronous speed
(c) rotor speed is equal to synchronous speed
(d) rotor may have any speed

ANSWERS

1. (a)	2. (d)	3. (c)	4. (b)	5. (d)	6. (d)
7. (a)	8. (c)	9. (a)	10. (c)	11. (b)	12. (b)
13. (a)	14. (a)	15. (c)	16. (b)	17. (c)	

❑❑❑

7
Inverters

7.0 INTRODUCTION

In Chapter 2 we have discussed methods to convert a.c. supply into d.c. supply. Some of the loads require ac supply for their operation. In the event of failure of mains a.c. supply, d.c. supply is converted into a.c. for driving such loads. This task is done by inverters. Thus we can define an inverter as a device which is used to convert d.c. power into a.c. power.

Oscillators are also used to produce sinusoidal waves which are ac in nature. However the a.c. supply with high voltage is not obtained from oscillators as it causes distortions at higher voltages. For obtaining a.c. supply from d.c. supply at higher voltages use of inverters is made. The output of inverters is generally not sinusoidal, it has generally other waveforms such as square, triangular etc. These wavefroms are approximated to sinusoidal waveform by reducing the width of the cycle of such wave forms so that the harmonic content in the output may be reduced. In what follows we shall have discussion on inverters.

7.1. WORKING PRINCIPLE OF AN INVERTER

It will be studied in next chapter that the speed of a d.c. motor or illumination of any lamp can be controlled by using a device called chopper. The chopper converts fixed d.c. into variable d.c. The output across the load is obtained during the period the switch is closed. When the switch is off the voltage does not appear acros the load. In inverter this OFF period is used to turn ON another chopper and output across the load appears for complete cycle. Thus an inverter is the combination of two choppers supplying power to a load to complement each other. The general circuit diagram of an inverter is shown in Fig 7.1 *(a)*.

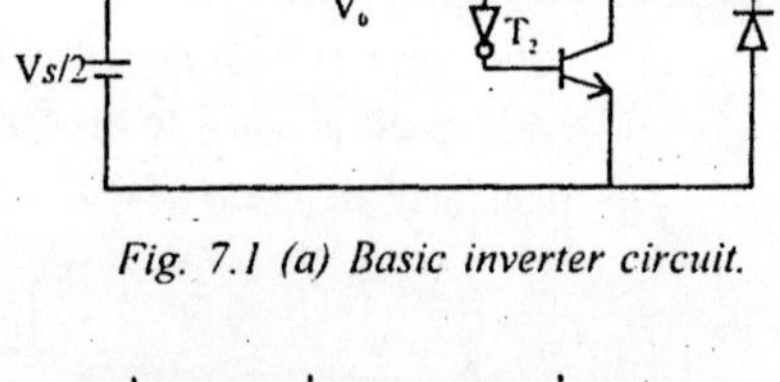

Fig. 7.1 (a) Basic inverter circuit.

From the ciruit we observe that when transistor switch T_1 is in conducting state, T_2 is OFF and vice versa. The number of cycles/sec generated by the pulse generator can be changed to chage the frequency of the output correspondingly. It changes from 50 to 400 cylces sec.

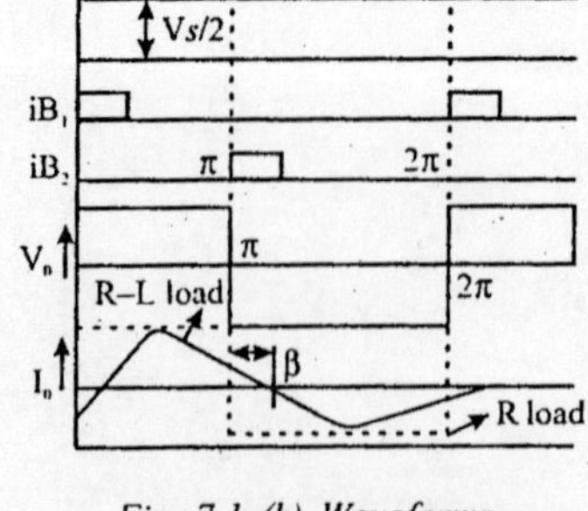

Fig. 7.1 (b) Waveforms.

When transistor T_1 is ON the current flows through the load in the direction shown through load-upper source $-T_1-$load. Let T_1 remains ON for a period t_1. As T_1 goes OFF and T_2 is turned ON, the voltage across the load follows the source voltage and goes to zero. However the load current may or may not be zero accordingly if the load is resistive or inductive. If the load is resistive as well as inductive, the energy stored in the inductor is released first through load, lower source and diode D_2 and then the transitor switch T_2 takes part in conduction and the current flows in the direction opposite to one shown in the Fig. 7.1. The path of the current is load-T_2-lower half voltage source. The polarity of the load voltage also reverses. This constitutes the negative half cycle. The waveforms are shown in Fig. 7.1 *(b)*. We observe from the waveforms that whenever any of the switch is open or closed the load voltage immediately follows the source voltage. However the behaviour of the current waveforms depends upon the type of the load. If load is purely resistive the load current and load voltage are in phase and rise and fall simultaneously as shown by dotted square wave. On the other hand, for R–L load the load voltage and load current are not in the same phase. Hence these do not rise and fall simultaneously as shown by the solid curve. The angle for which the current falls to zero is called extinction angle. From the wavefroms we observe that sinusoidal wavefrom is not available – square wave is obtained across the load. This is how a.c. supply is generated from d.c. supply by using inverters.

7.2. TYPES OF INVERTERS

Inverters are categorized below:

(a) *On the basis of method of commutation:* The inverter may be line commutated or forced commutated. A line commutated inverter feeds an a.c. system. The a.c. line voltage is used for commutation. Forcd commutation inverters use any method of forced commutation.

(b) *On the basis of connection of commutating elements:* In this category we have series inverter, parallel inverter and bridge inverter.

(c) *On the basis of power source:* Under this category voltage and current source inverters are included. A voltage source inverter is fed by a constant voltage source while a current source inverter is fed by a constant current source.

In the following sections we shall describe in detail some inverter circuits.

7.3. SINGLE PHASE BRIDGE INVERTER

It is called bridge inverter since the devices are connected in the form of the bridge. It is devided into two types:

1. Single phase half bridge inverter.
2. Single phase full bridge inverter

1. Single phase half bridge inverter

The circuit diagram is shown in Fig. 7.2 *(a)* and the corresponding waveforms in Fig. 7.2 *(b)*. The working of this circuit is the same as explained in section 7.1. When thyristor T_1 conducts the current flow through upper source-T_1-load. After interval T/2, T_1 is commutated while T_2 is set into conduction for the remaining interval T/2. For resistive load voltage and current have the same type of waveforms. However for R–L load, the voltage and current do not vary simultaneously. The waveforms will be as shown in Fig. 7.1 *(b)*. The diodes D_1 and D_2 are for feedback operation for R–L load. When the load is purely resistive D_1 and D_2 do not take part into conduction. The operation of thyristor is governed by the number of pulse at gate per second. If the no of pulse are f per second at two thyristors, the frequency of the output will be f. The frequency of output in radian will be $\omega = 2\pi f$ rad/sec. By changing the value of f an output of frequencies upto 400 Hz can be obtained.

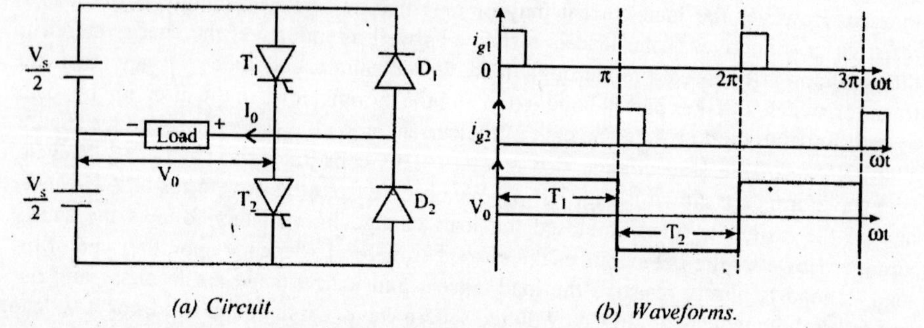

(a) Circuit. *(b) Waveforms.*

Fig. 7.2. 1 φ half bridge inverter.

By fourier series, output can be written as $V_0 = \sum_{n=1,3,5}^{\infty} \frac{2V_s}{n\pi} \sin n\omega t$

2. Single phase full bridge inverter

The circuit of single phase full bridge inverter is shown in Fig. 7.3 *(a)*. Four thyristor Th_1–Th_4 are used in the form of a bridge. For this reason it is called full bridge inverter. Diodes D_1– D_4 are for feedback operation for inductive or capacitive loads. The direction of voltage and current shown are assumed positive. A voltage source of e.m.f. + V_s has been considered in the circuit. When thyristors Th_1 and Th_2 are closed for angle 0 to π, the current flows in the circuit through load-Th_2-the source-Th_1-load. For angle π to 2π, thyristors Th_3 and Th_4 are turned ON. The current, in this case, flows through load. Th_4-source-Th_3-load in the direction opposite to one shown in the circuit. If the load is purely resistive the voltage and current vary in the similar way. The waveforms are shown in Fig. 7.3 *(b)*. For R–L load the current and voltage are not in the same phase. After first half cycle the freewheeling operation takes place through D_3-source, D_4-load while after second half cycle the freewheeling operation takes place through D_1-source-D_2 and load. The waveforms are shown in Fig. 7.3 *(b)*.

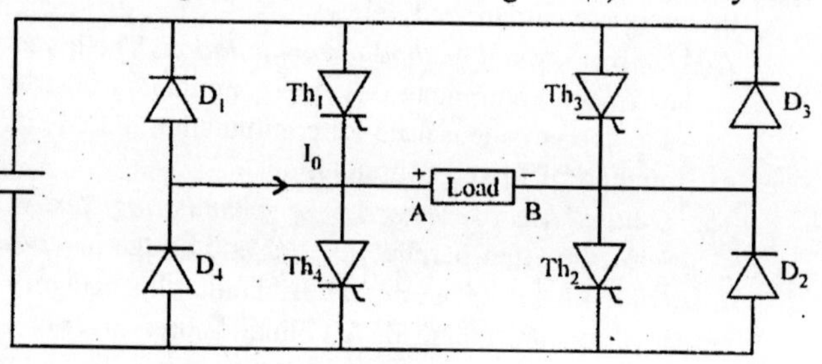

(a) Circuit.

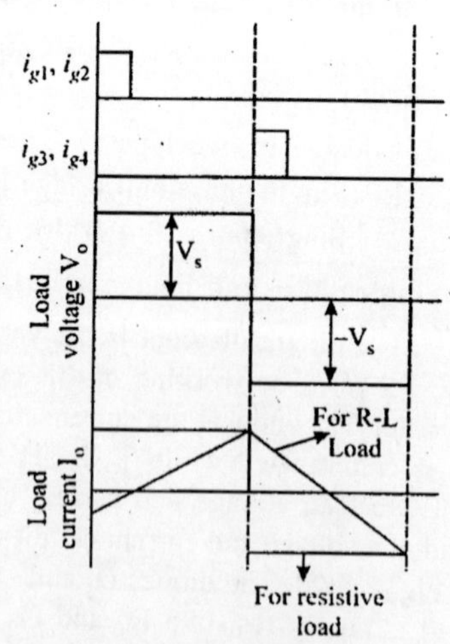

(b) Waveforms.

Fig. 7.3. 1 φ full bridge inverter.

Analysis of single phase full bridge inverter

As we have already studied that for inductive load the waveshape of voltage and current are not in phase. We consider the circuit shown in Fig. 7.4 to analyze the behaviour of load which are not purely resistive in nature. Voltage eqn by KVL is

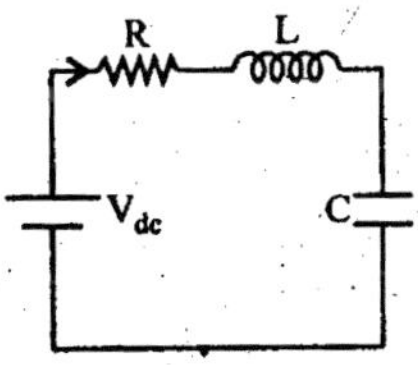

Fig. 7.4 (a)

$$V_{dc} = Ri + \frac{Ldi}{dt} + \frac{1}{C}\int idt \text{ for } 0 < t < T/2 \quad ...(1)$$

and $-V_{dc} = Ri + \frac{Ldi}{dt} + \frac{1}{C}\int idt$ for $T/2 < t < T$ ...(2)

We now consider different loads

(a) R–L load

For R–L loads the equation (1) and (2) are modifed to

$$V_{dc} = \frac{Ldi}{dt} + Ri \qquad 0 < t < T/2 \quad ...(3)$$

$$-V_{dc} = \frac{Ldi}{dt} + Ri \qquad T/2 < t < T \quad ...(4)$$

The solution of the equation of these type is given by

$$i = I_{dc}\left(1 - e^{-\frac{R}{L}t}\right) \quad 0 < t < T/2$$

$$i = -I_{dc}\left(1 - e^{-\frac{R}{L}t}\right) \text{ for } T/2 < t < T$$

The vriation of i with t is shown in Fig. 7.4 *(b)*. We observe that the current and voltage are not in same phase, the phase difference between voltage and current being given by

$$\theta = \tan^{-1}\frac{\omega L}{R}$$

Before time $t = 0$, the current is –ve which shows that thyristors Th_3 and Th_4 are conducting in this condition and direcion of load current is from B to A in Fig. 7.3 *(a)*. When Th_3 and Th_4 are turned OFF at $t = 0$ the inductance does not allow load current to fall to zero and feedback diodes D_1 and D_2 start conducting against battery. When Th_1 and Th_2 are turned ON at $t = 0$ they can not conduct until time $t = t_1$. At time $t = t_1$ due to reverse bias of diodes D_1 and D_2 thyritors Th_1 and Th_2 start conducting because they are forward biased at this instant. These keep conducting until t = T/2 when they are forward biased and will remain ON until t = T. This is how the cycle is completed.

Output voltage can be writtin as $V_0 = \sum_{n=1,3,5}^{\infty} \frac{4V_s}{n\pi}\sin n\omega t$

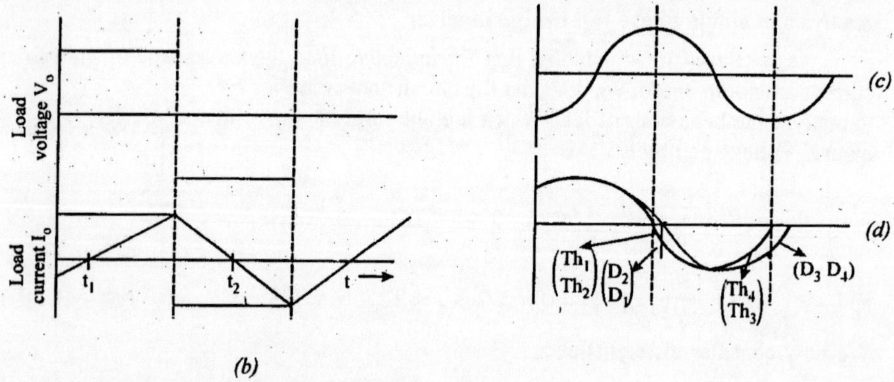

Fig. 7.4. Analysis of full bridge inverter.

(b) RLC Load

The solution of eqn. (1) and (2) are given by

$$i = c_1 e^{p_1 t} + c_2 e^{p_2 t} \quad (5)$$

where
$$p_1, p_2 = \frac{-R/L \pm \sqrt{(R/L)^2 - 4/LC}}{2} \quad (6)$$

Now we consider two different cases

Case 1: When $\left(\frac{R}{2L}\right)^2 > \frac{1}{LC}$, the roots p_1 and p_2 are real but unequal. The current is said to be overdamped. The waveshapes are shown in Fig. 7.4 *(c)*. The order of conduction is the same as that of R–L circuit. However, here the waveshape is sinusoidal.

Case 2: When $\left(\frac{R}{2L}\right)^2 < \frac{1}{LC}$. In this case the roots of the eqn. (5) are imaginary which indicates oscillatory nature with decaying amplitude. This is called R–L–C underdamped circuit. The waveforms are shown in Fig. 7.4 *(d)*. The shape of the waveforms is sinusoidal due to R–L–C components. As seen from the Fig., thyritors Th_1 and Th_2 stop conducting before Th_3 and Th_4 are turned ON. During this interval the current flows through diodes D_1 and D_2. Similarly during interval between turn ON of SCRs Th_1 and Th_2 and turn OFF of thyristors Th_3 and Th_4 diodes D_3 and D_4 conduct.

Example 7.1. A single phase half bridge inverter has a resistive load of 5 ohm. The source voltage is 40 V. Find *(a)* rms value of fundamental component of output voltage *(b)* output power *(c)* peak current in each thyristor *(d)* average current of each thyristor *(e)* peak reverse blocking voltage.

Solution : The output voltage is given as

$$v_0 = \sum_{n=1,3,5}^{\infty} \frac{2V_s}{n\pi} \sin n\omega t$$

$$v_{01} = \frac{2V_s}{\pi} \sin \omega t$$

rms value of fundamental component

$$v_{01}\text{ rms} = \frac{2V_s}{\pi}\times\frac{1}{\sqrt{2}}$$

$$= \frac{\sqrt{2}V_s}{\pi} = 0.45V_s = 0.45\times 40$$

$$= 18 \text{ volts.}$$

(b) Since output is a.c. power, we shall find first rms value of load voltage which is given as

$$v_{orms} = \left(\frac{1}{T}\int_0^T (V_0)^2\,dt\right)^{0.5}$$

$$= \left(\frac{2}{T}\int_0^{T/2} (V_0)^2\,dt\right)^{0.5},\quad V_0 = \frac{V_s}{2}$$

$$= \left(\frac{2}{T}\int_0^{T/2}\left(\frac{V_s}{2}\right)^2 dt\right)^{0.5}$$

$$= \left[\frac{2}{T}\left(\frac{V_s^2}{4}t\right)_0^{T/2}\right]^{0.5} = \left[\frac{2}{T}\frac{V_s^2}{4}T/2\right]^{0.5}$$

$$= \frac{V_s}{2} = \frac{40}{2} = 20\text{ V}$$

$$\text{Hence output power} = \frac{V_{orms}^2}{R} = \frac{(20)^2}{5} = 80\text{ watts}$$

(c) Peak current in each thyruter $=\dfrac{\text{Peak voltage}}{\text{resistance}}=\dfrac{20}{5}=4$ Amp.

(d) Each thyristor conducts for 50% of the time, Hence average current

$$I_{oav} = \frac{\text{Peak current}}{2} = \frac{4}{2} = 2\text{ Amp}$$

(e) Peak reverse block voltage = 2 × 20 = 40 volts.

Example 7.2. What will be the value of the parameters in example 7.1 for a full bridge inverter?

Solution. *(a)* For full bridge inverter o/p voltage is given by

$$V_0 = \sum_{n=1,3,5}^{\infty}\frac{4V_s}{n\pi}\sin n\omega t$$

$$V_{01} = \frac{4V_s}{\pi}\sin\omega t$$

(b) rms value of o/p voltage is

$$V_{orms} = \left[\frac{2}{T}\int_0^{T/2} V_s^2\, dt\right]^{0.5} = V_s\left[\frac{2}{T}\times\frac{T}{2}\right] = V_s$$

$$= 40 \text{ volts}$$

$$\text{Hence output power} = \frac{V_{orms}^{\;2}}{R} = \frac{(40)^2}{5} = \frac{1600}{5} = 320 \text{ watt.}$$

$$\text{(c) Peak current in each thyristor} = \frac{\text{Peak voltage}}{\text{resistance}} = \frac{40}{5} = 8 \text{ Amp}$$

$$\text{(d) Average current} = \frac{\text{Peak current}}{2} = \frac{8}{2} = 4 \text{ Amp.}$$

(e) Peak reverse voltage = Source voltage = 40 volt.

Example 7.3. A full bridge inverter has input voltage 200 V dc. The load is R–L load with R = 10Ω, L= 10 mH and inverter frequency is 100 Hz. Calculate *(a)* rms value of load current *(b)* output power *(c)* Peak value of load current *(d)* rms value of fundamental component of load current and total harmonic distortion, *(e)* average input current and peak value of thyristor current.

Solution. *(a)* For full wave bridge inverter output voltage is given by

$$V_0 = \sum_{n=1,3,5}^{\infty} \frac{4V_s}{n\pi}\sin n\omega t$$

$$\text{and load current } I_0 = \sum_{n=1,3,5}^{\infty} \frac{4V_s}{n\pi\left[R^2 + (n\omega L)^2\right]^{0.5}}\sin n\omega t$$

$$= I_{01} + I_{03} + I_{05} + \ldots$$

Hence $I_{orms} = \sqrt{I_{01rms}^2 + I_{03rms}^2 + I_{05rms}^2 + \ldots}$, neglecting higher order harmonics

$$\text{Hence} \quad I_{o1rms} = \frac{4V_s}{\pi\left[R^2 + (\omega L)^2\right]^{0.5}} \times \frac{1}{\sqrt{2}}$$

$$= \frac{4\times 200}{3.14\left[(10)^2 + (2\times 3.14\times 100\times 10\times 10^{-3})^2\right]^{0.5}} \times \frac{1}{\sqrt{2}} \text{ Amp.}$$

$$I_{03rms} = \frac{4V_s}{3\pi\left[R^2 + (3\omega L)^2\right]^{0.5}} \times \frac{1}{\sqrt{2}}$$

$$= \frac{4\times 200}{3\times 3.14\left[100 + (3\times 2\times 3.14\times 100\times 10\times 10^{-3})^2\right]^{0.5}} \times \frac{1}{\sqrt{2}}$$

$$I_{o5rms} = \frac{4 \times 200}{5 \times 3.14\left[100 + (5 \times 2 \times 3.14 \times 100 \times 10 \times 10^{-3})^2\right]^{0.5}} \times \frac{1}{\sqrt{2}}$$

Hence $I_{0rms} = 15.89$ Amp.

(b) rms load voltage

$$V_{0rms} = V_s = 200 \text{ V, for square wave}$$

$$\text{output power} = \frac{V^2_{0rms}}{R} = \frac{(200)^2}{10} = 4000 \text{ Watt}$$

(c) Peak value of load current = rms value of load current $\times \sqrt{2}$

(d) rms value of fundamental component of load current = 21.59 Amp

$$\text{Total harmonic distortion} = \frac{I_{01rms}}{I_{0rms}} = 0.96$$

(e) Average input current $= \dfrac{\text{Power output}}{\text{Input voltage}}$

$$= \frac{4000}{200} = 20 \text{ Amp.}$$

7.4. COMMUTATION IN SINGLE PHASE INVERTER

As we all are familiar with the fact that once a thyristor is turned ON it does not turn OFF even if the gating pulse is removed. The thyristor is turned OFF by using external circuits when it is fed by dc supply. The process of turn OFF of thyristor and changing the path of current is called commutation. The circuits used for this purpose are called commutation circuits. In what follows we shall discuss the inverters employing commutation methods.

1. Modified Mc Murray half bridge inverter : The circuit of Mc Murray half bridge inverter is shown in Fig. 7.5 *(a)* and the waveforms are shown in Fig. 7.5 *(b)*. In the circuit forced commutation is used. Thyristors Th_{a1} Th_{a2} C and L form the commutation network while Th_1 and Th_2 are main thyristors. Diodes D_1, D_2, D_3, D_4 provide the discharge path for the capacitor C. Load may be purely resistive, RL or RLC.

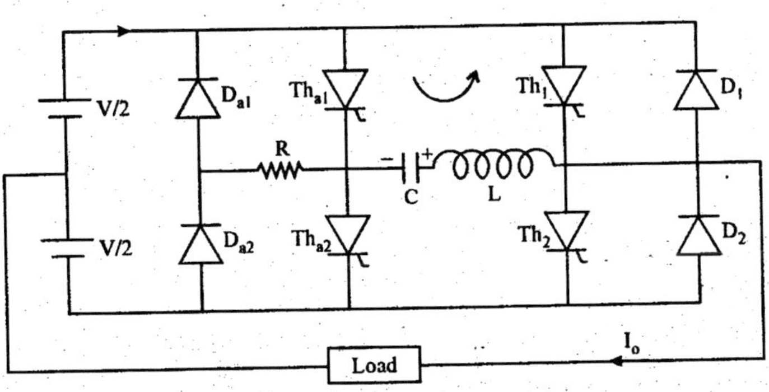

Fig. 7.5 (a) Modified Mc Murray half bridge inverter.

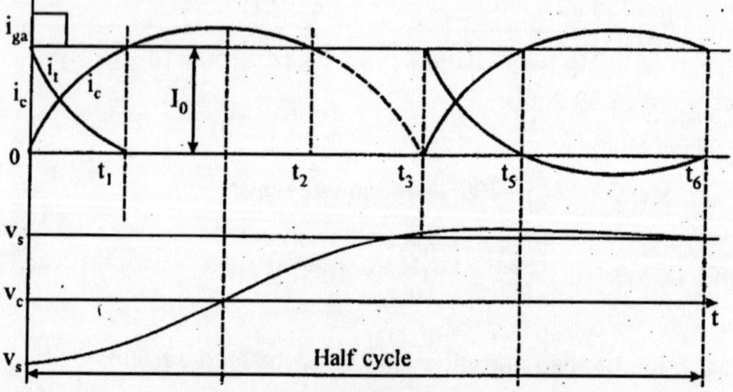

Fig. 7.5. (b) Waveforms for Fig. 7.5 (a)

Operation. We assume that a pulse is applied at the gate of SCR Th_1 at instant t=0. Prior to t = 0 SCR Th_1 is conducting with the capacitor charged with right plate positive.

As SCR Th_{a1} is turned ON the capacitor C starts discharging through Th_1, Th_{a1} and L in the direction shown. As a result the main thyrister Th_1 is reverse biased and hence turned OFF. The capacitor discharges now through diode D_1, SCR Th_{a1} and inductor L. Now capacitor consists of two parts–loads current and discharge current. As the potential across capacitor is reduced to zero it starts charging with left plate positive. When it is charged to supply voltage the SCR Th_{a1} is turned OFF and it is overcharged due to the energy stored in the inductive load.

2. **McMurray full bridge inverter.** Mc Muray full bridge inverter employs four main thyristors and commutation circuit as shown in Fig. 7.6 *(a)*. Thyristors Th_{a1}, Th_{a2}, C_1 and L form the commutation circuit for thyristor pair Th_1, Th_2. While the network consisting of Th_{a3}, Th_{a4}, C_2 and L form the commutation circuit for thyristor pair Th_3, Th_4. Diodes D_1–D_4 are for freewheeling operation. Load used may be purely resistive, R–L or RLC type load. In the interval $0 < t < T/2$ thyristor pair Th_1, Th_4 conduct while Th_2, Th_3 conduct during interval $T/2 < t < T$, the current flowing in direction from B to A. While drawing waveforms following assumptions are made.

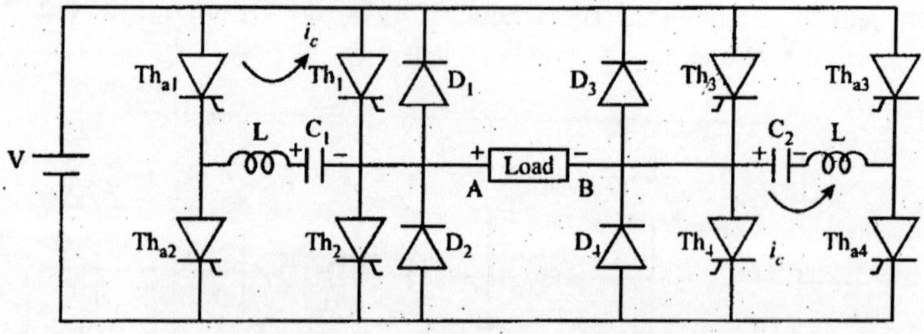

Fig. 7.6 (a) Mc Murray full bridge inverter.

(1) Load is inductive and load current is constant during commutation
(2) All elemets are ideal.

Operation. The operation of the circuit is divided into following modes.

Mode I $(0 \le t \le t_1)$. It is assumed that thyristor pair Th_1, Th_4 is conducting at time t=0 when auxiliary thyristors Th_{a1}, Th_{a4} are turned ON. This marks the beginning of Mode

1. Capacitors C_1 and C_2 start discharging through paths Th_1-Th_{a1}-C_1 and Th_4-Th_{a4}-L and C_2 respectively. The capacitor current starts rising and reaches the value $I_c = I_0$ at $t = t_1$. As the thyristor currents and capacitor currents are equal and opposite the thyristors Th_1, Th_4 are turned OFF at $t = t_1$.

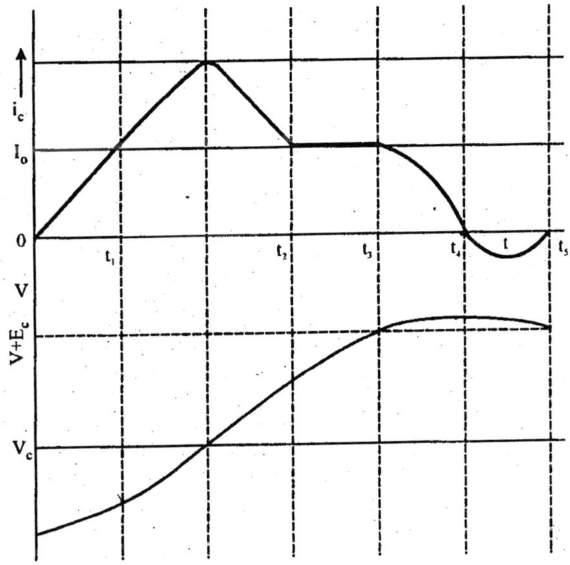

Fig. 7.6 (b) Waveforms.

Mode 2 $(t_1 \le t \le t_2)$. After time $t = t_1$ the capacitor currents start rising through paths consisting of L, C_1, D_1, Th_{a1} and L, C_2, D_2, Th_{a2}, sinusoidally. The capacitor voltage starts decreasing and reachs zero value when capacitor current reaches the peak value. After this point it starts charging in reverse direction and consequently current I_C starts falling. In interval $[t_1, t_2]$ the current I_C is more than load current. The excess current $I_C - I_0$ flows through the diodes D_1 and D_2. At the instant $t = t_2$ the diode current is reduced to zero and only load current flows through the capacitor. This marks the end of Mode 2.

Mode 3 $(t_2 \le t \le t_3)$. End of mode 2 marks the beginning of mode 3 at instant $t = t_2$. The capacitor starts recharging through the load. At $t = t_3$ the voltage across it reaches the value of supply voltage-V and it tends to overcharge due to energy stroed in inductors.

Mode 4 $(t_3 \le t \le t_4)$. At time $t = t_3$ the voltage across the capacitors is equal to supply voltage V. Due to energy stored in the inductors the capacitors charge to further higher voltage E_C. Diode D_2, D_3 become forward biased. At $t = t_4$ the capacitor is charged to maximum voltage and the current through it becomes zero.

Mode 5 $(t_4 \le t \le t_5)$. Once the capacitor gets charged to maximum voltage $V + E_c$ at the end of mode 4 it begins to discharge against the supply in Mode 5. As a result the capacitor voltage begins to fall and ultimately attains the value of supply voltage at instant $t = t_5$. This marks the end of Mode 5. This is how half of inverter O/P cycle is completed.

3. Modified Mc Murray full bridge inverter

Modified Mc Murray full bridge inverter is shown in Fig. 7.7. Additional diodes D_{A1}–D_{A4} and resistance R have been used in the circuit shown in Fig. 7.6 to remove excess charge across capacitor at the end of commutation.

After commutation of thysitor Th_1 the energy stored in capacitor C_1 is partly dissipated in resistance R and partly fed to the power supply through R, D_{a1}, power supply D_2.

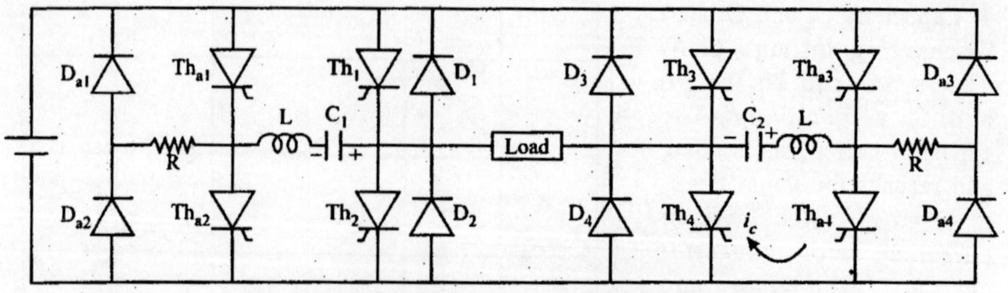

Fig. 7.7. Modified Mc Murray full bridge inverter.

In similar manner the capacitor C_2 supply energy to the source through D_3 source, D_{a4}, R and L. Modified bridge is used to remove the excess charge across the capacitors.

4. Mc Murray Bed ford half bridge inverter

In this inverter complementary voltage commutation is used . The circuit is shown in Fig. 7.8 *(a)*, Thyristors Th_1 and Th_2 commutate each other. C_1, C_2 and L form the commutation circuit. The operation of the circuit is divided into following three modes.

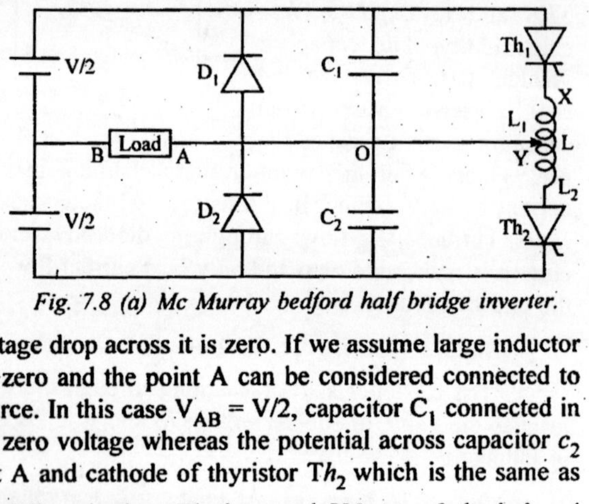

Fig. 7.8 (a) Mc Murray bedford half bridge inverter.

Mode 1. We assume that thyristor Th_1 is conducting at time $t = 0$ when the pulse is applied to thyristor Th_2. As Th_1 is conducting prior to $t = 0$ the voltage drop across it is zero. If we assume large inductor the voltage drop across it will be zero and the point A can be considered connected to positive terminal of upper half source. In this case $V_{AB} = V/2$, capacitor C_1 connected in parallel with Th_1 and L will be at zero voltage whereas the potential across capacitor c_2 will be the same as between point A and cathode of thyristor Th_2 which is the same as the source potential, *i.e.*, $V_{c_2} = V$. When thyristor Th_2 is turned ON at $t = 0$ the induced *e.m.f.* in the lower half of the coil will be V. Same e.m.f. will be induced in upper half of the coil with *x* and *y* point having same polarity. Hence effective e.m.f. induced at point $x = 2V$ which reverse biases the thyristor Th_1 and it is turned OFF. Load current is transferred to the lower inductor L_2. The capacitor C_2 discharges through L_2, so its voltage falls. The sum of V_{c_1} and V_{c_2} should be equal to supply voltage V, *i.e.*, $V_{c_1} + V_{c_2}$ = V. As voltage V_{c_2} falls V_{c_1} increases correspondingly.

Applying KVL for loop consisting of C_1, C_2 and source, we obtain

$$V = -\frac{1}{C}\int I_{c_2}\,dt + \frac{1}{C}\int I_{c_1}\,dt \quad ...(6)$$

L.H.S. of eqn. (6) is constant hence one term on R.H.S. should rise to compensate decrease is other term.

In other words $-\frac{1}{C}\int I_{c_2}\,dt + \frac{1}{c}\int I_{c_1}\,dt = \text{constant}$ ...(7)

All parameters in eqn. (7) having same value, we should have $I_{c_1} = I_{c_2}$ which implies that rate of discharge of capacitor c_2 is equal to rate of charging of capacitor c_1. By applying KCL at point 0 we get

$$I_{c_1} + I_{c_2} = I_{T_2} + I_0$$

or

$$I_{c_1} + I_{c_2} = I_{T_2} + I_0 \quad ...(8)$$

or $I_{c_1} = I_{c_2} = \dfrac{I_{T_2} + I_0}{2}$. At time $t = 0$ I_{T_2} is maximum having value I_0 hence the currents I_{c_1} and I_{c_2} have value at $t = 0$, given by

$$I_{c_1} = I_{c_2} = \frac{I_0 + I_0}{2} = I_0 \quad (9)$$

This mode ends at $t = t_1$ when voltage v_{c_2} becomes zero and v_{c_1} attains the value V. After this instant the capacitor c_2 starts charging in reverse direction which makes diode D_2 forward biased and it begins to conduct carrying current I_{T_2}. Current I_{c_2} falls to zero as $V_{c_1} = V$. Since v_{c_1} is continuously being applied across inductor L_2 its current keeps rising sinusoidally and at $t = t_1$ $I_{T_2} = I_m$ where I_m is the peak current of sinusoidal wave. We can write eq,. (10) as

$$I_{c_1} = I_{c_2} = \frac{I_0 + I_m}{2} \quad ...(10)$$

The equivalent circuit of mode 1 is shown in Fig. 7.8 *(b)*

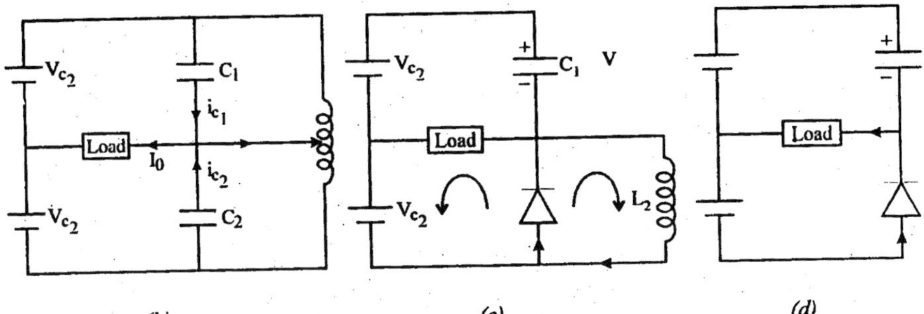

(b) *(c)* *(d)*

Fig. 7.8. Equivalent circuits.

Mode 2. Mode 2 begins at time $t = t_1$ when I_{c_1} and I_{c_2} are transferred to diode D_2. Diode current is equal to $I_0 + I_m$. Now energy stored in inductance L_2 is dissipated through L_2, T_2 and D_2. At time $t = t_2$; i_{T2} falls to zero and thyristor T_2 is turned OFF. This marks the end of mode 2. The equivalent circuit of mode 2 is shown in Fig. 7.8 *(c)*.

Mode 3. After instant $t = t_2$ only load current flows through the diode D_2. This feedback mode ends when load

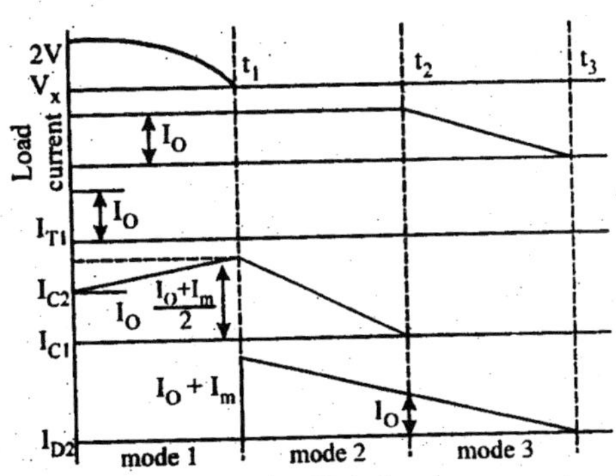

Fig. 7.9. Waveforms.

current falls to zero and reverse bias across Th_2 is automatically removed. As SCR is turned ON again since gating signal is applied continuously for 180°. This mode is shown in Fig. 7.8 *(d)*. The waveforms of the circuit are shown in Fig. 7.9.

7.5. THREE PHASE INVERTERS

The power delivered by single phase inverter to drive medium and larger load may not be suficient. Under such conditions three phase supply is required which is obtained by using three phase inverters. In what follows we shall discuss few methods to generate three phase a.c. power from a d.c source.

1. **Three phase bridge inverter.** Three phase bridge inverter is obtained by connecting three single phase inverters in parallel. The gating pulses to thyristors conducting for same polarity differ by 120°. The circuit is shown in Fig. 7.10. In the circuit six thyritors T_1–T_6 and six diodes D_1–D_6 are used. Diodes are used for feedback operation for R–L or RLC load. For purely resistive loads these do not conduct. Three phases A, B, and C have been obtained, each phase differing from each other by 120°. Mathematically each phase voltage can be written as

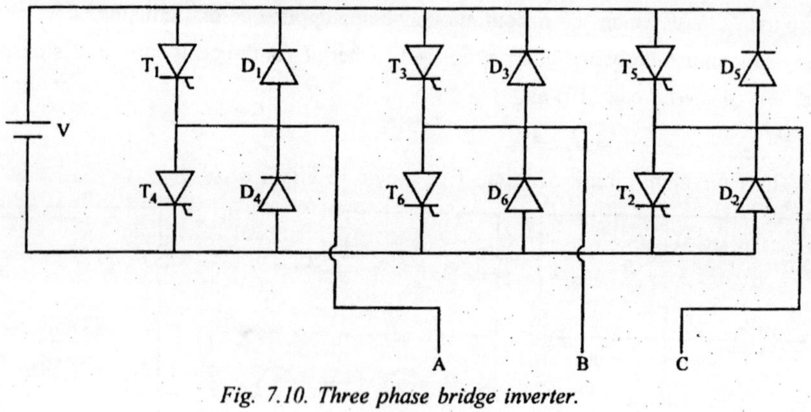

Fig. 7.10. Three phase bridge inverter.

$$V_{AN} = V_m \sin \omega t$$
$$V_{BN} = V_m \sin (\omega t - 120°)$$
$$V_{CN} = V_m \sin (\omega t - 240°)$$

Graphical representation of above phase voltages is given in Fig. 7.11.

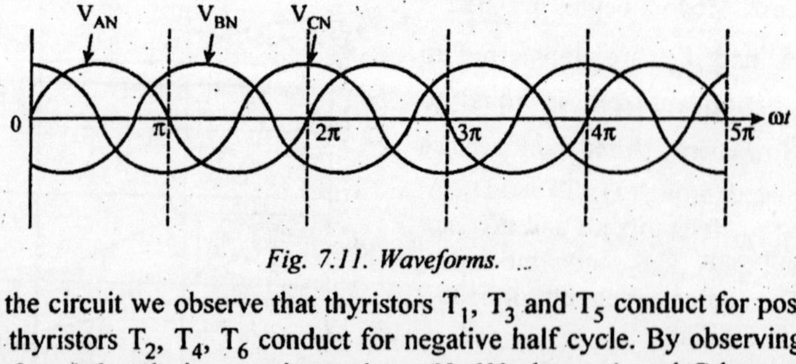

Fig. 7.11. Waveforms.

From the circuit we observe that thyristors T_1, T_3 and T_5 conduct for positive half cycle while thyristors T_2, T_4, T_6 conduct for negative half cycle. By observing the two figures it is found that during any interval say 0°–60° phases A and C have +ve value while phase B has –ve value and the thyristors which take part in conduction are T_1,T_5 and T_6. Similarly for any other interval different -2 devices take part in conduction.

Three phase inverter can have two different modes of operation–one is 180° conduction scheme and second is 120° conduction scheme. These schemes are discussed below.

(a) **180° conduction scheme.** Under this scheme every thyristor conducts for 180° and three thyristors conduct at one time as is evident from Fig. 7.11. The sequence of operation of thyristors is as under

Interval	Turn on Instant	Device turned ON	Devices conducting
I 0°–60°	0°	T_1	T_1, T_5, T_6
II 60°–120°	6°	T_2	T_1, T_2, T_6
III 120°–180°	120°	T_3	T_1, T_2, T_3
IV 180°–240°	180°	T_4	T_2, T_3, T_4
V 240°–300°	240°	T_5	T_3, T_4, T_5
VI 300°–360°	300°	T_6	T_4, T_5, T_6

As is clear from above table that during any interval three thyristors conduct at one time. If the load is star connected and purely resistive and between each phase and netural N a resistance R is connected, the equivalent circuit can be drawn as follows :

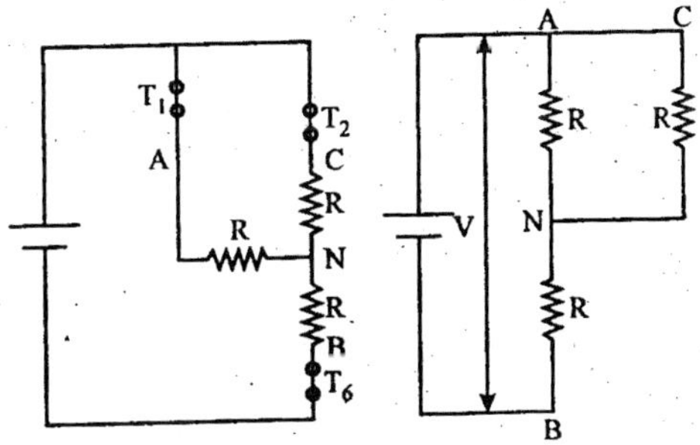

Fig. 7.12. Equivalent circuit.

During interval I thyristors T_1, T_5, T_6 are conducting, hence the circuit shown in Fig 7.10 can be reduced to the circuit shown in Fig. 7.12. The phase voltages V_{AN} and V_{CN} from the Fig. are equal, *i.e.*,

$$V_{AN} = V_{CN}$$

Equivalent resistance between pulse A and B is

$$R^1 = R11R+R = \frac{R}{2}+R = \frac{3}{2}R$$

Hence current in the circuit

$$I_0 = \frac{V}{R^1} = \frac{V}{\frac{3}{2}R} = \frac{2V}{3R}$$

$$V_{AN} = V_{CN} = R \times \frac{I_0}{2} = \frac{R}{2} \times \frac{2V}{3R}$$

$$= \frac{V}{3}$$

$$V_{NB} = R \times I_1 = \frac{2V}{3}$$

$$V_{BN} = -\frac{2V}{3}$$

In similar manner the phase voltage in any interval can be found out. Any two devices having same polarity in two phases will have same phase voltage and $\frac{V}{3}$ each. The third device in third phase will have phase voltage $\frac{2V}{3}$ and opposite in sign to the other two.

For example we consider interval II. The devices conducting in this interval are T_1, T_2 and T_6. T_1 conduct in +ve cycle while T_2,T_6 conduct in negative half cycle. T_1, T_2 and T_6 are connected in phases A, C and B respectively. Hence from above discussion phase voltage V_{BN} and V_{CN} will have the same value of $-\frac{V}{3}$ each while phase voltage V_{AN} will have the value of $\frac{2V}{3}$. Carrying out above discussion, we can write

During interval (III) $120° \leq \omega t \leq 180°$

$$V_{AN} = V_{BN} = \frac{V}{3},\ V_{CN} = -\frac{2V}{3}$$

For interval IV $180° \leq \omega t \leq 240°$

$$V_{AN} = V_{CN} = -\frac{V}{3},\ V_{BN} = \frac{2V}{3}$$

For interval V $240° \leq \omega t \leq 300°$

$$V_{BN} = V_{CN} = \frac{V}{3},\ V_{AN} = -\frac{2V}{3}$$

For interval VI.$300° \leq \omega t \leq 360°$

$$V_{AN} = V_{BN} = -\frac{V}{3},\ V_{CN} = \frac{2V}{3}$$

The waveforms are shown in Fig. 7.13. the firing angle for each thyristor is taken zero i.e. $\alpha = 0$. In the figure turn ON instants and conduction periods of all six thyristors have been shown and corresponding phase voltages have been drawn. For example thyristor T_1 is turned ON at $\omega t = 0°$ and in the interval $0° \leq \omega t \leq 60°$ thyrostors T_1, T_5 and T_6 conduct. hence phase voltage $V_{AN} = V_{CN} = \frac{V}{3}$ and $V_{BN} = -\frac{2V}{3}$

These voltages are shown in the Fig. 7.13. Similarly for other intervals phase voltages have been drawn which constitute voltage waveforms for each phase. The waveforms are not sinusoidal. These can be approximated to sinusoidal nature by decreasing duration of each interval. For purely resistive loads diodes D_1–D_6 do not conduct. However for inductive load these diodes conduct for freewheeling operation. If firing angle for the thyristors is α, waveforms shown in Fig. 7.13 are shifted by α.

(b) **120° conduction Scheme.** We again consider the phase voltages represented graphically by Fig. 7.11. Each thyristor conducts for 120° in a cycle. Let for simplicity, the firing angle be 0°. We observe the waveform at $\omega t = 0°$ when phase voltage V_{AN} starts appearing across the load which corresponds to firing of thyristor T_1. We divide again a cycle into six interval each interval of 60° duration. From 0–60° interval only V_{AN} and

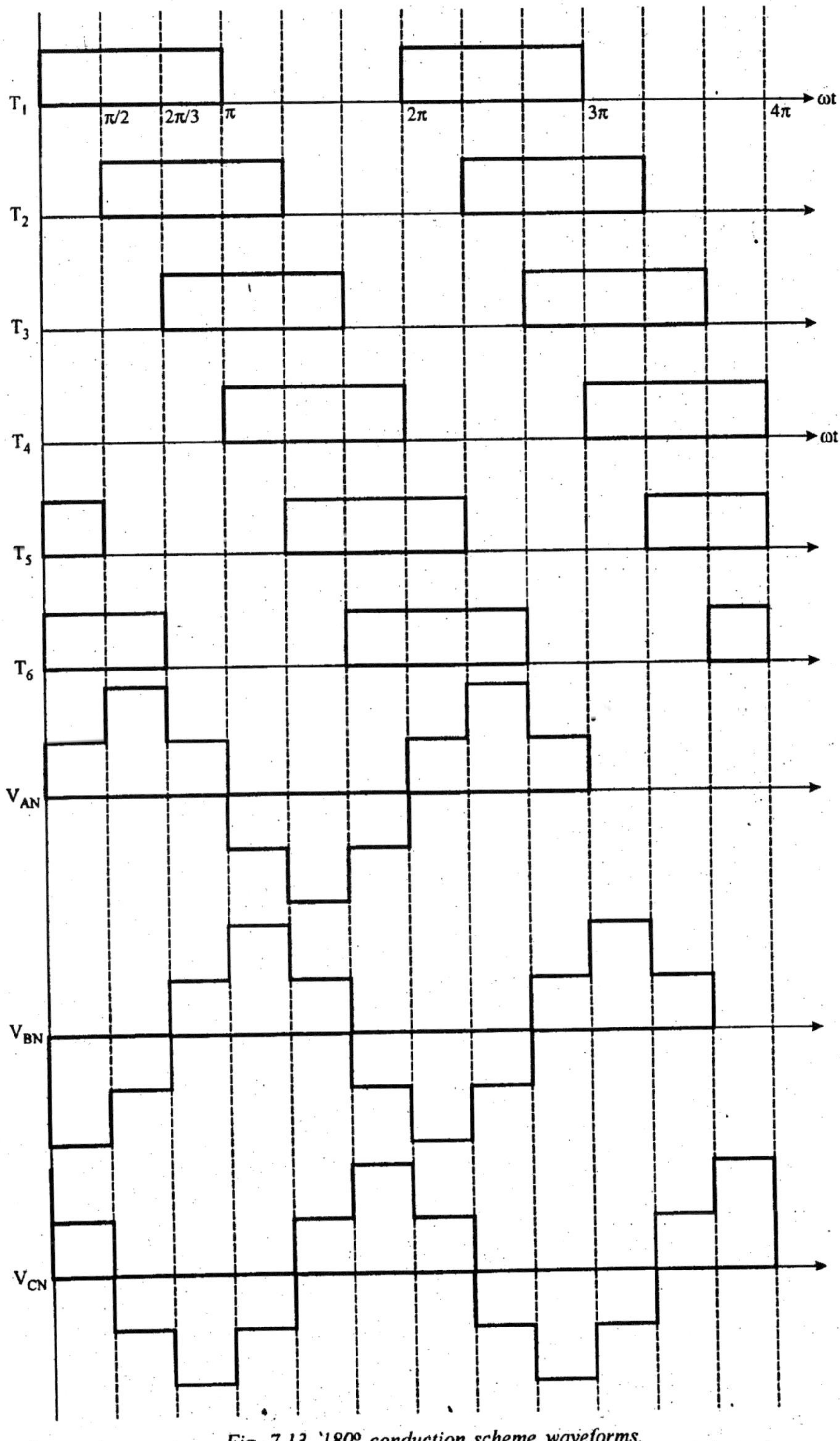

Fig. 7.13. '180° conduction scheme waveforms.

$-V_{BN}$ will appear across the load since V_{CN} has already appeared across the load for 120° duration. We conclude therefore that only two thyristors conduct at one time. The sequence of operaion of the thyristor is given in the following table.

Interval	Turn on Instant	Device turned ON	Devices conducting
I 0°–60°	0°	T_1	T_1, T_6
II 60°–120°	60°	T_2	T_1, T_2
III 120°–180°	120°	T_3	T_2, T_3
IV 180°–240°	180°	T_4	T_3, T_4
V 240°–300°	240°	T_5	T_4, T_5
VI 300°–360°	300°	T_6	T_5, T_6

Analysis

During interval I the thyristors T_1 and T_6 are conducting while all others are OFF the equivalent circuit is shown in Fig. 7.14. For purely resistive loads, from the figure

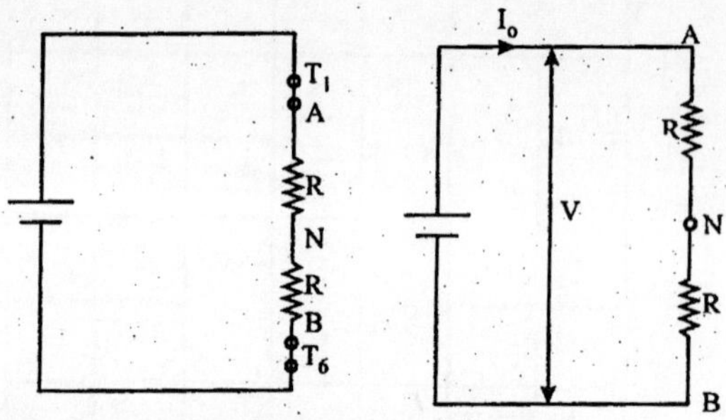

Fig. 7.14. Equivalent circuits for 120° scheme.

$$\text{Current } I_0 = \frac{V}{R+R} = \frac{V}{2R}$$

Phase voltage $V_{AN} = I_0 \times R = \frac{V}{2R} \times R = \frac{V}{2}$

$$V_{NB} = I_0 \times R = \frac{V}{2R} \times R = \frac{V}{2}$$

$$V_{BN} = -\frac{V}{2}$$

Similarly we can draw the equivalent circuit in other intervals and the phase voltages are given as below

Internval II 60° ≤ wt ≤ 120°

Equivalent circuit shown in Fig. 7.15.

The load current $I_0 = \frac{V}{2R}$

phase voltage $V_{AN} = \frac{V}{2R} + 1 = \frac{V}{2}$

$$V_{NC} = \frac{V}{2R} + R = \frac{V}{2}$$

$$V_{CN} = -\frac{V}{2}$$

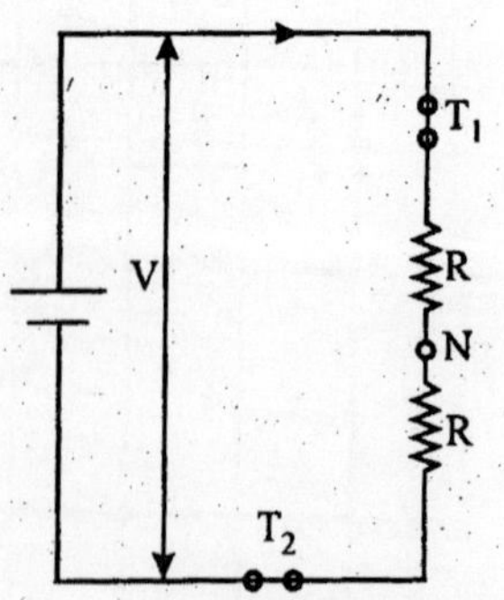

Fig. 7.15. Equivalent circuits for 120° scheme.

Interval III $120° \leq wt \leq 180°$

Equivalent circuit shown in Fig. 7.16.

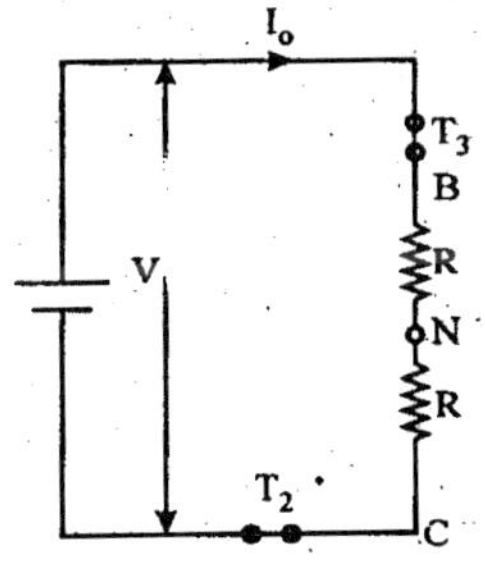

Fig. 7.16. Equivalent circuits for 120° scheme.

Load current $$I_0 = \frac{V}{2R}$$

phase voltage $$V_{BN} = \frac{V}{2R} \times R = \frac{V}{2}$$

$$V_{CN} = -\frac{V}{2R} \times R = -\frac{V}{2}$$

Interval IV $180° \leq wt \leq 240°$

Equivalent circuit shown in Fig. 7.17.

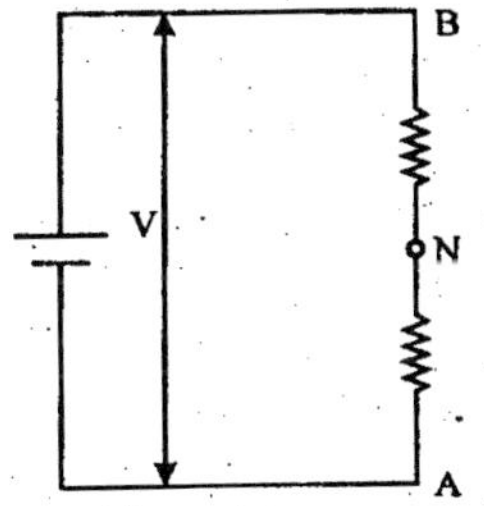

Fig. 7.17. Equivalent circuits for 120° scheme.

phase voltage are $$V_{AN} = -\frac{V}{2}$$

$$V_{BN} = \frac{V}{2}$$

Interval V $240° \leq wt \leq 300°$

See the circuit in Fig. 7.18.

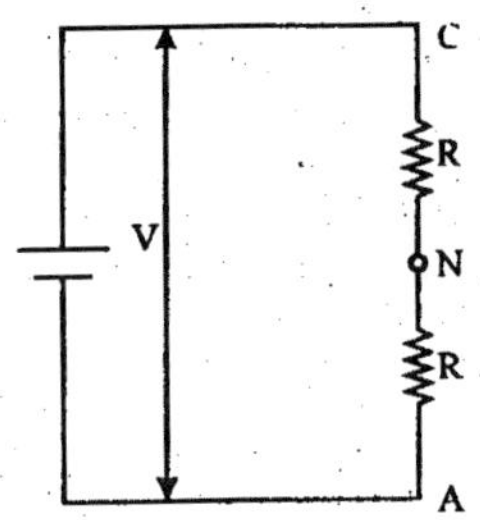

Fig. 7.18. Equivalent circuits for 120° scheme.

$$V_{AN} = -\frac{V}{2}$$

$$V_{CN} = \frac{V}{2}$$

Internal VI. The circuit is shown in Fig. 7.19.

$300° \leq wt \leq 360°$

The phase voltages are

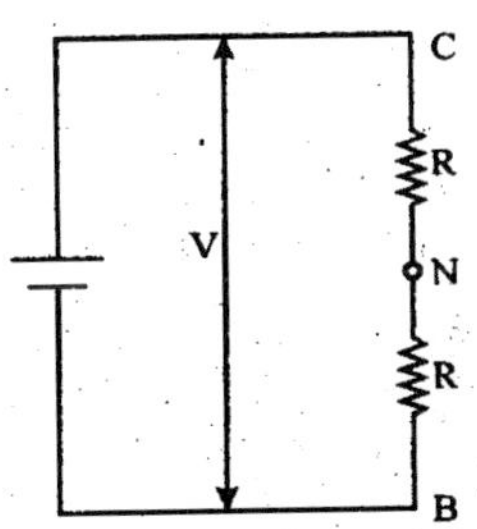

Fig. 7.19. Equivalent circuits for 120° scheme.

$$V_{CN} = \frac{V}{2}$$

$$V_{BN} = -\frac{V}{2}$$

120° conduction scheme is used for delta and star connected load. The curve forms are shown in Fig. 7.20.

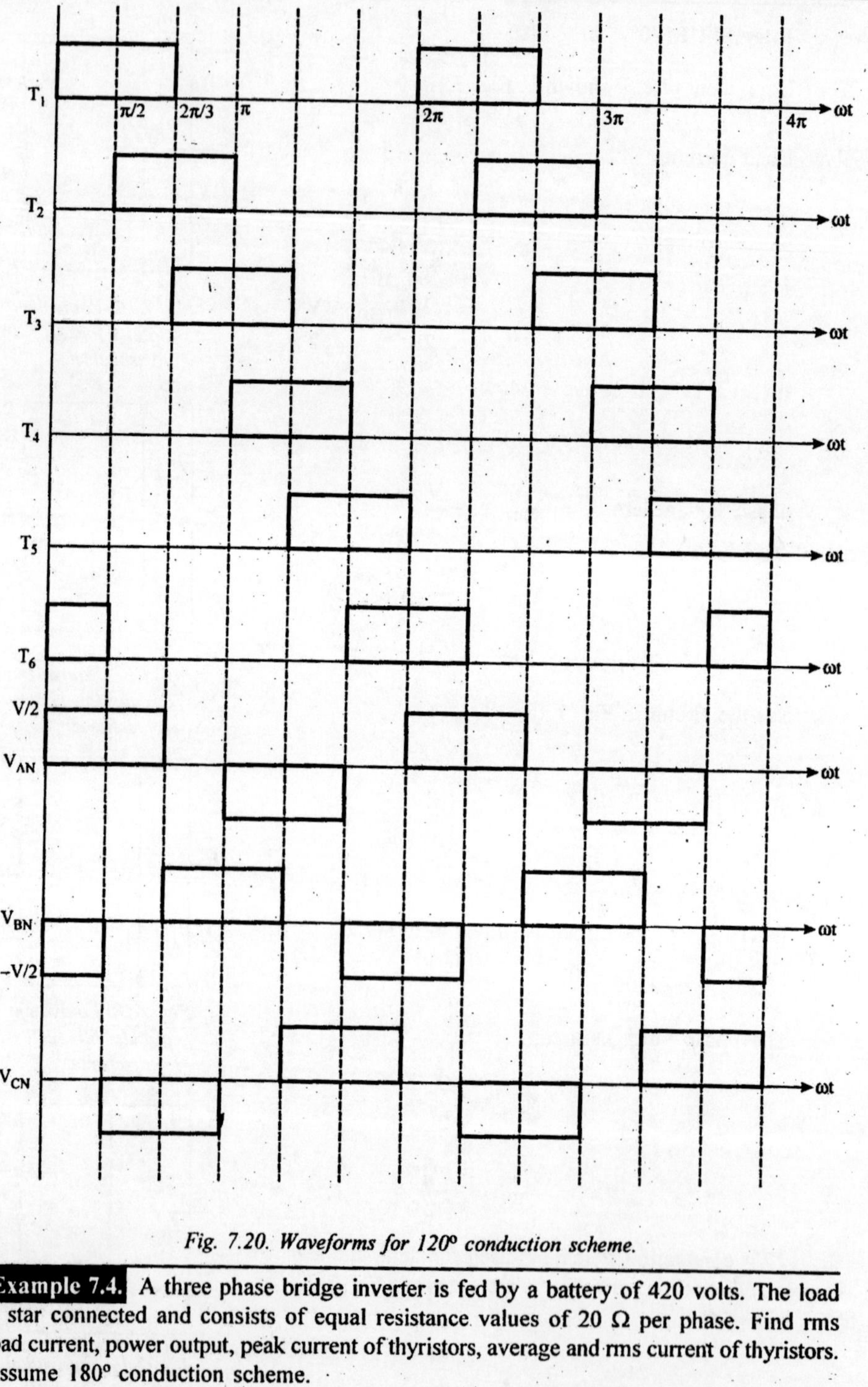

Fig. 7.20. Waveforms for 120° conduction scheme.

Example 7.4. A three phase bridge inverter is fed by a battery of 420 volts. The load is star connected and consists of equal resistance values of 20 Ω per phase. Find rms load current, power output, peak current of thyristors, average and rms current of thyristors. Assume 180° conduction scheme.

Solution. Refer to Fig. 7.12. The equivalent resistance in the load is

$$\text{R}\|\text{R} + \text{R} = 1.5\text{ R}$$
$$= 1.5 \times 20 = 30\Omega.$$

The load current has peak value during $\frac{1}{3}$ cycle and half value for $\frac{2}{3}$ cycle. The rms current will be

$$I_{0rms} = \left[\frac{1}{2\pi}\left\{\int_0^{\frac{2\pi}{3}} I_0^2 d(\omega t) + \int_{\frac{2\pi}{3}}^{2\pi} I_0^2 d(\omega t)\right\}^2\right]^{\frac{1}{2}}$$

$$I_A = I_B = I_C = I_0 = \frac{V}{1.5R} = \frac{420}{30} = 14 \text{ Amp}$$

$$I_{0rms} = \left[\frac{1}{2\pi}\int_0^{\frac{2\pi}{3}} 196 d(\omega t) + \int_{\frac{2\pi}{3}}^{2\pi} 49 d(\omega t)\right]^{\frac{1}{2}}$$

or $$I_{0rms} = \left[196 \times \frac{2\pi}{3} + 49\frac{4\pi}{3}\right]^{0.5}$$

$$= \frac{1}{\sqrt{2\pi}}\left[\frac{2\pi}{3}(196 + 98)\right]^{0.5} = \frac{1}{\sqrt{2\pi}}\left[\frac{2\pi}{3} \times 294\right]^{0.5}$$

$$= (98)^{0.5} = 9.98 \text{ Amp.}$$

Output power for three phases $= I_{orms}^2 \times R \times 3$

$$P_{out} = 90 \times 20 \times 3$$
$$= 5880 \text{ watt}$$

Peak current of thyristors $= I_0 = 14$ Amp.

Avarage current of each thyristor: Each thyristor carries a current of 14 Amp for 60°, Amp for 120° and no current for 180°. Hence the average current is

$$I_{Thav} = \frac{1}{2\pi}\left[\int_0^{\frac{\pi}{3}} 14 d(\omega t) + \int_{\frac{\pi}{3}}^{\pi} 7 d(\omega t)\right]$$

$$= \frac{1}{2\pi}\left[\frac{14\pi}{3} + 7 \times \frac{2\pi}{3}\right] = \left[\frac{14}{3} + \frac{14}{3}\right]\frac{1}{2} \text{ Amp}$$

$$= \frac{14}{3} = 4.67 \text{ Amp.}$$

rms current of each thyristor

$$I_{Thrms} = \left[\frac{1}{2\pi}\left\{\int_0^{\frac{\pi}{3}} I_0^2 d(\omega t) + \int_{\frac{\pi}{3}}^{\pi}\left(\frac{I_0}{2}\right)^2 d(\omega t)\right\}\right]^{0.5}$$

$$= \left[\frac{1}{2\pi}\left\{\frac{196\pi}{3}+\frac{49}{3}\times 2\pi\right\}\right]^{0.5}\left(\frac{294}{3\times 2}\right)^{0.8}$$

$$= (49)^{0.5} = 7 \text{ Amp.}$$

Example 7.5. Repeat example 7.4 for 120° conduction scheme.

Solution. In 120° conduction scheme load resistance of two phases are in series at any time. Hence load current

$$I_0 = \frac{v}{2r} = \frac{420}{2\times 20} = 10.5 \text{ Amp.}$$

The load current has full value during $\frac{2}{3}$ of cycle and zero value for $\frac{1}{3}$ cycle. Hence the rms load current is.

$$I_{0rms} = \left[\frac{2}{2\pi}\int_0^{2\pi/3} I_0^2 (d\omega t)\right]$$

$$= \left[\frac{1}{2\pi}(10.5)^2\left(\frac{2\pi}{3}\right)\right]^{\frac{1}{2}} = \left[\frac{(10.5)^2}{3}\right]^{\frac{1}{2}}$$

$$= 6.08 \text{ Amp.} \times 1.41 = 8.652 \text{ Amp}$$

A thyristor conducts for 120° in a complete cycle.

The peak current of thyristors is $I_{pth} = I_0 = 10.5$

$$\text{Avarage thyristor current} = \frac{1}{2\pi}\int_0^{\frac{2\pi}{3}} I_0 d(\omega t)$$

$$= \frac{1}{2\pi}[10.5\omega t]_0^{\frac{2\pi}{3}}$$

$$= \frac{1}{2\pi}\times 10.5\times\frac{2\pi}{3}$$

$$= \frac{10.5}{3} = 3.5 \text{ Amp.}$$

$$\text{rms value of thyristor current} = \left[\frac{1}{2\pi}\int_0^{\frac{2\pi}{3}} I_0^2 d(\omega t)\right]^{0.5}$$

$$= \left\{\frac{1}{2\pi}\left[(10.5)^2\times\frac{2\pi}{3}\right]\right\}^{0.5}$$

$$= 6.08 \text{ Amp}$$

$$\text{output power } p_{out} = I_{orms}^{\;2}\times R + 3 \text{ (for three phase)}$$

$$= (8.652)^2\times 20\times 3 \text{ watt}$$

$$= 4491.42 \text{ watt}$$

7.6. CURRENT SOURCE INVERTERS

In the inverters discussed so far we have used voltage sources which provide constant voltage irrespective of what the value of load is. In present section we shall discuss current source which supply fixed current to the circuit under consideration. When a voltage source has large inductance in series, it behaves like a current source. A single phase current source inverter is shown in Fig. 7.21. Th_1–Th_4 are the thyristors. D_1–D_4 are the diodes which prevent discharging of the capacitors C_1 and C_2. A variable voltage source is used in the circuit.

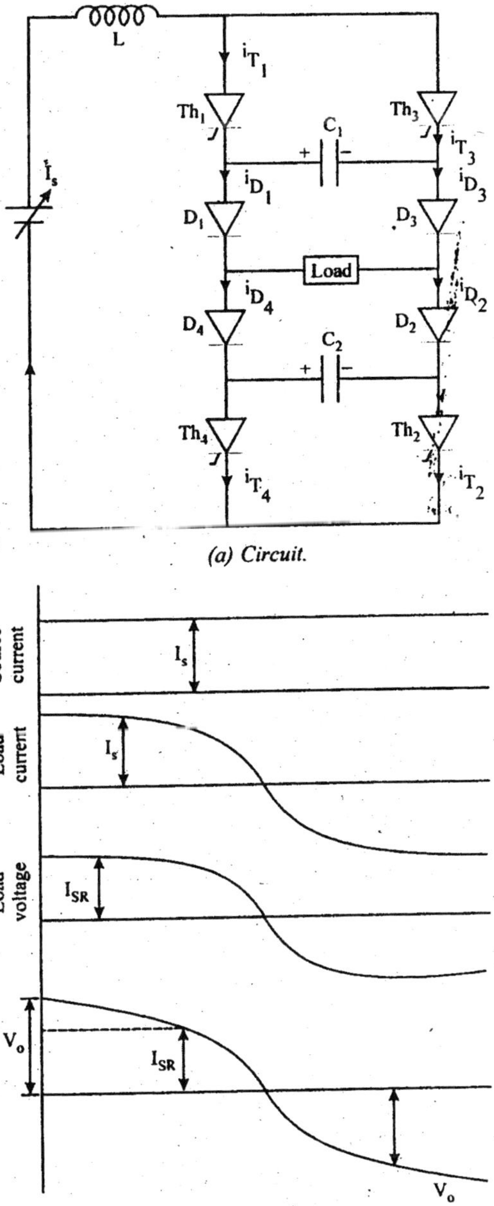

(a) Circuit.

(b) Waveforms.

Fig. 7.21. Current source inverter.

Let Th_1 and Th_2 be conducting. The thyristor current I_{T1} flows partially through capacitor C_1 to D_3 and partially D_1 and load and through D_4 – C_2. All these fractions constitute the current $I_{T2} = I_{T1}$. When the capacitors C_1 and C_2 are fully charged, whole the current flows through the path Th_1 – D_1 – load – D_2 – Th_2 – Source

The discharging of the capacitor C_1 is prevented by diode D_3 and that of C_2 is prevented by diode D_4, when the capacitors are charged with the polarity shown. Now let the pulse is applied to thyristors Th_3 and Th_4 at instant $t = 0$. The capacitor C_1 starts discharging through Th_1 and consequently it is turned OFF by impulse commutation. In a similar way Th_2 is commutated by capacitor C_2.

The waveforms of a current source inverter are shown in Fig. 7.21 *(b)*.

7.7. SERIES INVERTERS

In series inverters commutating elements L and C remain connected in series with load. Basic series inverter circuit is shown in Fig. 7.22 *(a)*. The value of L, R and C are selected such that it forms an underdamped circuit, *i.e.*,

$$R^2 < \frac{4L}{C}$$

Circuit Operation. The operation of the circuit is divided into following three modes.

Mode 1. The capacitor has initial voltage E_c. When the thyristor is turned ON at $t = 0$, the current i starts rising sinusoidally. Since the circuit is underdamped the current attains peak value when capacitor voltage reaches a value almost equal to V. The current starts falling, whereas the voltage across capacitor still keeps rising and finally reaches the value $V + E_c$ when current falls to zero value and thyristor Th_1 is turned OFF.

Mode 2. During this mode thyristors Th_1 and Th_2 are OFF. The voltage across the capacitor is still $V + E_C$. The time between *a* and *b* should be more than turn OFF time t_{off} of the thyristor.

Mode 3. This mode begins at *b* when thyristor Th_2 is turned ON. The capacitor starts discharging through L and R. The reverse current attains the peak value when the capacitor voltage falls to V. Afterwards the current starts falling, attains zero value at point c when thyristor Th_2 is turned OFF.

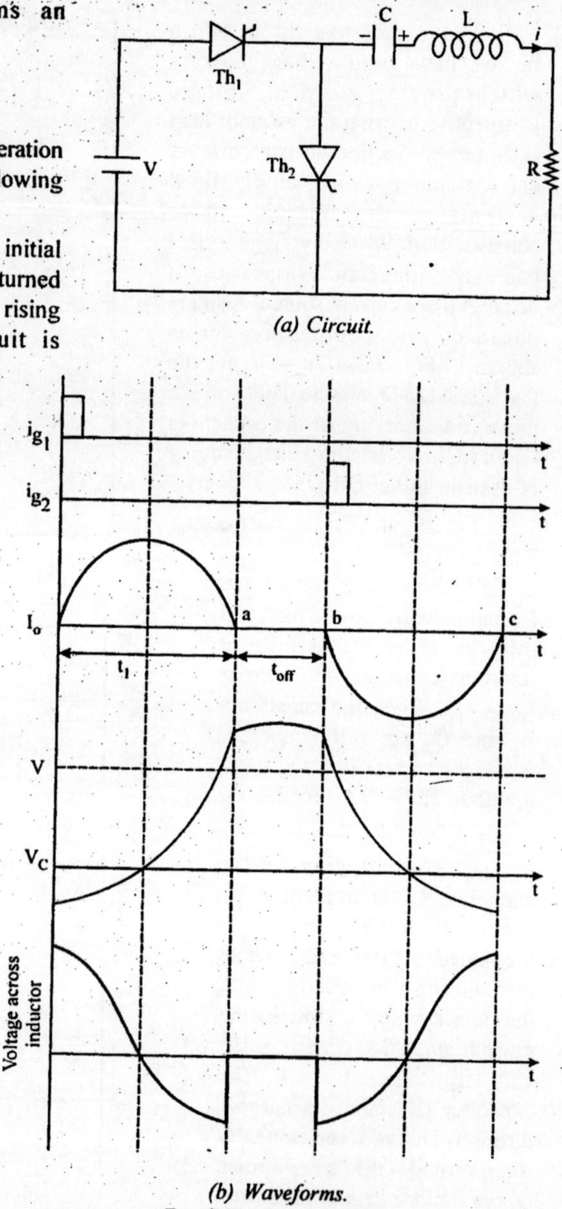

(a) Circuit.

(b) Waveforms.

Fig. 7.22. Series inverter.

This process is repeated again with the firing of thyristor. Th_1. The frequency of the output voltage is given by $f = \frac{1}{T} = \frac{1}{2(t_1 + t_{off})}$. Hz

Analysis of the Circuit

Applying KVL to the mesh containing source, Th_1, C, L and R, we get

$$V + E_c = \frac{Ldi}{dt} + Ri + \frac{1}{c}\int idt \quad ...(11)$$

Laplace transformation gives

$$\frac{V + E_c}{s} = LsI(s) + RI(s) + \frac{1}{C_s}I(s)$$

or

$$I(s) = \frac{(V + E_c)}{s\left\{Ls + R + \frac{1}{Cs}\right\}} = \frac{V + E_c}{L\left(s^2 + \frac{Rs}{L} + \frac{1}{LC}\right)}$$

The characteristic eqn. is

$$s^2 + \frac{R}{L}s + \frac{1}{LC} = 0 \quad ...(12)$$

The roots of eq. (12) are s_1 and s_2 given by

$$s_1, s_2 = \frac{-\frac{R}{L} \pm \sqrt{\left(\frac{R}{L}\right)^2 - 4 \times \frac{1}{LC}}}{2}$$

$$= -\frac{R}{2L} \pm \sqrt{\left(\frac{R}{2L}\right)^2 - \frac{1}{LC}} \quad ...(13)$$

For underdamped case, second term in eq. (13) should be imaginary for which one should have

$$\left(\frac{R}{2L}\right)^2 < \frac{1}{LC}$$

or

$$R^2 < \frac{4L}{C}$$

Hence we can write eqn. (13) as

$$s_1, s_2 = -\frac{R}{2L} \pm j\sqrt{\frac{1}{LC} - \left(\frac{R}{2L}\right)^2}$$

Eqn. (12) can be written as

$$\left(s^2 + \frac{R}{L}s + \frac{1}{LC}\right) = (s - s_1)(s - s_2)$$

$$= \left[s - \left\{\frac{R}{2L} - j\sqrt{\frac{1}{LC} - \left(\frac{R}{2L}\right)^2}\right\}\right]\left[s - \left\{\frac{R}{2L} + j\sqrt{\frac{1}{LC} - \left(\frac{R}{2L}\right)^2}\right\}\right] \quad ...(14)$$

Let us put $\frac{1}{LC} = \omega_n^2$ and $\varsigma = \frac{R}{2\sqrt{L/C}}$ in eqn. (14) and get the eqn. as

$$= \left[s - \left\{ -\frac{\frac{R}{LC}}{2\frac{LC}{C}} + j\sqrt{\frac{1}{LC} - \left(\frac{\frac{R}{2LC}}{C}\right)^2} \right\} \right] \left[s - \left\{ -\frac{\frac{R}{2LC}}{C} - j\sqrt{\frac{1}{LC} - \left(\frac{\frac{R}{2LC}}{C}\right)^2} \right\} \right]$$

$$= \left[s - \left\{ -\varsigma\omega_n + j\sqrt{\omega_n^2 - \varsigma^2\omega_n^2} \right\} \right] \left[s - \varsigma\omega_n + j\sqrt{\omega_n^2 - \varsigma^2\omega_n^2} \right]$$

$$= \left[(s + \varsigma\omega_n) - j\sqrt{\omega_n^2 - \varsigma^2\omega_n^2} \right] \left[(s + \varsigma\omega_n) + j\sqrt{\omega_n^2 - \varsigma^2\omega_n^2} \right]$$

$$= (s + \varsigma\omega_n)^2 + \omega_n^2 - \varsigma^2\omega_n^2$$

$$= s^2 + 2\varsigma\omega_n s + \varsigma^2\omega_n^2 + \omega_n^2 - \varsigma^2\omega_n^2$$

$$= s^2 + 2\varsigma\omega_n s + \omega_n^2 \qquad \text{...(15)}$$

Hence, we get

$$I(s) = \frac{V + E_c}{L\left(s^2 + 2\varsigma^2\omega_n s + \omega_n^2\right)},$$

$$= \frac{V + E_c}{L(s^2 + \varsigma^2\omega_n^2 + 2\varsigma\omega_n s + \omega_n^2 - \varsigma^2\omega_n^2)}$$

$$= \frac{V + E_c}{L(s^2 + \varsigma^2\omega_n^2 + 2\varsigma\omega_n s + \omega_n^2(1 - \varsigma^2)}$$

$$I(s) = \frac{(V + E_c)\omega_d}{\omega_d \, L\left[(s + \varsigma\omega_n)^2 + \omega_d^2\right]} \qquad \text{...(16)}$$

where $\omega_d = \omega_n\sqrt{1 - \varsigma^2}$ is called damped natural frequency.

Taking inverse laplace transform of eqn. (16), we obtain

$$i(t) = \frac{V + E_c}{\omega_d L}\left(e^{-\varsigma\omega_n t} \sin \omega_d t\right) \qquad \text{...(17)}$$

The load current given by eqn. (17) is shown in Fig. 7.22 *(b)*. The current reaches the value zero at time t_1 such that $\omega_d t_1 = \pi$ or $t_1 = \dfrac{\pi}{\omega_d}$.

However this inverter suffers from following drawbacks.

1. Major drawback of the circuit is that if thyristor Th_2 is turned ON before Th_1 is completely turned OFF, the source may be short circuited and hence damaged.
2. Due to time t_{off} this output is not purely sinusoidal which introduces harmonics in the output.
3. Maximum load current depends upon the load resistance. Therefore output regulation of the circuit is poor.
4. Circuit should be underdamped. Hence for large inverter the values of L, C, and R should be high.

The above drawbacks (1, 3) are removed by modifying the circuit of the inverter which is discussed below.

7.8. MODIFIED SERIES INVERTER CIRCUIT

The circuit diagram is shown in Fig. 7.23 *(a)*. Two identical inductors L_1 and L_2 have been used such that L_1, C, R and L_2, C, R form underdamped circuits. Let the capacitor has initial voltage E_c with upper plate negative. When thyristor Th_1 is fired at time $t = 0$, e.m.f. is induced in inductor L_1 with the polarity shown. This in association with capacitor voltage turns thyristor Th_2 OFF. The voltage rising across the load reaches peak value. The analysis of the circuit is exactly the same as that discussed in section 7.7. The advantage of this circuit is that thyristor Th_2 (or Th_1) can be turned ON even when thyristor Th_1 (or Th_2) are in ON condition without short circuiting the source. The reason for this is the presence of inductors. When thyristor Th_2 is turned ON while Th_1 also is ON the voltage across inductors will be non zero. The result wil be that the source will be avoided from short circuit even under the condition that $t_{off} = 0$. The frequency of output voltage can be increased more than the resultant frequency. The expression for the frequency becomes.

$$f = \frac{1}{2t_1}$$

The waveforms are shown in Fig. 7.23 *(b)*.

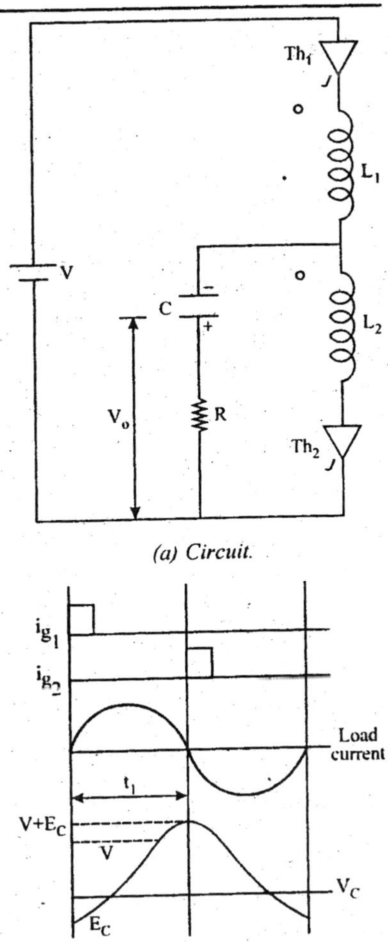

(a) Circuit.

(b) Waveforms.

Fig. 7.23. Modified series inverter.

Modified series inverter can also be designed by using two capacitors as shown in Fig. 7.24 *(a)*. Let the voltage across capacitor C_2 is E_c with the polarity shown. Then by applying KVL in the mesh containing C_1, C_2 and V the voltage across C_1 will be $V + E_C$ with the polarity shown. Let a pulse is applied to thyristor Th_1 at time $t = 0$. The current starts flowing through the path V, Th_1, L_1, R and C_2. Another current flows from upper plate of capacitor C_1 through Th_1, L_1 and load. Since $V_{c_1} \pm V_{c_2} = V$, a constant. As a result increase in V_{c_1} should be accompanied by decrease in V_{c_2}. The voltage across capacitor C_2 reaches the value $V + E_C$ with upper plate positive when the current I_L reaches zero. The capacitor C_1 attains voltage E_c with lower plate positive. Now the pulse is applied to thyristor Th_2. Again the same process occurs. The current flows through the load in the direction opposite to one shown in the figure. The load current flows through source, C_1, R and Th_2. Another current is due to discharging of capacitor C_2 through C_2, R, L_2 and Th_2. Thus in this type of inverter power from the source is drawn in both the half cycles. The waveforms are shown in Fig. 7.24 *(b)*.

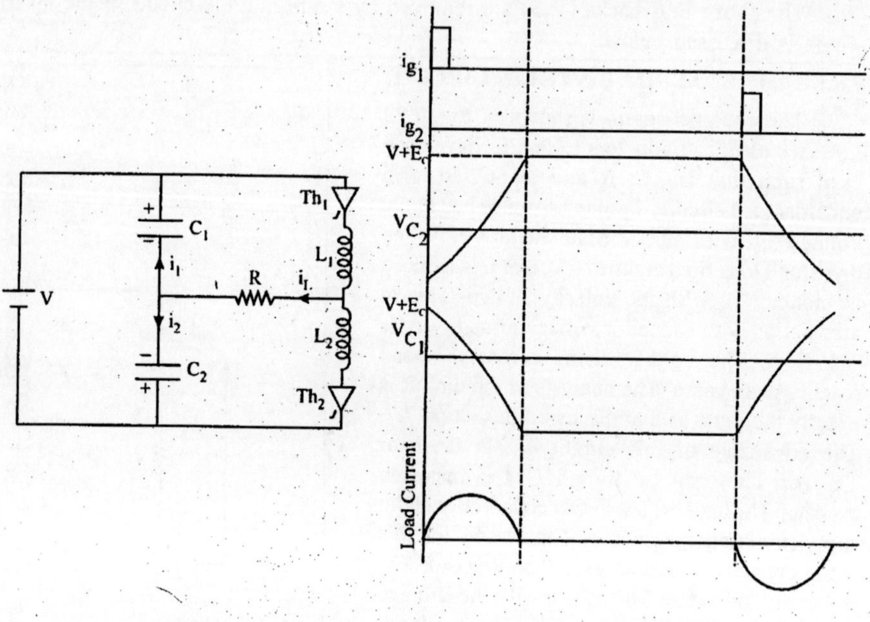

(a) Circuit *(b) Waveforms*

Fig. 7.24. Modified series inverter.

7.9. BIDIRECTIONAL SWITCH IN SERIES INVERTER

The circuit is shown in Fig. 7.25 *(a)*. In the circuit, an antiparallel diode D is connected across the thyristor T. When T is turned ON the current flows through underdamped circuit. After sometime the thyristor is self commutated. The voltage developed across the capacitor is more than the supply voltage V. After T is turned OFF the capacitor discharges through the load connected through transformer. A part of the energy is fedback through diode D. Waveforms are shown in Fig. 7.25 *(b)*.

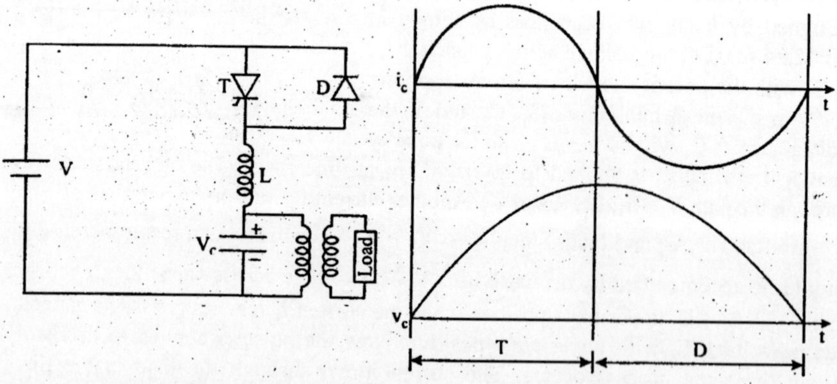

Fig. 7.25. Series inverter using bidirectional switch.

7.10. THREE PHASE SERIES INVERTER

Three phase inverter is the parallel combination of three single phase inverters. The circuit is shown in Fig. 7.26. The capacitors C and inductors are used to turn off the thyristors as in case of single phase inverters. The firing sequence of the six thyristors is 1, 2, 3, 4,5, 6. It is clear that for one output cycle six pulses are to be applied. The capacitors C_1 and C_2 are chosen larger in values so that constant neutral can be provided.

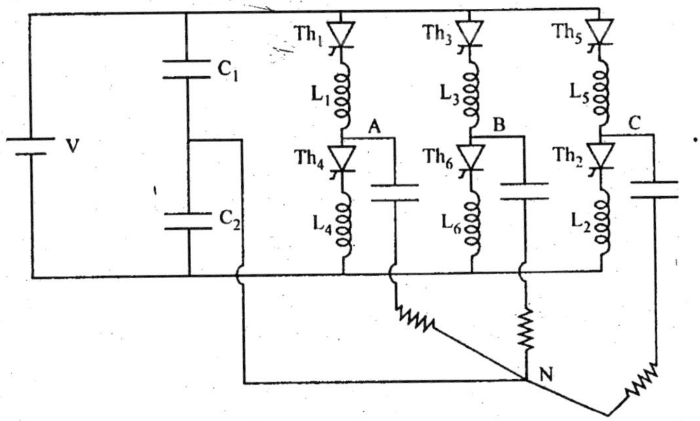

Fig. 7.26. Three phase series inverter.

7.11. HIGH FREQUENCY SERIES INVERTER

The limitations of the thyristors discussed above is that the frequency of output has maximum value of resonant frequency. These limitations on the frequency of output voltage can be overcome by using high frequency series inverter. The circuit is shown in Fig. 7.27. When pulse is applied at Th_1, the current flows through V_1, P_1, Th_1, L_1 and C in the direction shown. When Th_6 is fired, the current flows through V'_1, C, L_6, Th_6, and P_2 which is opposite to the current through P_1. Thus a cycle is completed. The thyristors are turned OFF by self commutation. Similarly other thyristors are turned ON.

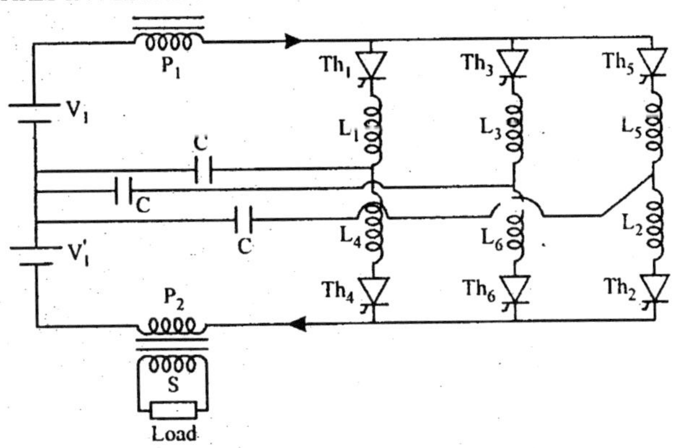

Fig. 7.27. High frequency inverter.

Example 7.6. A series inverter is fed by 2000 volts. The value of circuit components are C = 10 μF, L = 20 mH R = 80Ω. If turn OFF time of thyristor is 6 ms, what is the frequency of output voltage?

Solution. The time of duration of half cycle is given by $t_1 = \dfrac{1}{2f} = \dfrac{\pi}{\omega_d}$

where $\omega_d = \sqrt{\dfrac{1}{LC} - \dfrac{R^2}{4L^2}}$

Hence
$$t_1 = \frac{\pi}{\sqrt{\frac{1}{LC}-\frac{R^2}{4L^2}}} = \frac{3.14}{\sqrt{\frac{1}{10\times10^{-6}\times20\times10^{-3}}-\frac{80^2}{4\times20\times10^{-6}+20}}}$$

$$= \frac{3.14}{\sqrt{\frac{1}{2\times10^{-7}}-\frac{6400}{16\times10^{-4}}}} = \frac{3.14}{\sqrt{5\times10^{+6}-4\times10^{6}}}$$

$$= 3.14\times10^{-3}\text{ Sec}$$

Frequency of output voltage

$$f_o = \frac{1}{2(t_1+t_{off})} = \frac{1}{2(3.14\times10^{-3}+6\times10^{-6})}\text{Hz}$$

$$= \frac{10^3}{18.28}\text{ Hz.} = 54.7\text{ Hz.}$$

Example 7.7. In the above example the value of damping factor is increased by 10%. What will be the value of damped frequency?

Solution. Previously the value of damping factor is

$$\varsigma = \frac{R}{2\sqrt{L/C}} = \frac{80}{2\sqrt{\frac{20}{10}\times10^3}} = \frac{40}{\sqrt{20}} = 0.89$$

10% increase in value gives the total damping factor

$$= 0.89+0.089$$

$$= 0.979$$

$$\text{damped frequency} = \omega_d = \omega_n\sqrt{1-\varsigma^2}$$

$$= \frac{1}{\sqrt{20\times10^{-3}\times10\times10^{-6}}}\sqrt{1-(0.979)^2}$$

$$= 229.6\text{ Hz.}$$

7.12. PARALLEL INVERTER

Circuit diagram of parallel inverter is shown in Fig. 7.28 *(a)*. Capacitor C is in parallel with the transformer primary and load while load resistance is in parallel with its secondary. L and C are commutating elements in the circuit. The number of turns in the primary of transformer is twice those in secondary. Midpoint O of the primary MN is connected to the battery through inductor L. As T_1 is turned ON the current flows through part OM of the primary.

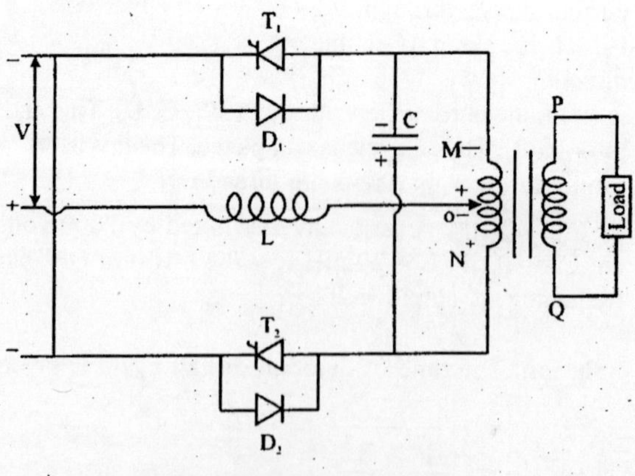

Fig. 7.28 (a) Parallel inverter.

Induced e.m.f. is developed in it with point O positive. By transformer action same potential is developed in the half ON with N end being positive. Hence the total potential between MN and the capacitor connected across it is V+ (V) = 2V with the polarity shown in the figure. The negative plate of the capacitor is connected to the anode of SCR T_1.

Immediately the pulse is applied to thyristor T_2, the capacitor starts discharging through T_2, source and D_1 after T_1 is turned OFF. After the capacitor is fully discharged it will again charge to the voltage 2V with the upper plate positive. The lower negative plate connected to the anode of thyristor T_2 turns it OFF. The pulse is again applied to thyristor T_1. This process is continuous. The thyristors are gated in alternate manner. The waveforms are shown in Fig. 7.28 *(b)*. The current flowing in the lower and upper half of the primary is the same as the current in the secondary and hence the load. When the current flows through part ON, the current in the load flows from bottom to top and from top to bottom when the current in the primary is from O to M. Thus we get a.c. output across the load.

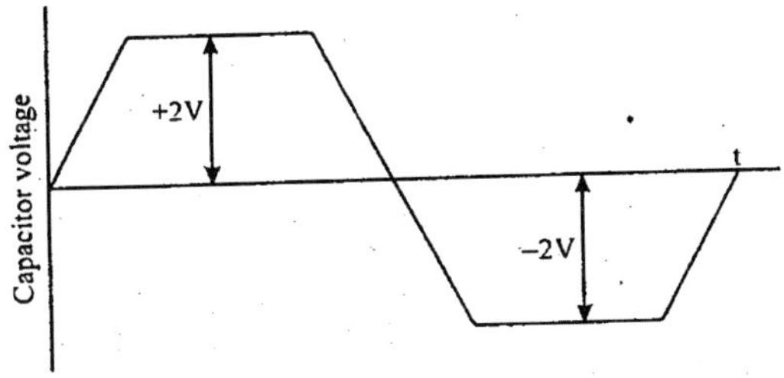

Fig. 7.28 (b) Waveforms.

Example 7.8. A parallel bridge inverter is fed by a d.c. source of 200 V. The thyristors are fired at an interval of 5 m sec. Find r.m.s. load current and rms power developed across the load. The load resistance is 24 Ω.

Solution. Assuming load voltage waveforms rectangular, we obtain the rms current

$$I_{orms} = \sqrt{\frac{1}{T}\int_0^T I_0^2\, dt}$$

The secondary voltage = source voltage = V

$$\text{Hence load current} = \frac{V}{R} = \frac{200}{20} = 10 \text{ Amp.}$$

$$I_{Orms} = \sqrt{\frac{1}{(5+5)\times 10^{-3}} \int_0^{10\times10^{-3}} 10^2\, dt}$$

$$= \sqrt{10^3\left[10\times 10^{-3}\times 10\right]} \text{ Amp.} = 10 \text{ Amp. } \textbf{Ans.}$$

$$\text{rms power} = I_{orms}^2 \times R$$

$$= 10^2\times 20 = 2\text{kW}$$

SUMMARY

An inverter is used to convert dc into ac. For this purpose various circuits have been described in this chapter. The circuits we have studied in the present chapter include bridge inverter - single and 3ϕ both, series inverter, parallel inverter, Mc Murray inverter, modified Mc Murray inverter, Mc Murray Bedford inverter, current source inverters etc.

EXERCISE

1. What is an inverter? Describe working principle of an inverter.
2. Describe single phase half bridge and full bridge inverters. Compare the harmonic contents of the two.

3. Explain the operation of modified Mc Murray half bridge inverter.
4. Describe Mc Murray Bedford half bridge inverter. Draw the waveforms.
5. Why are three phase inverters used? Describe working principle of three phase bridge inverters.
6. Explain 180° conduction schemes in three phase bridge inverter.
7. What is the use of 120° conduction scheme in three phase inverter. Explain its working.
8. Write a note on current source inverter
9. What is meant by series inverter. Derive expression for the load current in series inverter.
10. How can the frequency of the output be increased by modifying series inverter? Explain the related circuit.
11. Describe in detail parallel inverters.
12. Explain series inverter using bidirectional switch.

NUMERICAL PROBLEMS

1. A single phase half bridge inverter has resistive load of 5 ohms. The d.c. input voltage is 50 V. Find *(a)* rms value of fundamental component of output voltage *(b)* output power *(c)* peak and average thyristor current.
2. A single phase full bridge inverter has resistive load of 5Ω. The source voltage is 50 V. Find *(a)* rms value of load current *(b)* output power *(c)* peak and rms value of thyristor current.
3. A full bridge inverter has input voltage of 200V. The load is an RL type with R = 10 Ω and L = 20 mH, the operating frequency is 60 Hz. Find *(a)* peak and rms value of fundamental component of load current *(b)* find peak load currnt considering harmonics upto 5th level.
4. A three phase full bridge inverter feeds a star connected load having 10 Ω resistance in each phase. Input voltage is 150V. find rms load current, power output, peak, rms and average thyristor current for 120° operation.
5. Repeat problem 4 for 180° operation.
6. A series inverter has R = 60Ω, L = 6 mH, C = 1μf. Find maximum output frequency.
7. A series inverter has R = 100Ω, L = 10 mH and C = 1.3 μf. If turn off time is 6μs, find output frequency.
8. A single phase full bridge inverter has a load resistance of 5 ohms and dc input voltage of 50V. Find *(a)* output power *(b)* peak and average thyristor current *(c)* peak inverse voltage.

OBJECTIVE TYPE QUESTIONS

1. An inverter is a

   (a) d.c. to a.c. converter *(b)* d.c.-d.c. converter

   (c) fixed d.c. to variable d.c. converter *(d)* none of the above
2. In a series inverter

   (a) RLC form an underdamped circuit

   (b) RLC has no role to play

   (c) power supply may be short circuited

   (d) highest frequency may be equal to resonant frequency
3. In single phase full bridge inverter

   (a) The output has no harmonic

   (b) The output has harmonics less then those in half bridge inverter

(c) The output voltage is given by $v_0 = \sum_{n=1,3,-5}^{\infty} \frac{4V_s}{n\pi} \sin n\omega t$

(d) both (b) and (c). (e) none of tthe above.

4. In modified Mc Murray half and full bridge inverters :

(a) Natural commutation is used (b) Forced commutation is used

(c) Current commutation is used (d) No commutation is required

5. In Mc Murry Bedford inverter

(a) Current commutation is used

(b) complementary voltage commutation is used

(c) Natural commutation is used (d) Auxiliary voltage commutation is used

6. In 120° conduction scheme

(a) Two devices conduct at one time

(b) Two or three devices conduct at one time

(c) Every device conduct for 180° (d) Three devices conduct at one time

7. In a series inverter

(a) commutating elements and load are in series

(b) commutating elements and thyristors are in series

(c) Capacitor may be in parallel with L but in series with load

(d) none of the above

8. Frequency of output of series inverter is given by

(a) $f = \frac{1}{T}$ (b) $f = \frac{1}{2(t_1 + t_{off})}$

(c) $\frac{1}{2(t_{off})}$ (d) $\frac{1}{2\pi\sqrt{LC}}$

9. In series inverter

(a) source supplies power for half cycle

(b) source supplies power for both half cycles

(c) The frequency has maximum value of reasonance frequency

(d) there is no limit in frequency.

10. In modified series inverter

(a) the output frequency may be higher than in case of series inverter

(b) there is no change in output frequency

(c) source supplies power for half cycle only

(d) a natural commutation occurs.

ANSWERS

1. (a)	2. (a), (d)	3. (d)	4. (b), (c)	5. (b)	6. (a)
7. (a)	8. (b)	9. (a), (c)	10. (a)		

8

Induction And Dielectric Heating

8.0. INTRODUCTION

Among various methods of heating electrical heating is preferable because of its fast response, automatic control, cleanliness, high range of heating etc. Electrical heating is categorized as (1) Resistance heating (2) induction heating and (3) dietectric heating. The first method is very old method and is generally used when source is d.c. However later two methods viz. induction heating and dielectric heating require a.c. source for heating as the heat dissipated depends upon the frequency. For heating, high frequency is required. These two processes of heating are therefore known as high frequency heating.

8.1. RESISTANCE HEATING

Suppose a current I amp flows through a resistance R Ω. The power dissipated as heat is I^2R. The resistances used may be metal conductors, carbon tubes, liquids etc. The metal conductors may be heated by placing them in an oven and connecting to d.c source. For maintaining high temperature upto 200° F, carban tube on heating elements are used. As high current passes through these tubes, the heat is dissipated.

For heating liquids, like electrolyte two electrodes are placed in the liquid and rated current is passed between them. The liquid possessing resistance generates heat I^2R. The electrodes generally used are of carbon or graphite. The method is used in chemical and metallurgical furnaces. This method, however, has certain limitations for its use in high temperature heating.

8.2. INDUCTION HEATING

The induction heating is based on the princple of electromagnetic induction. When time varying current flows through any coil, it produces time varying flux which produces eddy currents in a metallic slab subjected to it. The eddy current produces heat in the specimen. Induction heating is sometimes called eddy currents heating. If the specimen is magnetic, hysteresis also takes place which further leads to heat dissipation called hysteresis loss. Thus these two types of heating viz. eddy current heating and hysteresis loss are due to e.m.f induced. Induction heating is used in melting, forging, surface hardening, annealing etc.

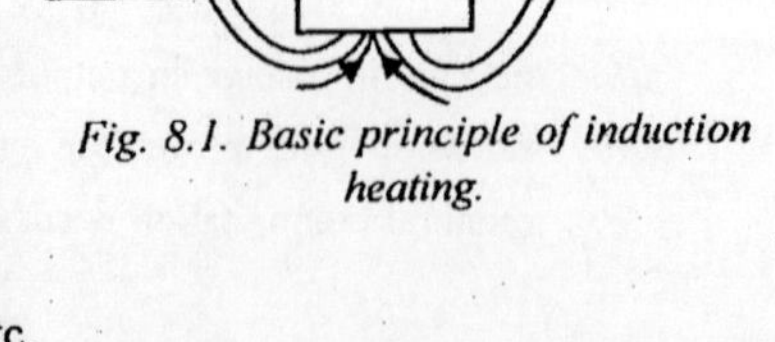

Fig. 8.1. Basic principle of induction heating.

We consider Fig. 8.1. Current carrying coil is wrapped around the specimen called the charge. As a.c current flows through the coil, it produces time varying magnetic flux which is cut by the specimen. As the flux through the specimen changes e.m.f. is induced which generates eddy currents through it. This current generates heat in the specimen which is in the form of eddy current losses. The eddy current loss is given by the expression

$$W_e = Kf^2 B_m^2 V \quad ...(1)$$

where K is a constant,

f = frequency of input supply, cycles/sec.

B_m = max flux density, wb/m^2

V = volume of specimen, m^3.

From above expression one can conclude that heating of the speciman can be affected in many ways. For induction heating, therefore, high frequency input is more useful as it depends upon square of the frequency. However, if the specimen is a magnetic material, in addition to eddy current losses, hyisteresis loss also takes place. Such materials can be heated to higher degrees as compared to non magnetic materials, under the same conditions. The hysteresis loss is given by

$$W_h = K' B_m^{1.6} fV \quad ...(2)$$

Adding eqn. (1) and (2), we obtain expression for total heat generated in magnetic material as

$$W_T = w_e + w_h = Kf^2 B_m^2 V + K' B_m^{1.6} fV \quad ...(3)$$

From eqn. (3) following points may be noted.

1. All other parameters such as frequency, flux density, volume etc being given, the heat input depends upon the permeability of the material. Magnetic materials get heated up faster than the non-magnetic materials.

2. The depth of heat penetration can be controlled by changing the frequency.

3. Flux density affects the heat input to the material which can be changed by changing the number of turns in the coil around the specimen.

4. The heat input to the specimen depends on its resistance. The heat input is higher at higher resistance.

5. Upto Curie point the magnetic materials heat up faster than nonmagnetic materials. Above Curie point both are heated equally as only eddy current loss occurs above Curie point.

6. For different purposes different ranges of frequency of input can be used. For example for melting purpose mains frequency is used, but for forging, annealing higher frequency ranging from 0.5 to 4 kHz is used.

For producing high frequency inputs various methods are used. Some of the devices used to convert frequency include motor-generator set (960, 3000, or 9600 Hz), Mercury pool frequency converter for 1 kHz, spark gap converter in the range from 50 Hz to 200 kHz and electric heater in range from 300 kHz to 1 MHz.

The working of spark gap generator is based on the principle of oscillating condenser. The condenser discharges through the spark gap. For very high frequency current heating, however, electronic heaters are used.

Induction heating has proved to be very advantageous, particularly in metal industries. Some of the advantages of the process are as under:

1. Very high heating rates can be attained.

2. These can be used to heat up selected part of the specimen without heating whole material.

3. Wastage of heat is not much.

4. Automatic control of temperature can be attained.

5. Continuous furnance temperature is not required. The furmnce may be energized at the moment of heating.

6. Automatic stirring effect results in thorough mixing of alloys in the melting process.

7. Absence of ash, flue gas, etc in the process provides better working condition. However, the system is costly as it requires a ferquency converter. The efficiency of the system, however, is quite low.

Example 8.1. In an induction heating process following parameters were used $f = 200$ Hz, $B_m = 1$ wb/m², K = 0.83, V = 2 m³. Calculate the value of heat produced due to eddy current losses in two hours.

Solution. Eddy current loss is given by

$$\begin{aligned} W_e &= K.B_m^2 f^2 V \text{ watt} \\ &= 0.83 \times 1 \times 1 \times (200)^2 \times 2 \text{ watt} \\ &= 0.83 \times 8 \times 10^4 \text{ watt} \\ &= 6.64 \times 10^4 \text{ watt} = 66.4 \text{ kw} \end{aligned}$$

$$\begin{aligned} \text{Electrical energy} &= 66.4 \times 2 \text{ kwh} = 132.8 \text{ kwh} \\ \text{Heat produced} &= 132.8 \times 3.6 \times 10^5 \text{ Joule} \\ &= \frac{132.8 \times 3.6 \times 10^5}{4.18} \\ &= 1.14 \times 10^5 \text{ k cal.} \end{aligned}$$

Example. 8.2. In an induction heating process the following data were obtained.

$K = 0.80, f = 500$ Hz, $B_m = 1.5$ Wb/m², $V = 2$ m³, $K' = 500$

Calculate total heat losses in half an hour.

Solution. Eddy current losses

$$\begin{aligned} W_e &= K\, Bm^2 f^2 v \text{ watt} \\ &= 0.80 \times (1.5)^2 \times (500)^2 \times 2 \text{ watt} \\ &= 0.80 \times 2.25 \times 25 \times 10^4 \times \text{ watt} \\ &= 0.80 \times 450 \times 25 \times 10^4 \text{ watt} \\ &= 9000 \times 10^4 \text{ watt} = 9 \times 10^4 \text{ kw} \end{aligned}$$

$$\begin{aligned} \text{Electrical energy} &= 9 \times 10^4 \times \frac{1}{2} \text{ kwh} \\ &= 4.5 \times 10^4 \text{ kwh} \end{aligned}$$

$$\begin{aligned} \text{Hysteresis loss} \quad W_h &= k^1 B_m^{1.6} FV \\ &= 500 + (1.5)^{1.6} \times 500 \times 2 \text{ watt.} \\ &= 5 \times 10^3 \times (1.5)^{1.6} \text{ watt} \\ &= 5 \times 10^3 \times 1.91 = 9.56 \times 10^3 \text{ watt} \\ &= 9.56 \text{ k watt} \end{aligned}$$

$$\text{Electrical energy} = 9.56 \times \frac{1}{2} \text{ kwh} = 4.78 \text{ kwh.}$$

$$\text{Total electrical energy} = 4.5 \times 10^4 + 4.78 \text{ kwh} = 45000 + 4.78 \text{ kwh}$$

$$= 45004.78 \text{ kwh}$$

$$\text{Total heat} = (860 \times 45004.78) \text{ kCal} = 3.87 \times 10^7 \text{ k cal}$$

Example 8.3. In the above problem what percentage of total heat is contributed by the hysteresis ?

Solution. Contribution to heat by hysteresis loss

$$= 4.78 \times 860 \text{ kCal}$$

$$\text{Total heat} = 45004.78 \times 860 \text{ kCal.}$$

$$\% \text{ of hysteresis loss} = \frac{4.78}{45004.78} \times 100\% = 0.01\%$$

8.3. OSCILLATOR CIRCUITS USED IN INDUCTION HEATERS

Objects of medium sizes can be heated up at frequency about 50 kHz. However for very small objects high frequencies are needed to heat them up. The high frequency signals are produced by electric generators which are basically oscillators. We shall discuss oscillators to produce very high frequcncy outputs.

1. Electronic generators. The circuit diagram of an oscittator is shown in Fig. 8.2. Capacitor C_T and inductor L_T form the tank or tune circuit which is connected to the plate of triode valve through blocking capacitor C_b which blocks dc supply. Any transient present in the supply E_b will pass through capacitor C_b and the tank circuit $C_T - L_T$ will be set into oscillation generating a frequency given by $f = \frac{1}{2\pi\sqrt{C_T L_T}}$. A part of the oscillator output is fedback to the grid of the tube in phase with the input signal. This results in an increase in positive grid potential and larger current flows between plate and cathode through the tank circuit which further increases the grid potential. In this way amplitude of oscillator output keeps increasing until it reaches a saturation value. Beyond this value amplitude does not increase. $R_G - C_G$ network provides negative biasing to the grid circuit. Loss of power across the resistance R_T in series with L_T is compensated by the battery E_b. The block diagram of an electronic generator is shown in Fig. 8.3.

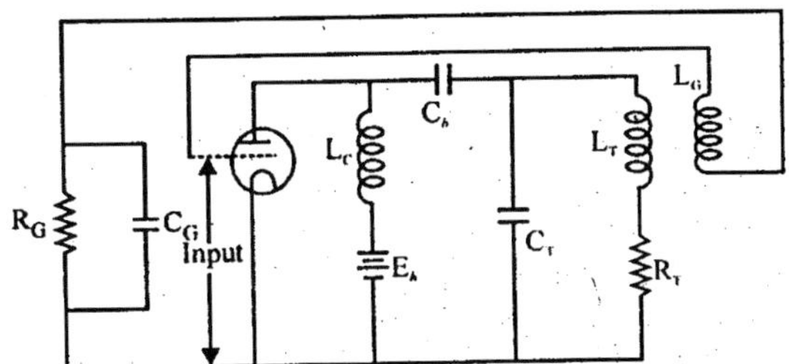

Fig. 8.2. Basic circuit of an oscillator.

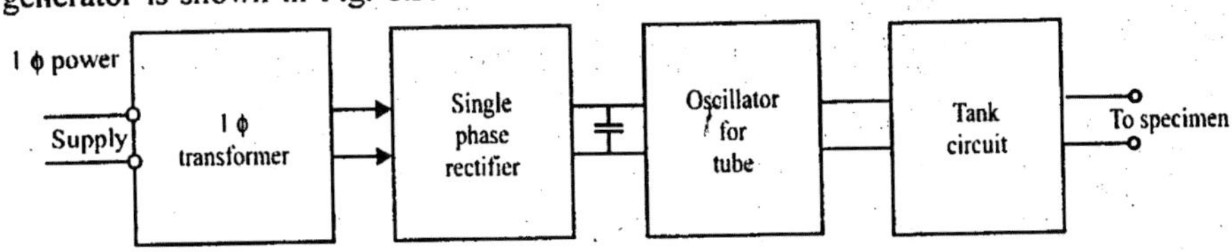

Fig. 8.3. Block diagram of an electronic generator.

A complete electronic generator is shown in Fig. 8.4. A step up transformer has been used to increase the voltage which is rectified by bridge rectifier and smoothened by choke coil. It is used as plate potential and biases the triode valve and given to the oscillator through the valve. As the amplitude of oscillation by tank circuit increases the plate current increases which further increases the amplitude of oscillations. This process continues untill the amplitude of oscillation reaches a saturation value. This high frequency a.c. voltage available at the heater is used to heat up the specimen.

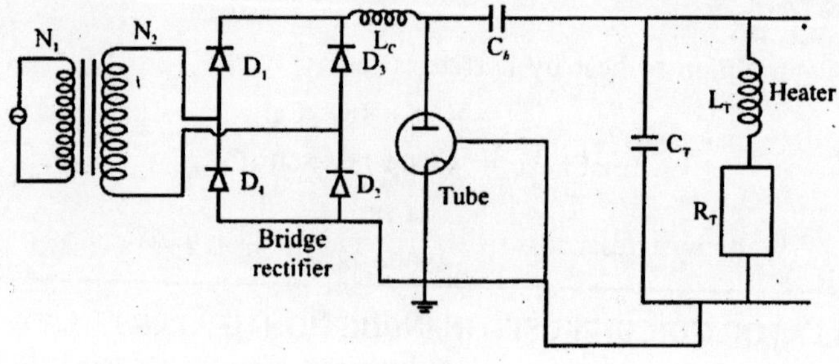

Fig. 8.4. Circuit of an electronic generator.

Electronic generators are used to generate voltage at very high frequencies. However if the requirement of frequency is not so stringent we can use spark gap generator as discussed under :

2. Spark gap generator. This is basically an oscillator which produces signal at frequencies lower than those produced by electronic generators, in the range from 50 Hz to 200 kHz. The circuit diagram of a spark gap generator is shown in Fig. 8.5. L_2 is the work coil connected in the tank circuit formed by L_1 - C_2. It acts like a primary of the transformer while the specimen as the secondary. Capacitor C_1, called blocking capacitor protects the coil L_2 against high dc voltages. The frequency of the tank circuit can be adjusted by changing inductor L_1. Spark gaps are formed by using discs of non-linear resistance materials such as tungston and are either air cooled or water cooled or both. The work coil can carry the current as high as 1000 A and large amount of heat is produced. The coil is cooled by taking it in the form of hollow copper tube and the water is allowed to pass through it . By changing different parameters such as frequency, magnetic flux density and in fact the resistance of the material the amount of heating can be changed.

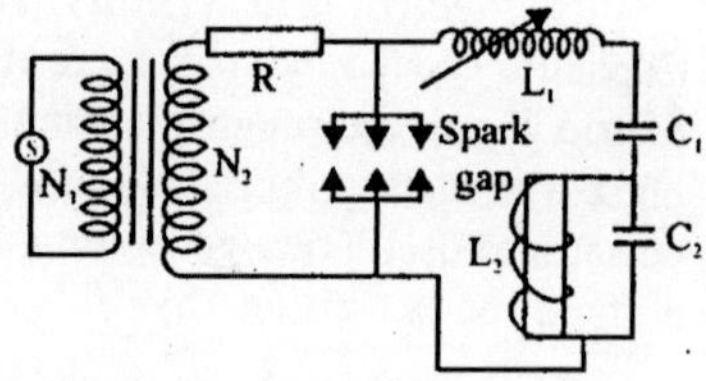

Fig. 8.5. Spark gap generator.

As a large range of frequencies can be produced by spark gap generator it can be used for different processes such as annealing, deep surface hardening in the frequency range of 500 Hz to 3000 Hz. The conversion efficiency of these convertors is about 50% and the power varies from 5 kw to 50 kw.

In other method the secondary of a transformer is shorted through the job (specimen). Due to large secondary current the job is heated.

8.4. MERCURY ARC CONVERTER AND MOTOR GENERATOR SETS

The requirements of these frequency converter are in different frequency ranges. Mercury arc generator is used to generate frequency in the range of 1 kHz to about 1.5

kHz while motor generator set is used to generate frequency in the range of 2 kHz to 10 kHz with efficiency from 60 to 65 percent. The effeciency of mercury arc converter is about 90 percent.

8.5. CIRCUIT ARRANGEMENT FOR INDUCTION HEATING

An electronic heater employed for induction heating is shown in Fig. 8.6. The oscillator is excited by a constant dc which is obtained by filtering the output of the rectifier through L – C filter. The oscillator frequency is changed to change the amount of heating of the specimen. The heating can be done in two ways. In first method the inductor of the tank circuit is tapped to the job coil as shown in Fig. 8.7. (*a*).

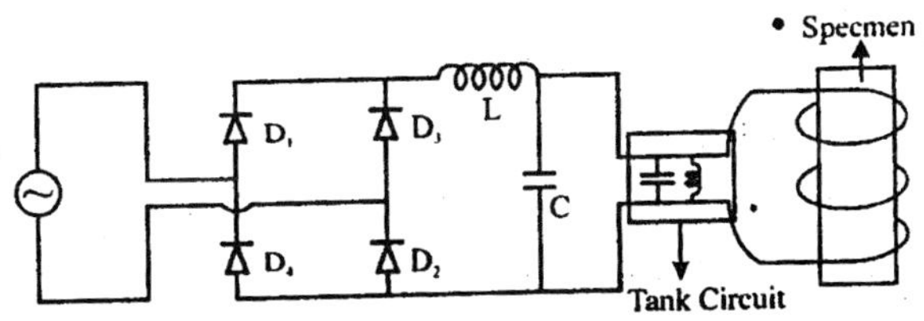

Fig. 8.6. Schematic of induction heating.

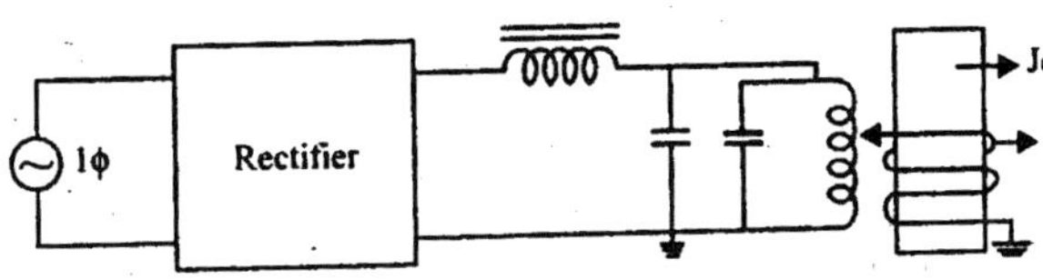

Fig. 8.7 (a) Schematic of induction heating.

In the second method the coupling through transformer action is used. Related circuit is shown in Fig. 8.7 (*b*). Job coil acts as primary and the specimen acts as secondary of the transformer. By changing the frequency of the oscillator the heating of the specimen can be controlled. The oscillator circuit may be Hartley oscillator, Calpit's oscillator or tuned plate tuned grid oscillator. In what follows we shall discuss Hartley oscillator used for induction heating.

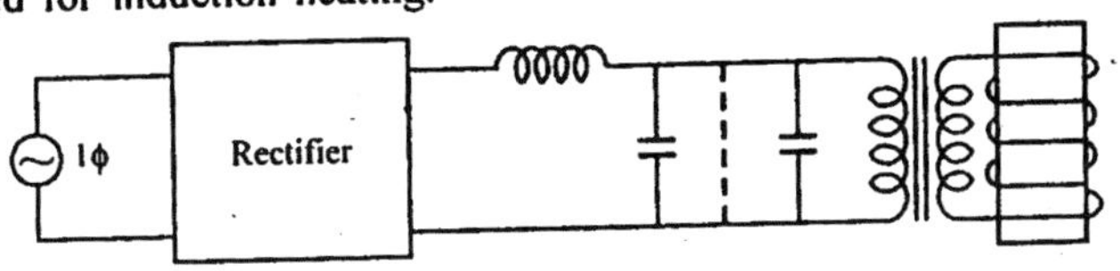

Fig. 8.7 (b) Induction heating.

1. Hartley Oscillator : The circuit is shown in Fig. 8.8.

AC supply is enhanced by using step up transformer. The transformer output is rectified by using bridge rectifier which is fed to the oscillator circuit formed by power transistor Q and tank circuit consisting of inductor L_1 and L_2 and capacitor C. Choke coil L_c is used as dc coupler which acts as a filter and removes ripples present in the rectifier's

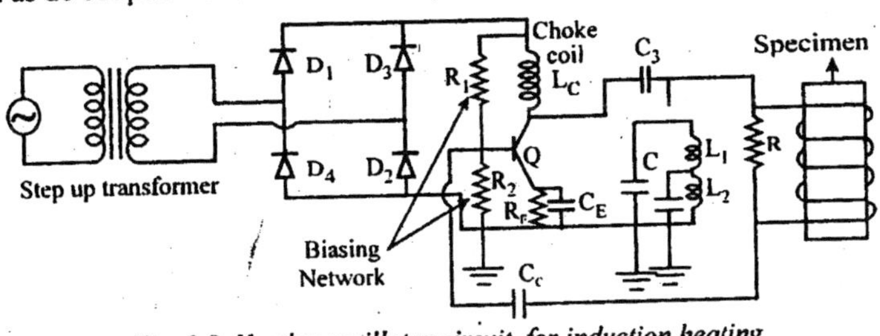

Fig. 8.8. Hartley oscillator circuit for induction heating

output. R_1, R_2 form the biasing network, R_E is used to stabilize the gain while C_E is used to bypass a.c to avoid negative feedback to the transistor. The oscillator output is fed-back to the base of transistor by coupling capacitor C_C. The phase difference between input and transistor output is 180°. Another 180° phase difference is introduced by inductor L_2. Thus a total of 360° phase difference occurs between input and output which is the condition for sustained oscillations. Capacitor C_3 blocks the dc to the tank circuit.

The frequency of oscillation is given by $f = \frac{1}{2\pi\sqrt{C(L_1 + L_2 + 2M)}}$, where M is the coefficent of mutual inductance between the two coils. By changing L_1, L_2 or C the frequency of oscillator and hence the heating of the material can be changed.

The output of the oscillator is coupled to job coil through resistance R as shown in the Fig. The job coil induces eddy current which results in the heating of the meterial. If material is magnetic, hysteresis loss also is used to heat up the material..

If the oscillations of varibale amplitude are required diodes D_1-D_4 in the bridge rectifier are replaced by thyristors. The output of the bridge now can be controlled by changing the firing angles of thyristors.

The transistor may be replaced by power MOSFETs for reducing power loss and it has high input impedance causing reduction in loading effect. The drawback of MOSFET is that it is not available for high power rating where vacuum tubes are most suitable.

2. Push-Pull Oscillators : For handling large power simply Hartley and Colpitt's oscillators are not suitable as they work at low power. For high power applications, push pull oscillators are used.

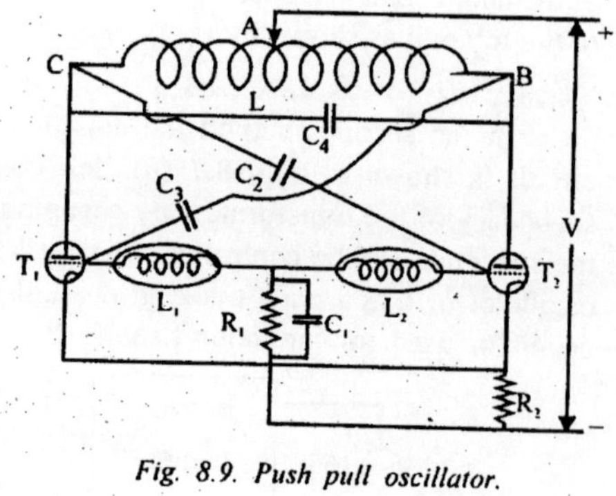

Fig. 8.9. Push pull oscillator.

The circuit of a push pull oscillator is shown in Fig. 8.9. T_1 and T_2 are two vacuum triode valves which constitute the push pull arangement. Large coils L_1 and L_2 are used to protect the grid against short circuit when RF signal is applied. R_1-C_1 network is used to provide bias stability at the grid while resistance R_2 is safety bias resistor. The capacitor C_2, C_3 are used to provide feedback to the grid from the tank or tune circuit constituted by L-C_4. The reactance of C_2 and C_3 should be very low at RF frequencies to ensure distribution of grid voltage between grid and earth only.

Circuit Operation : The voltage at point B is fedback to grid of tube T_1 making it more positive relative to cathode. As a result more electrons move from cathode to plate or more current flows from plate to cathode which in turn increase the feedback to the tube and hence the conduction. At this stage point B is more negative relative to point A while point C is more positive relative to point A. Consequently no feedback is available to the tube T_2 and it remains in the cutoff state while T_1 is in full conduction state. After the current in T_1 has reached its peak value, it starts falling making point C negative and point B positive with respect of point A. As a reselt tube T_1 starts pulling out of conduction while T_2 starts conducting gradually. A stage is reached when T_1 is in off state while T_2 is in full conduction state. This process of one tube in cutoff and another in conduction state is called push pull action. The frequency of the current through inductor L is governed by $L - C_4$ parameters in tank circuit. The output is taken from tank coil as shown in Fig. 8.10.

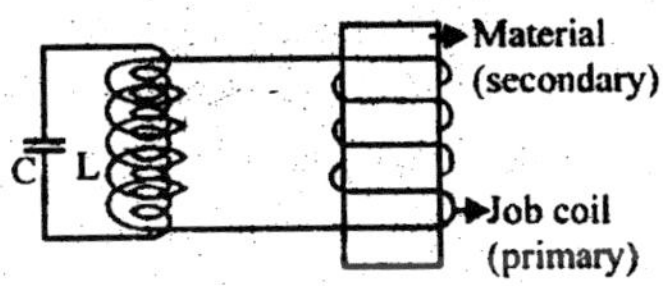

Fig. 8.10.

By changing the frequency of tank circuit the desired induction heating can be achieved.

Triode vacuum tubes can be replaced by transistors with some modifications for push pull action. The push-pull oscillator using transistors is shown in Fig. 8.11. The circuit is a multivibrator. The output of Q_1 is fed to the base of Q_2 and vice versa. When Q_1 is off, its output is maximum and Q_2 is in conducting state. When Q_2 is off it pulls Q_1 in conduction state. Thus the process continues. The frequency of oscillation is given by

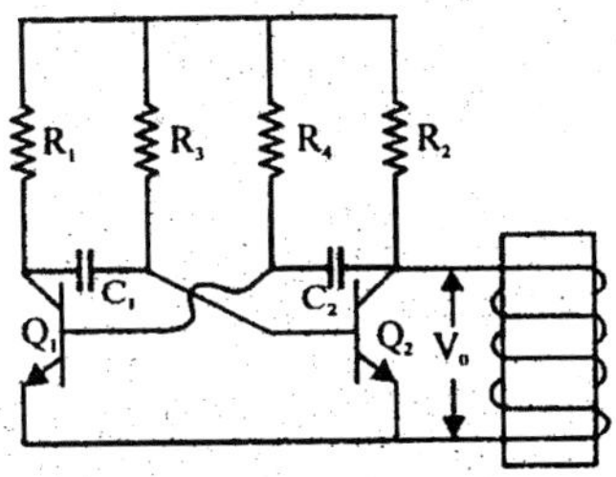

Fig. 8.11. Push pul oscillator.

$$f = \frac{0.72}{RC}$$

where $R_L = R_1 = R$, say.

Thus by changing the values of RC the heating of the material can be controlled.

3. Static converters used for induction heating : Static generators are used for producing output in frequency range of 200 Hz to 10 kHz. The generated block diagram of static converter or static generator is shown in Fig. 8.12.

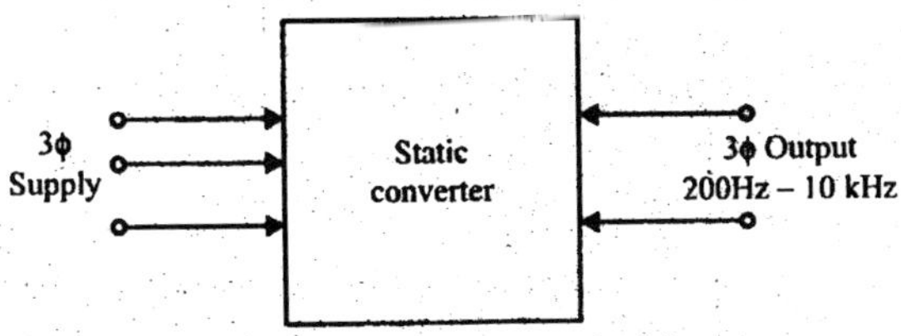

Fig. 8.12. Static converter.

8.6. POWER SUPPLIES USED IN INDUCTION HEATING

The requirement of the power supply is governed by the limit on the voltage output of the rectifier. Power supplies using simple diodes have already been used in different circuits. The drawback of such type of power supply is that desired output voltage cannot be obtained from such supplies. This difficulty is overcome by replacing diodes by thyristors in the supplies. The circuit is shown in Fig. 8.13. A.C input from mains supply is rectified by a fullwave bridge converter. The output of the converter has ripples which are removed by filter circuit constituted by L and C. The rectifier output can be controlled to desired value by changing the firing angles of the thyristors Th_1 - Th_4 and is fed as input to the inverter circuit formed by four thyristors $Th_5 - Th_8$. The frequency of output of inverter circuit may be changed by changing the switching frequencies of the thyristors. By changing the duration of firing pulses the output frequency is adjusted. Normal range of frequency is chosen from 500 Hz to 50 kHz. The output voltage of inverter can also be changed by changing turn ON period of the pulse applied to the gates of thyristors. Thus the required voltage can be obtained by changing the firing angle or switching frequencies of thyristors in rectifier or inverter bridge or in both the bridges. The inverter's output is fed to the load.

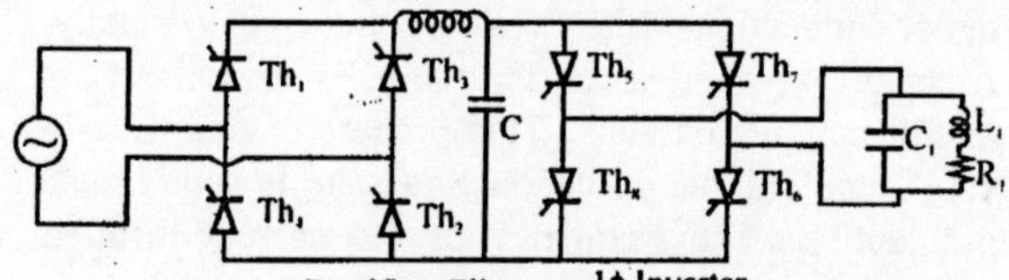

Fig. 8.13. Power supply for induction heating.

First the desired frequency is obtained by changing the firing sequences of the bridges. The temperature in the furnace is maintained constant by automatic control. The inductor L_1 connected in series with load resistance R_1 maintains the current through the load constant. The capacitor C_1 connected in parallel with L_1 R_1 improves the power factor. The R_1 L_1 C_1 combination itself acts as commutation circuit for the inverter.

8.7. DIELECTRIC HEATING

When a non conducting material is subjected to ac electric field, dielectric loss takes place inside the material which appears as heat. This is the principle of dielectric heating. For given field the value of heat produced changes with dielectric strength of the material. By this method the heat supplied is constant throughout the body. Dielectric materials such as wood, tobacco, jute etc. are dried by this method. Also this is of immense use in textile, chemical, rubber and food industries.

In dielectric heating the material to be heated is placed between two electrodes which are connected to a.c supply having frequency about 1 MHz. The electrodes act as two plates of a capacitor with non-conducting material as the dielectric material between them. For uniform heating of the material the dimension of electrods should be greater then those of the dielectric material. Also separation between them is less than their lengths. If V is the applied voltage and I the current flowing through the capacitor, the capacitive reactance X_c is given by

$$X_c = \frac{V}{I}\Omega$$

Power across the material = dielectric loss

$$P = \frac{V^2}{X_c}\tan\delta$$

$$= V^2 \omega C \tan\delta \qquad \text{... (4)}$$

where, ω = frequency of input supply.

C = capacitance of the capacitor

$\delta = 90 - \theta$

θ = phase angle of the dielecite, *i.e.*, the phase difference between voltage and current.

From eqn. (4) we conclude that

1. By increasing the supply voltage the dielectric heating can be changed but there is always an upper limit to the voltage applied. If the voltage applied is more than 15 kV, corona and arcing effect occur.

2. Capacitance should be high. For this purpose the air gap between dielectric and electrodes should not exist.

3. This is very efficient method to heat up non-conducting materials uniformly.

4. Degree of heating is dependent upon the frequency of the input supply.

The generalized block diagram for dielectric heating is shown in Fig. 8.14.

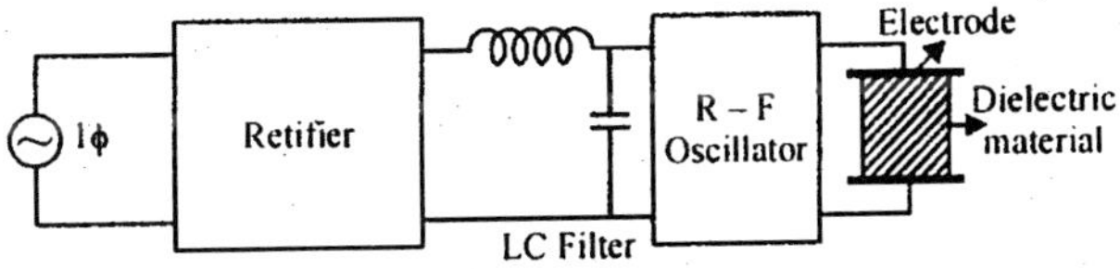

Fig. 8.14. Schematic of dielectric heating.

When there is no material to be heated the air gap between the electrodes acts as dielctric medium. Corresponding frequency of the tank circuit used is adjusted so that the maximum power may be transferred to the air. It C is the capactance of the tuning circuit and L the inductance used in the oscillator, the resonance frequency is written as

$$f_r = \frac{1}{2\pi\sqrt{LC}} \quad ...(5)$$

Now the air is replaced by dielectric material to be heated. As the dielectric constant of the material is not equal to that of the free space, the capacitance will have different value. Let the capacitance of the tuning circuit when dielectric is introduced is C', then the resonance frequency for maximum heating will be given by

$$f'_r = \frac{1}{2\pi\sqrt{LC'}} \quad ...(6)$$

From eqn. (5) and (6) it is concluded that f'_r is not equal to the resonant frequency of the tank circuit. As a result the maximum power will not be transferred to the material. Hence returning of the oscillator is needed. To avoid returning a coupling capacitor of proper value is used to automatically adjust the change in value of working capacitor.

For frequency generation different kinds of oscillators such as push-pull oscillator, tuned plated tuned grid oscillator etc. are used. After the selection of the frequency another factor to be taken into account for uniform heating of the materials is the electric field. For uniform distribution of electric field parallel plate electrodes are generally used. When a high frequency ac voltage is applied across these electrodes, a uniform electric field is generated within the material to be heated. However, if the frequency is higher than a particular value the electric field will not be uniformly distributed. As a result different parts of the materials will be heated by different amounts. Voltage gradiant is another problem in dielectric heating. If the voltage gradiant is not constant, the heating will not be uniform. Some air gap exists between specimen and the electrodes for a number of reasons such as (1) irregular shape of the dielectric material (2) keeping the provison for the movement of the dielectric body between the electrodes. For overcoming such diffculty the shape of the electrodes should be designed such that each part of the body irrespective of its shape is heated uniformly.

Applications : Dielectric heating has wide range of applications which are discribed as under.

1. Used for drying textiles such as rayon, nylon, terrylene etc.

2. Processing of chemicals.
3. Drying, curing of wood.
4. Heat treatment of plastics.
5. Used in rubber industries and in curing resins adhesive.
6. Used in sterilization of food and medical supplies
7. Used in tobacco industries as drier.

Example 8.4. In a dielectric heating process a voltage of 230 V is applied at 30 kHz. If the electrodes used have area of 4 cm^2 separated by 8 cm, what is the dielectric loss filled between the electrodes? Assume phase angle of dielectric = 30° and dieleetric constant as 10.

Solution. The dielectric loss is given by

$$P = V^2 \omega c \tan \delta$$
$$= V^2 \omega c \tan (90 - \theta)$$
$$= V^2 \omega c \cot \theta.$$

where
$$C = \frac{\varepsilon_o \varepsilon_r A}{d} \text{ farad}$$

$$= \frac{8.85 \times 10^{-11} \times 10 \times 4 \times 10^{-4}}{8 \times 10^{-2}} \text{ farad}$$
$$= 1.16 \times 40 \times 10^{-14} \text{ farad}$$
$$= 46.40 \times 10^{-14} \text{ farad}$$
$$= 0.47 \text{ pF}$$

Hence
$$P = (230)^2 \times 2\pi \times 30 \times 10^3 \times 0.47 \times 10^{-12} \times \cot 30°$$
$$= 8.1 \times 10^{-3} \text{ watt.}$$

SUMMARY

At high frequencies induction and dielectric heatings are used as resistance heating does not serve the purpose of high degree of heating. In the present chapter various schematic of induction heating, dielectric heating, oscilator circuits used to produce power at high frequencies have been studied. Power supplies required for various applications of these heatings have been discussed in detail.

EXERCISE

1. Explain the process of heating a material by induction method.
2. How do induction and dielectric heating depend on the frequency of input supply ?
3. Draw the circuit of electronic generator. Explain its working.
4. What is the working principle of spark gap generator ? Explain related circuit.
5. Draw Hartley oscillator circuit used for induction heating. What is the limitation of this circuit ?

6. Explain push-pull oscillator circuit using transistors. How is the drawback of this oscillator overcome by replacing transistors with triode vacuum tube ?
7. Describe the power supply used for induction heating.
8. Explain delectric heating. Draw the circuit used for heating a non conducting material.
9. Derive expression for dielectric loss. Descirbe complete procedure for heating a dielectric material.
10. Enumerate the applications of dielectric heating.
11. Describe methods to connect a load to oscillator circuit

NUMERICAL PROBLEMS

1. A specimen is subjected to induction heating for 2 hours. The following data were obtained. K = 1.00, Maximum flux density 6.0 wb/m^2, supply frequency 20 kHz. The dimension of the specimen is 3 cm × 2 cm × 1.5 cm. Calculate the heat produced due to eddy current losses in the specimen.
2. In the above example if K′ = 3.68. Find the hystersis loss and the total loss in the material.
3. An inductive coil having resistance of 2 ohms and inductance 30 mH is subjected to a supply of 240 V at 20 kHz. The no of turns per unit length in the coil are 10. The specimen to be heated has a volume of 25 cm^3. Find the ratio of resistance heating to induction heating.
4. In induction heating supply voltage of 200 V at 100 kHz is used. Maximum magnetic flux density of 2.5 web/m^2 is used for heating the material having volume of 10 cm^3. If eddy current loss is 1.2 × 10^5 kW, find the value of K.
5. In dielectric heating an input supply of 300 V at 100 kHz is used. The capacitance of the capacitor formed when the dielectric material is inserted between parallel plate electrodes is 0.01 μF. If phase angle of dielectric is 45°, calculate dielectric loss.
6. A dielectric material having dielectric constant 100 is introduced between two parallel plate electrodes. Effective area of each electrode is 6 cm^2 and distance between them is 10 cm. If the input supply of 200 V at 1 MHz is used and the phase angle of dielectric is 30°, calculate dielectric loss.

OBJECTIVE TYPE QUESTIONS

1. Induction heating requires

   (*a*) AC input (*b*) high frequency ac input

   (*c*) mains frequency ac input (*d*) none of the above.

2. Induction heating in non magnetic materials is due to.

   (*a*) eddy current loss (*b*) hysteresis loss

   (*c*) (*a*) and (*b*) above (*d*) none of the above.

3. Above curie point

   (*a*) Magnetic and nonmagnetic materials are heated equally

   (*b*) magnetic materils are heated faster

   (*c*) non magnetic materials heat up faster

   (*d*) all the above.

4. For the purpose of heating an electrolyte the electrodes are generally made of

   (*a*) germanium or sillicon (*b*) Copper and nickel

   (*c*) Carbon or graphite (*d*) insulators.

5. Induction heating is used for

   (*a*) melting (*b*) annealing

   (*c*) forging (*d*) all the above

   (*e*) none of the above.

6. A spark gap generator can generate a frequency of

   (*a*) 50 Hz to 200 kHz (*b*) 20 Hz to 20 kHz

   (*c*) 100 Hz to 10 kHz (*d*) 50 Hz to 1 kHz

7. For small objects the induction heating is done generally at a frequency of.

   (*a*) 10 kHz (*b*) 50 kHz

   (*c*) 10 mHz (*d*) 100 kHz

8. Spark gap generators have power rating ranging from

   (*a*) 1 kW to 4 kW (*b*) 100 kW to 500 kW

   (*c*) 5 kW to 50 kW (*d*) 100 W to 3 kW.

9. Efficiency of spark gap generator is approximately

   (*a*) 50 percent (*b*) 10 percent

   (*c*) 90 percent (*d*) 30 percent.

10. A load in an induction heating is coupled to the oscillator circuit in

    (*a*) 2 ways (*b*) 3 ways

    (*c*) 4 ways (*d*) 6 ways.

11. In case of high power requirement oscillator circuits is developed using

    (*a*) field effect transistor (*b*) power transistor

    (*c*) UJT (*d*) tube device.

12. Number of transistors used in push-pull oscillator is

    (*a*) two (*b*) three

    (*c*) five (*d*) seven.

13. For dielectric heating the supply requires

    (*a*) low frequency (*b*) very low frequency

    (*c*) high frequnecy (*d*) very high frequency.

14. In dielectric heating process the supply voltage should not exceed.

    (*a*) 15 kV (*b*) 33 kV

    (*c*) 6 kV (*d*) 1 kV.

ANSWERS

1. *(b)*	2. *(a)*	3. *(a)*	4. *(c)*	5. *(d)*	6. *(a)*
7. *(b)*	8. *(c)*	9. *(a)*	10. *(a)*	11. *(d)*	12. *(a)*
13. *(d)*	14. *(a)*				

❑❑❑

9

Introduction to Instrumentation

9.0. INTRODUCTION

Measurement is the act or the result of comparison between a given quantity and a standard quantity chosen as a unit. The device used for comparing unknown quantity with the unit of measurement or standard quantity is called measuring instrument. The result of the device used for measurement is the ratio of unknown quantity to the standard quantity chosen as unit. Measurement of any physical quantity may involve more than one object. In what follows we shall discuss how more than one object can be used to measure any quantity.

We can define an instrumentation system as a physical system which is collection of physical objects connected in such a way as to give a desired output. For example an electronic amplifier uses various components to amplify a given input. Thus it is an instrumentation system. From process industry or plant we can cite numerous examples. From other fields also many examples can be cited.

Physical objects are not ideal. The physical system, therefore, cannot be treated as an ideal system. We have, therefore, to make idealizing assumptions for analysis and synthesis of the system. An idealized physical system is called physical model. Depending upon the problem a physical system can be modelled in a number of ways. Once physical model of the system is obtained, the relationship between variables is expressed by using mathematical equations. The mathematical representation of the physical model by making use of physical laws is called mathematical model. A mathematical model may be a set of differential equations or linear equations or any other form, depending upon the choice of variables and co-ordinate system used. For example a network can be solved by different methods to get the same result.

We can classify an instrumentation system as analog system and digital system.

An analog function may be defined as a continuous function. For example Voltage and current w.r.t. time. A system which is used for measurement of information in analog form is called analog system.

A digital quantity may be defined as discrete or discontinuous pulses whose time relationship contains information regarding magnitude or nature of the pulse. The system used to measure information in digital form is called digital system. In the following sections the role of instrumentation will be dealt with clearly.

9.1. MEASUREMENT METHODS

Suppose we want to measure any physical quantity such as temperature, pressure, velocity etc. or electrical quantity such as current, voltage, resistance etc. Measurement of any quantity can be done by two methods-direct and indirect. In direct method of measurement the quantity under test is directly measured such as measurement of current by ammeter, measurement of voltage by voltmeter, measurement of temperature by thermometer etc. In indirect method of measurement the unknown quantity is calculated in terms of other quantities related to unknown quantity. For example the resistance of any conductor can be calculated by measuring the voltage across the conductor and current flowing through it. The resistance is calculated by using ohms law viz. $R = \frac{V}{L}$ ohms. Another example is the horizontal component of earth's magnetic field by tangent galvanometer. In this method the horizontal component of magnetite field at any point on the earth is proportional to the tangent of deflection is tangent galvanometer, *i.e.,* $H \alpha \tan\theta, \theta$ is the deflection.

Direct method of measurement can be classified as

1. Deflection method
2. Comparison method

In deflection method the quantity under measurement is determined by means of measuring instrument having a scale graduated to the quantity under measurement such as measurement of voltage by voltmeter.

In comparison method the unknown quantity is measured in terms of given standard of quantity. Such as measurement of e.m.f. of a cell by comparison with standard e.m.f..

Direct methods of measurement are most commonly used as they are simple and least time consuming. The accuracy of such instruments is in the range of 0.2 to 10%. Direct methods of measurement have small errors in comparison to indirect measurements. Comparison instruments are used where higher accuracy in measurements is required.

9.2. ELEMENTS OF MEASUREMENT SYSTEM

It is essential for an instrument to maintain functional relationship between physical properties of variables and it should have means to communicate to the observer. The capability of the instrument to maintain the functional relationship between variables is associated with static calibration while its performance is given in terms of static and dynamic characteristics.

The Operation of a measurement system can be explained is terms of the functional elements of the system. Block diagrammatic representation of a measurement system is shown in Fig 9.1.

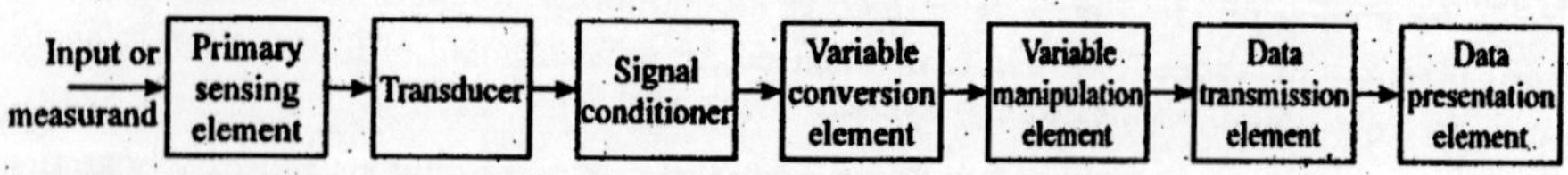

Fig. 9.1. A typical instrumentation system.

Various elements of instrumentation system are described below.

1. Primary sensing Element : The input or measurand makes its first contact with the primary sensing element. The measurand is disturbed by primary sensing element which should be minimized. If the input and output of the primary sensing elements are non-electrical such as manometer, bourdon tube, capsules etc., the sensor output is fed to transducer. Transducer is a device which converts one form of energy into another. In this particular case the output of transducer is electrical energy in the form of voltage or current. If primary sensing element has non-electrical input and electrical output such as

LVDT, thermo couple, etc., it is a transducer. Thus primary sensing element and transducer are separate elements for non-electrical input and non electrical outputs. However for nonelectrical input and electrical output primary sensing element itself is a transducer. The transducer output may be distorted one due to the presence of external sources. For removing distortion some linear processes such as amplification, attenuation, integration, differentiation, addition and subtraction or non-linear processes such as modulation, sampling, filtering, chopping etc. are required. These processes are called signal conditioning and the circuit used for this purpose is called signal conditioner. Signal conditioning will be discussed in sec 9.3. The signal conditioner follows transduder or primary sensing element as the case may be.

2. Variable conversion element : The output of the transducer may or may not be in form suitable to the instrument. In case the output signal of the signal conditioner is not suitable to the instrument, it is converted into a suitable form to the instrument. This conversion is carried by variable conversion element. For example if the electrical signal is in analog form and the next stage accepts the signal in digital form only, an A to D converter is required which converts analog signal into digital signal suitable for the instrument.

3. Variable manipulation element : The function of this element is to manipulate the signal while preserving the original nature. For example an amplifier amplifies the signal while its nature remains unchanged. It is thus a variable manipulation element.

Data transmission element is needed when elements of the instrument are physically separated.

4. Data presentation element : The measurand is communicated by the instrument to the user. This task is carried by data presentation element or read out device Simplest form of this element is panel meter. Other forms include a printer, tracer, recorder etc. Depending upon the need the input signal can be observed by the user in any form. In case a control device is to be used for final measurement stage, some feedback is necessary to the input stage from output stage.

9.3. SIGNAL CONDITIONING

Signal conditioning is used to present the transducer output to the user in measurable form. The output of the transducer may be too small or it may be distorted one, which is different to be observed. To remove this difficulty signal conditioning is used. Two types of signal conditioning are used-DC signal conditioning and AC signal conditioning. shown, respectively in Fig 9.2 and 9.3.

In DC signal conditioning system the excitation source is required if passive transducer is used. For example if strain gauge is used an excitation source is required. The bridge output is amplified by D.C. amplifier. However for active transducer excitation source is not required. For example a solar cell needs no exciatation source. Its output is fed to the amplifier. The ripples present in the amplifier output are filtered by low pass filter. The dc output of the filter is available for the user.

In case of ac signal conditioning system excitation source is a carrier oscillator. The oscillator output is modulated by the transducer output if passive transducer is used. However for active transducer no oscillator is required. For resistive and inductive transducer a carrier frequency between 50 Hz and 20 kHz is used while for capacitive transducer carrier frequency of about 0.5 MHz is required. The bridge consisting the transducer as one arm (or transducers as more than one arm) is excited by carrier oscillator.

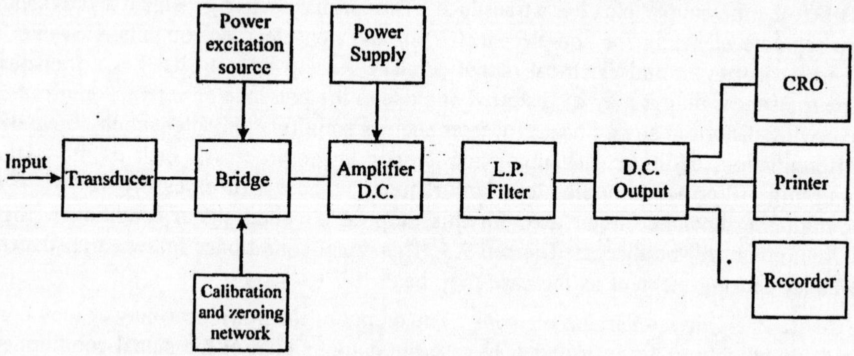

Fig. 9.2. D.C. Signal conditioning

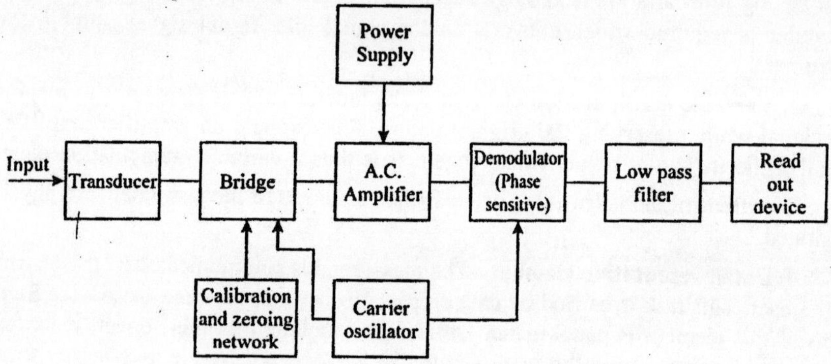

Fig. 9.3. A.C. Signal conditioning

The modulated signal being small is amplified by a.c. amplifier and demodulated by phase sensitive demodulator. Demodulator output is filtered by high and low pass filter which allows only the data signal to pass through and the sprious signals are removed. Phase sensitive demodulator filters out the carrier frequency from the data Signal. For reliable working of the circuit, carrier frequency should be at least five times the highest signal frequency. For example a carrier frequency of 5 kHz limits the signal frequency to 1 kHz

9.4. EXAMPLE ON INSTRUMENTATION SYSTEM

We Consider example on measurement of temperature of an electric oven. Two cases are possible (1) the temperature of the oven is to be measured (2) Temperature of the oven is to be maintained at set value. In first case the temperature sensing device such as thermocouple (transducer) inside the oven generates corresponding voltage which is fed to signal conditioner and displayed on the front panel of the oven. The block diagram is shown in Fig 9.4 *(a).*

In the second case the temperature inside the oven is to be maintained at desired value. For this purpose a fraction of the output of the system is fedback to input for automatic control. The block diagram is shown in Fig 9.4 *(b).*

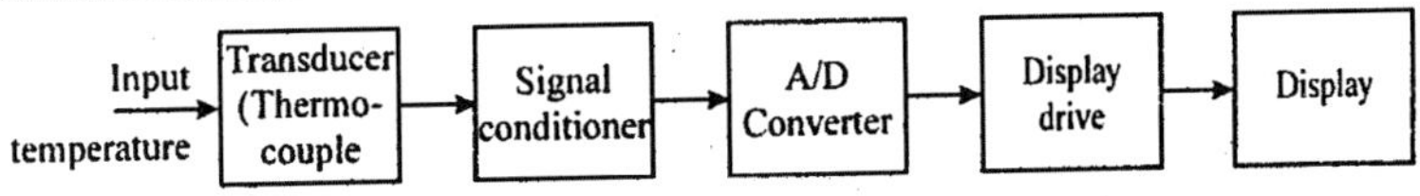

(a)

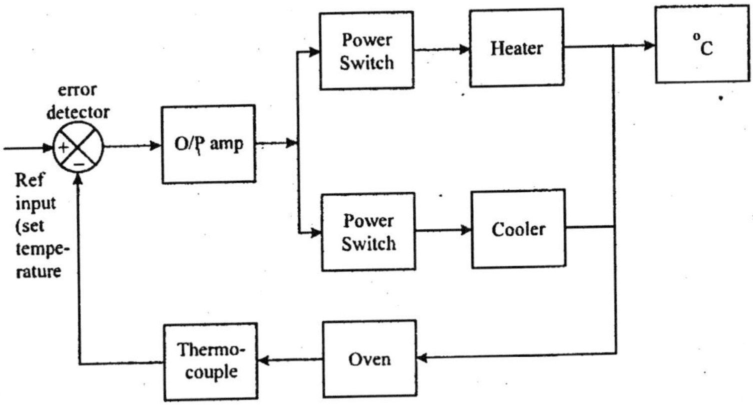

(b)

Fig. 9.4. Automatic temperature control.

One should know first the temp-voltage characteristics of the thermocouple in use. Once thermocouple output voltage per °C change in temperature is known, the reference input has the same value as the thermocouple output at given temperature. The temperature of the oven is set at the desired value by linear potentiometer on front panel, graduated in °C. Correspsonding reference input is set. So far as error signal is non zero the amplifier output is non-zero which produces heat and hence temperature is increased. With increase in temperature the thermocouple output increases and finally reaches the value of reference signal. Under this condition the amplifier output is zero and the oven temperature is maintained at the set value. When the temperature of the oven is lowered the cooler starts working and continues in ON state until the temp of the oven attains the set value.

9.5. INSTRUMENT CLASSIFICATION

Instruments can be classified on the basis of various criteria. These criteria are very useful in assessing the instruments in terms of various parameters. A broad classification of instruments follows as under:

9.5.1. Active/Passive Instruments

An active instrument generates output in accordance with the input to it. On the other hand a passive instrument generates the output by modifying the magnitude of the power drawn. Let us consider an example.

Temperature of a system can be measured by using either resistance thermometer or themocouple. In resistance type thermometer, the resistance of the sensing element depends upon the input temperature and its output draws the energy from the power source in accordance with the temperature. Thus it is a passive element. On the contrary if a thermocouple is used, it generates its own output according to the value of the input temperature - in this case no external power source is required. This is an example of active instrument.

9.5.2 Null/Deflection Type Instruments

Take an example of pressure gauge. In this instrument the pressure applied is indicated by the corresponding deflection of the pointer on the scale. Such an instrument is called deflection type instrument. In yet another type of pressure measurement technique dead weight gauge is very popular. In this gauge a weight is placed on a pan attached to a piston to bring it to the null (datum) point. The amount of weight required to bring the piston to null level is the measure of pressure in the vessel in which the piston is fitted. Such instruments are known as null type instruments. Null type instruments are, in general, more accurate than the deflection type instruments. Another example of null type instrument is Kelvin Double bridge which is used to measure low resistances.

9.5.3. Control /Monitoring Instruments

The instruments which give just the indication of physical quantities being measured are called the monitoring instruments. For example a strain gauge when subjected to input force just gives the indication of input quantity. On the other hand a control instrument is included in the feedback path to obtain the desired output. For example, a tachometer is an instrument which can measure the rotary speed and at the same time can be used to control the speed of that particular system.

9.5.4. Analog/Digital Instruments

An analog instrument gives the output which vary continuously in accordance with the input quantity to be measured. Thus it can give infinite output values within the instrument's range. A deflection type instrument is a good example of it. Another example is the measurement of force by a resistive transducer.

A digital instrument, however, gives output which varies in discrete steps and can have finite values in given range of input being measured by the instrument.

9.6. CHARACTERISTICS OF INSTRUMENTS

An instrument is characterized by (1) static characteristics and (2) Dynamic characteristics as described below.

9.6.1. Static Characteristics

Steady state response of an instrument is attributed to various parameters as defined below.

9.6.1.1. Accuracy

Accuracy is the extent by which a reading deviates from actual value. Thus the word Accuracy has a –ve meaning *i.e.*, it is the amount by which the result is wrong. It is expressed as percentage of full scale reading. For example if we measure the temperature by a thermometer of range 0-100° C having an accuracy of ±2% then for an input temperature of 50° C (say) it will have an error of 1°C. Thus it will indicate the temperature either 49 or 51°. The error due to instrument's accuracy can be reduced by using the instrument of the range which is suitable for given input.

Example 9.1. A pressure meter has the range 0-20 bars. If it has accuracy of ±2% fs, calculate the instrument reading for input pressure of 12 bars.

Solution. $$2\% \text{ full scale} = \frac{2}{100}\times 20 = \pm\ 0.4 \text{ bars for 20 bars}$$

For 12 bars input pressure, the accuracy (error) will be equal to

$$\frac{0.4}{20} \times 12 = \pm\ 0.24 \text{ bars}$$

Hence the reading indicated by the instrument will be

$$12 - 0.24 = 11.76 \text{ bars for +ve accuracy}$$

$$= 12 + 0.24 = 12.24 \text{ bars for −ve accuracy} \quad \textbf{Ans.}$$

9.6.1.2. Precision/Repeatability/Reproducibility

Precision of an instrument defines its immunity to random errors. Such errors may be caused by any source. Thus if the readings of a quantity is taken by high precision instrument a number of times, the spread in its output will be very small.

The term precision has meaning different to that of accuracy. It is not necessary that an instrument with high precision has high accuracy. For an instrument having poor accuracy, the precision may be good. Let us consider that an ammeter reads the input current of 2 Amp as 1.8 amp every time. In this case the accuracy of the instrument is poor but the precision is very good. Thus even if the instrument gives inaccurate results, these results should not vary with the same input.

Repeatability of an instrument defines closeness of its output values when same input is applied over a short period of time. Every time the reading is taken, the same measurement conditions, instrument, observer and same conditions of use and locations are to be maintained.

The reproducibility of the instrument defines closeness in its output for the same input when all above conditions may change. It means the instrument, observer, measurement conditions location, conditions of use, time of measurement etc. can change.

9.6.1.3. Range or Span

It defines maximum and minimum values of input quantity which the instrument is designed to measure. Consider a strain gauge which can be used to measure pressure between P_1 and P_2. Then the range of the instrument is from P_1 to P_2.

9.6.1.4. Linearity

Linearity of an instrument is a very important static characteristics. It defines the degree to which the output follows input to the instrument, linearly. The input output relationship is given by straight line of the type $y = mx + c$. The linearity of the instrument is expressed in terms of the deviation of instrument's input-output curve from the straight line behaviour as shown in Fig. 9.5. This deviation or non-linearity is usually expressed as percentage of full scale reading.

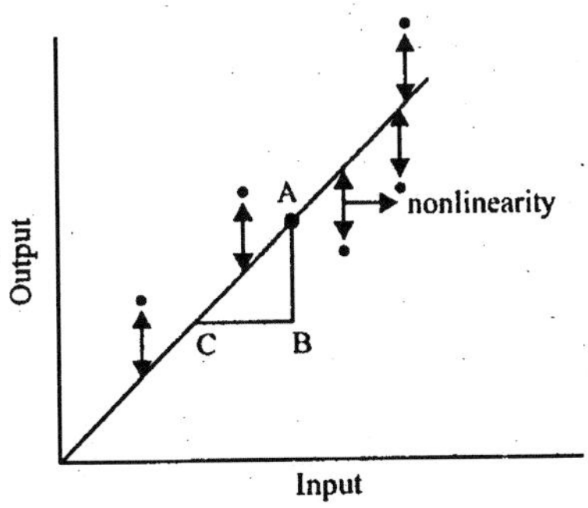

Fig. 9.5. Illustration of Linearity

9.6.1.5. Sensitivity of Measurement

Sensitivity of the instrument is defined as the ratio of change in its output to the change in input to it. Thus.

$$\text{Sensitivity} = \frac{\text{Change in output}}{\text{Change in input}}$$

From Fig. 9.5, the sensitivity is the slope of straight line. For a change from C to B in the input, the output changes from its value at C to the value at A. Thus

$$\text{sensitivity, } S = \frac{AB}{BC}$$

9.6.1.6. Sensitivity to Disturbance

The specification of any instrument are defined under controlled environmental conditions. With a variation in the environmental condition, the instrument characteristics change. This change in the instrument characteristics is the measure of its sensitivity to disturbances. It is further of two types.

1. Zero Drift and 2 Sensitivity Drift

Zero drift is the change in the zero of the instrument per unit change in the environmental parameter. For example the zero of an operational amplifiers output changes with the change in temperature. It is expressed as mV/°C. If more than one external parameters affect the zero value, the total zero drift is the resultant of all zero coefficients due all the parameters.

Sensitivity drift defines the amount by which an instrument's sensitivity varies with variation in the environmental conditions. It is expressed as sensitivity drift coefficients. An instrument can have any of the above drifts or both.

9.6.1.7. Hysteresis

Hysteresis is the non-linearity in the input-output relationship of an instrument. The phenomenon of hysteresis is shown in Fig. 9.6 between input x and output y of the instrument. We can observe from the curve that for increasing value of x the output has value $y = y_1$, for $x = x_1$ while $y = y_2$ for $x = x_1$ when x is decreasing. Thus for increasing value of input, the output decreases while for decreasing values of x, y increases. This kind of non-coincidence between input and output is known as hysteresis. The hysteresis of an ideal instrument should be zero.

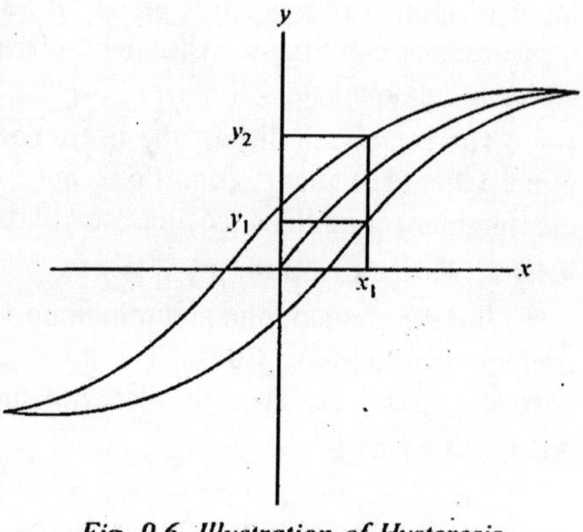

Fig. 9.6. Illustration of Hysteresis

9.6.1.8. Dead space

It is a non-linear element and is defined as the range of input values for which the instrument gives no output as shown in Fig. 9.7.

9.6.1.9. Resolution

It defines the minimum change in the input which produces an observable change in the output of the instrument.

9.6.1.10. Threshold

Threshold of an instrument defines the certain minimum level of input above which the output of the instrument becomes detectable.

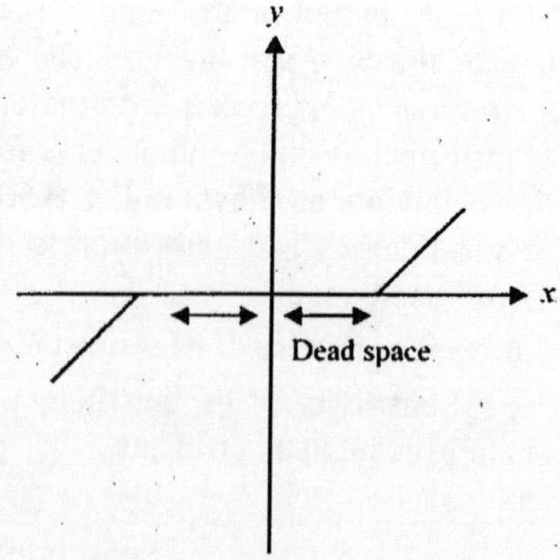

Fig. 9.7. Dead Space

9.6.2. Dynamic Characteristics of Instruments

The behaviour of an instrument due to variation in input to it is described by dynamic characteristics. As the input to the instrument is changed or applied, its output shows transient behaviour before attaining steady state. This transient behaviour is defined by dynamic characteristics of the instrument which are described as under:

We should know that the behaviour of an instrument can be governed by linear time invariant differential equation produced below

$$a_n \frac{d^n y}{dt^n} + a_{n-1} \frac{d^{n-1} y}{dt^{n-1}} + \ldots a_1 \frac{dy}{dt} + a_0 y = b_m \frac{d^m x}{dt^m} + b_{m-1} \frac{d^{m-1} x}{dt^{m-1}} + \ldots b_0 x \ldots \quad \ldots(9.1)$$

In the above equation y is the output for input x to the instrument. Certain cases are as follows:

9.6.2.1. Zero Order Instruments

A zero order instrument is described by the equation

$$a_0 y = b_0 x \quad \text{or} \quad y = \frac{b_0}{a_0} x \quad \ldots(9.2)$$

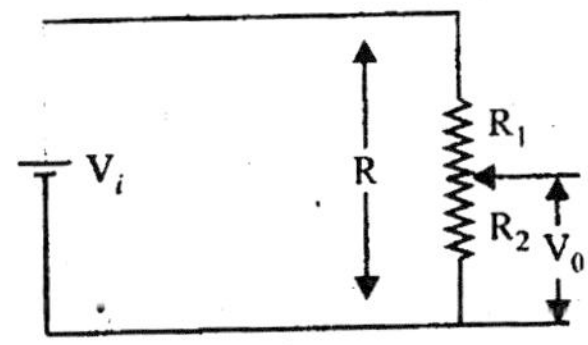

Fig. 9.8. Zero Order Instrument

$\frac{b_0}{a_0}$ is called instrument sensitivity. As per above equation, the output changes instantaneously as the input changes. As an example we consider the circuit shown in Fig. 9.8. The output voltage is given by

$$V_0 = \frac{R_2}{R_1 + R_2} V_i = \frac{R_2}{R} V_i \quad \ldots(9.3)$$

$\frac{R_2}{R}$ is called instrument sensitivity.

9.6.2.2. First Order Instruments

A first order instrument is one which can be defined by the equation

$$a_1 \frac{dy}{dt} + a_0 y = b_0 x$$

or

$$\left(\frac{a_1}{a_0} D + 1\right) y = \frac{b_0}{a_0} x$$

or

$$y = \frac{(b_0/a_0) x}{(1 + a_1/a_0 D)} \quad \ldots(9.4)$$

In the above equation b_0/a_0 = static sensitivity of the instrument and a_1/a_0 is called time constant which is denoted by greek letter τ (tau).

A very familiar example of first order instrument is thermocouple. When a thermocouple is inserted in hot water its temperature rises which follows first order system's behaviour. The behaviour of thermocouple is given in terms of time constant.

9.6.2.3. Second Order System

A second order system is governed by the equation

$$a_2 \frac{d^2y}{dt^2} + a_1 \frac{dy}{dt} + a_0 y = b_0 x$$

or
$$\frac{y}{x} = \frac{b_0/a_0}{\left(\frac{a_2}{a_0} D^2 + \frac{a_1}{a_0} D + 1\right)}; \frac{d}{dt} = D$$

In the above equation

$$\frac{b_0}{a_0} = K, \text{ static sensitivity}$$

$$\frac{a_2}{a_1} = \frac{1}{\omega}; \ \omega \text{ is undamped natural frequency of oscillations.}$$

and
$$\frac{a_1}{2a_0a_2} = \zeta \text{ the damping factor.}$$

Above equation can be written as

$$\frac{y}{x} = \frac{k\omega^2}{D^2 + 23\zeta D + \omega^2} \quad \text{...(9.5)}$$

Various parameters associated with the above equation are deterministic of instrument's behaviour to the dynamic inputs.

9.7. INTELLIGENT INSTRUMENTATION

In modern instrumentation system, a microcomputer is used as an integral part of it. It has various advantages over the instrumentation system without microcomputer such as higher accuracy, self diagnosis, higher immunity to external disturbances and noise, multiuser facility, transmission over long distance etc. Such an instrumentation system is called smart or intelligent instrumentation system. Block diagram of such a system in shown in Fig. 9.9.

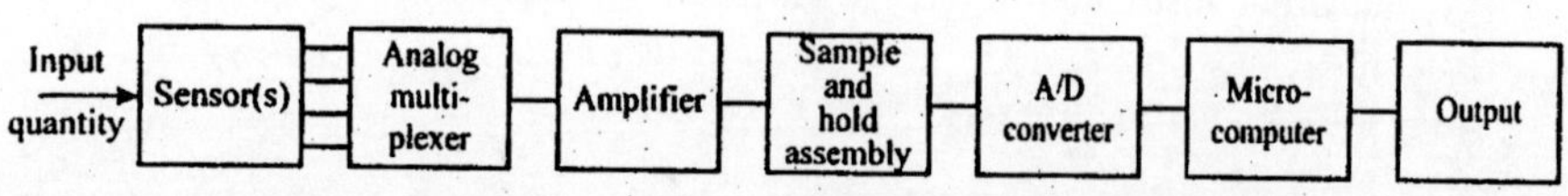

Fig. 9.9. Generalized block diagram of Intelligent Instrumentation.

From the figure, we observe that a microcomputer has been included which gets its input through an A/D converter in digital format. A microcomputer in an intelligent instrumentation has added many advantages. Various developments are taking place in this field in recent times.

SUMMARY

An instrumentation system is used to serve a specific purpose. This fact has been explained in this chapter by considering one or two examples. For automatic control of a given quantity a feedback system is required which is considered in the chapter. By studying this chapter one should have a clear understanding of an instrumentation system.

EXERCISE

1. Define measurement and instrument. Draw generalized block diagram to an instrumentation system.
2. Explain the working of each block of an instrumentation system.
3. What do you understand by the word "signal conditioning"? Draw block diagram of dc and ac signal conditioning systems.
4. Explain clearly the measurement methods.
5. Give an example of instrumentation system using automatic control.
6. What do you understand by static and dynamic characteristics of instruments ? Clarify.
7. Define various static characteristics of an instrument.
8. Distinguish between :
   (a) Accuracy and precision
   (b) Repeatibility and reproducibility.
9. Write a note on intelligent instrumentation system.

❑❑❑

10
Transducers

10.0. INTRODUCTION

As we have observed in last chapter that an instrumentation system is represented by the block diagram shown in Fig. 10.1 The measurand is in contact with the primary sensor which gives non-electrical output for non-electrical input. For example bourden tube, capsules, etc. are primary sensor for pressure, bimetallic strip for temperature etc. The output of the sensor is converted to electrical form by a device called transducer. Thus tranducer is defined as a device which converts one form of energy into another. The transducer output is processed by a circuit called signal conditioner and finally communicated to the user through output device.

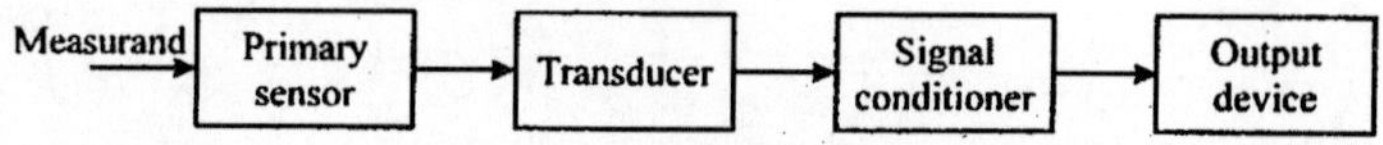

Fig. 10.1. Instrumentation system.

In this chapter various transducers and their applications in the measurement of different physical quantities will be discussed.

10.1. TRANSDUCER

Transducer or pick up is a device which converts one form of energy into another. For example a mechanical transducer converts any form of energy into mechanical energy, optical transducer converts any form of energy into light energy. In earlier times such types of transducer were in much use. Although the accuracy of these tranducers is very high, the transducer output can not be transmitted to remote locations. To overcome these difficulties electrical transducers are used. In this chapter the emphasis will be on electrical transducers.

Example : We consider a parallel plate capacitor with the provision of change in separation between the plates. As the separation between the plates changes the capacitance given by $C = \frac{\varepsilon_0 A}{d}$ also changes correspondingly. If the capacitor is connected in a circuit, the voltage across the capacitor changes with change in capcitance which in turn is due to change in displacement. Thus the capacitor converts the displacement into electrical quantity. Hence it is an example of a transducer.

10.2. CLASSIFICATION OF TRANSDUCERS

Transducers can be classified on the following basis:

1. On the basis of excitation source

(a) Passive transducers

(b) Active tranducers.

2. On the basis of nature of output

(a) Analog

(b) Digital tranducer

3. On the basis of input-output relationship

(a) Primary transducers

(b) Secondary transducers.

4. Inverse transducers

Description of the transducers is given below.

1. Passive transducers : Passive transducers require power source for their working. Such transducers perform their task by controlling the energy from the power source. For example, resistive, capacitive, inductive transducers. Let us consider LVDT. It is excited by a.c. power source connected to its primary. When the core moves between primary and secondaries; the output of the device changes correspondingly. If a.c. power source is removed, we can not get output. It is thus an example of a passive transducer. In another example consider a strain gauge connected to a d.c. power supply. Fig. 10.2 shows that when force is applied on the gauge its resistance and hence output voltage changes. In the above examples power sources are required for working of the the transducer. These are passive transducers.

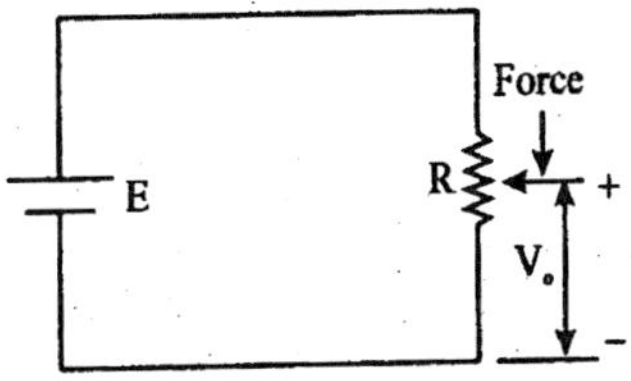

Fig. 10.2. A passive transducer.

2. Active transducers : Active transducers give electrical output in the absence of power source. Such transducers are therefore called self generating. The examples are tachogenerator (for velocity) thermocouple (for temperature) solar cell (for light intensity) etc. The output of such transducers is generally small and needs amplification.

3. Analog transducers : Analog transducers generate, for any given physical input, an analogous output which is continuous function of time. For example LVDT, strain gauge, thermocouple, capacitive transducers etc are analog transducers as their output is continuous function of time corresponding to any input to them.

4. Digital tranducers : Digital transducer convert any input quantity into an output which is in the form of pulses. The amplitude of the pulses are represented by two states. HIGH and LOW or 1 and 0 respectively. Punch card is an example of digital transducer.

5. Primary transducer : When input signal is directly sensed by the transducer and the physical phenomenon is directly converted into electrical form, the transducer is called primary transducer. For example when LVDT is used to measure displacement, the LVDT core movement is directly converted into electrical output. Thus it is a primary transducer.

6. Secondary transducer : When input quantity is sensed by some sensor and its output which is different from input signal is given to a transducer for conversion into electrical form, the transducer is called secondary transducer. In the example of LVDT cited above if it is used for measurement of force, the force is sensed by a senser coupled to the core of the LVDT. Corresponding to force there will be core movement which is different from input. Thus the force is measured in terms of core displacement. In this case LVDT is a secondary transducer.

7. Inverse transducer : Inverse transducer is one which converts electrical quantity into non-electrical quantity. For example when a piezoelectric crystal is subjected to electric field, a mechenical stress is developed in it. It is thus an example of inverse transducer. An inverse transducer is used in feedback measuring systems.

10.3. REQUIREMENTS OF A TRANSDUCER

A transducer should respond only to the measurand for which it is designed . It is necessary, therefore, to know the relationship between input and output. Basic requirements of a transducer are given as under.

1. Linearity : The input output characteristics should be linear and it should produce these characteristics in symmetrical way for any range of input.

2. Repeatability : The transducers should give the same output for same input applied again and again under fixed environmental conditions.

3. Reliability and stability : For any change in the environmental conditions it should give the same output for given input.

4. Hysteresis : There should be no hysteresis between input and output when input is changed from low value to high value or vice sersa.

5. Good dynamic response : For varying inputs the transducer should respond quickly, *i.e.,* variation in input should be reflected in the output.

6. High signal to noise ratio : The ratio of output signal to noise should be very high .

7. Ruggedness : It should be capable of sustaining overloads.

8. No deformation : There should be no deformation left after the removal of the input, *i.e.,* it should be perfectly elastic.

A perfect transducer should have the properties as close as possible to the above said ones.

10.4. SELECTION OF A TRANSDUCER

In measurement system first element to interact with the measurand is the transducer which converts the measurand into electrical form. A transducer can be used to measure more than one physical quantities or more than one transducer can be used to measure the same physical quantity. Suitability of a transducer for a particular measurand is determined by considering the following points.

1. Range : Amplitude and frequency range of the transducer should be large enough so that it can encompass all the amplitudes and frequencies in input.

2. Sensitivity : The sensitivity of transducer should be high i.e. for a change in input a measurable change should take place in the output .

3. Electrical output characteristics : The electrical characteristics of transducer viz. output impedance, frequency response etc. of the transducer should match rest of the measuring system.

4. Environmental conditions : The transducer should be immune to environmental conditions such as change in temperature, humidity, vibration and shocks etc.

5. Error : Errors present due to environmental conditions or those inherent in the transducer itself should be so small that these can be controlled.

6. Accuracy : Accuracy of the transducer should be within the specified range so that input in the specified range can be reliably measured.

Typical applications of transducers in the measurement of various physical quantities will be discussed in the following section.

10.5. RESISTIVE TRANSDUCERS

This is a passive transducer. In this transducer resistance between the output terminals of transducer varies according to the measurand. Both ac and dc power source can be used with such transducers.

The resistance of a conductor of length l is given by

$$R = \rho \frac{l}{A} \quad ...(1)$$

where R = resistance of the conductor, ohms

ρ = resistivity of the material of condutor ohm-meter

A = cross sectional area of conducting wire, m^2.

It is clear from eqn. (1) that the resistance of the conductor can be varied by changing any one of ρ, l and A . Variation in ρ can be used to measure temperature while displacement is measured by changing length l of conductor in potentiometer. Strain gauges are used to measure pressure and force. Resistance of some devices changes with change in light intensity. This effect is called photoconductive effect. Resistance of some devices changes with change in magnetic field. This effect called magnetoresistive effect is used to measure intensity of magnetic field. Typical range of resistance values for resistive transducer is between about 100 Ω and 100 kΩ.

Some of the resistive transducers with their application will be discussed below.

10.6. POTENTIOMETRIC RESISTANCE DEVICES

Such a transducer is used to measure displacement and position in its simplest form. The simplest form of potentiometer is a single length of wire with a slider in contact with the wire. The position of slider determines the effective length of the conducting wire and a change in this position is related to the change in resistance or voltage drop across the potentiometer. Other forms of the potentiometer are wire wound potentiometer, carbon film potentiometer, thin metal film potentiometer etc.

Wire wound potentiometer has a number of turns in given length of the wire so that the sensitivity and voltage drop across the device is increased due to large potentiometer resistance. In carbon film type potentiometer a carbon film is employed on potentiometer with a wiper in contact with the film . The position of the wiper determines the resistance and voltage across the potentiometer. The resolution of carbon film potentiometer is higher compared to that of wire wound potentiometer.

The potentiometer having rotary motion are called rotary potentiometer. They may have full scale angular displacement as small as 10 and as much as 60 full turns.

In additon to above, wiper in some potentiometers has translational as well as rotary motion. Such type of potentiometers are called helipots. All the three types of potentiometers are shown in Fig. 10.3.

From Fig. 10.3 *(a)*, we observe that the output voltage is given by

$$V_0 = \frac{R_2}{R_1 + R_2} V.$$

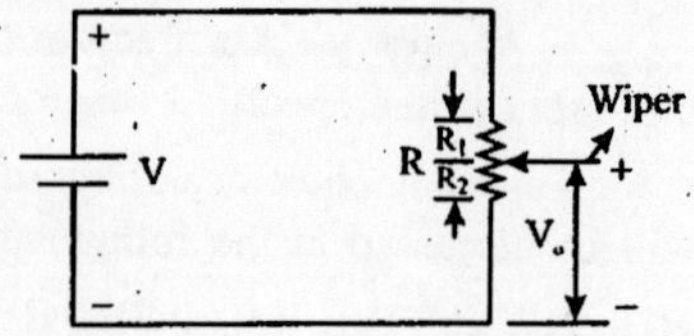

Fig. 10.3 (a) Linear potentiometer.

By changing the position of the slider, the resistance of the potentiometer in output circuit can be changed which results in change in the output voltage. Similarly for rotary potentiometer the rotation

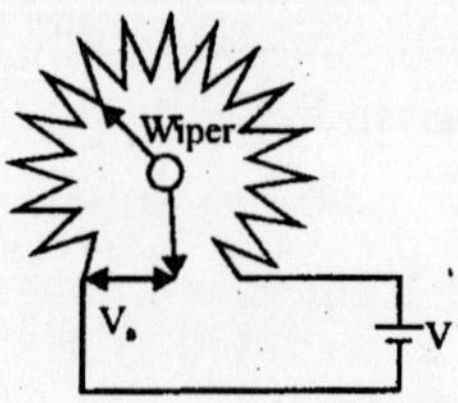

(b) Rotary potentiometer

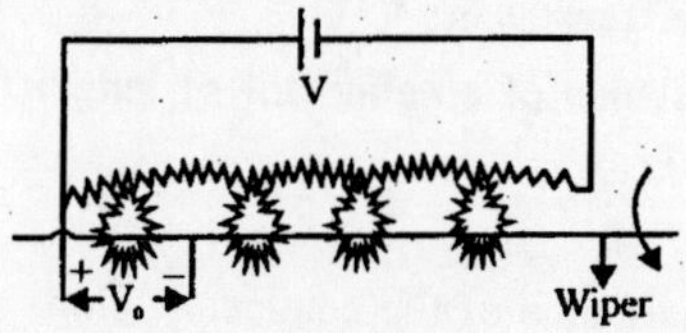

(c) Helipot

Fig. 10.3.

of the wiper results in change in the output voltage. In helpot both types of motion-linear as well as rotary result in change in output voltage.

Example 10.1. A resistive transducer has a length of 10 cm and cross sectional area of the wire is 5 mm. It is excited by a power of 10 V dc.

If total resistance of the wire is 200 ohm, what is its resistivity?

Solution : The resistance of the wire is given by

$$R = \rho \frac{l}{A} \text{ ohms}$$

Here $R = 200\ \Omega$

$\rho = ?$

$l = 10$ cm

$A = \pi r^2 = 5$ mm²

Putting the values of parameters, we obtain

$$200 = \rho \times \frac{10 \times 10^{-2}}{5 \times (10^{-3})^2} \text{ ohm} = \rho \times \frac{105}{5}$$

or
$$\rho = \frac{200 \times 5}{10^5} \text{ ohm-meter}$$

$$= 10^{-2} \text{ ohm-meter}$$

$$= 10 \text{ miliohm-meter}$$

10.7. APPLICATIONS OF POTENTIOMETER

Potentiometer can be used for measurment of various physical quantities. Some of the applications are described as under.

1. Measurement of Linear displacement : The circuit for measurement of linear displacement using potentiometer is shown in Fig. 10.4. It works on the principle that when the moving part of the system under test is attached to the wiper of the potentiometer, it moves along the length of the potentiometer as displacement in the moving system occurs. With the movement of the wiper the resistance of the potentiometer between output terminals changes.

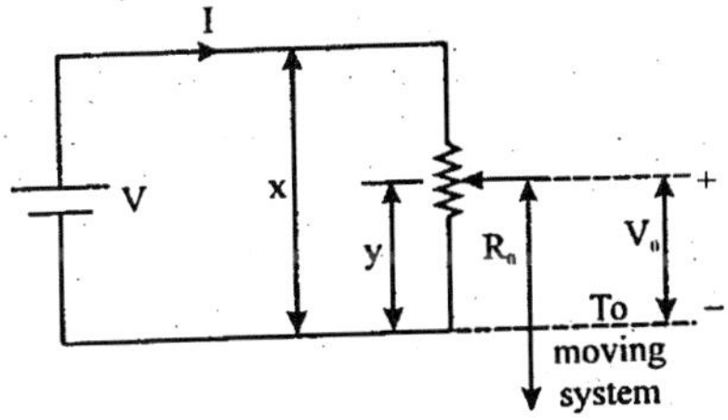

Fig. 10.4.

It results in a change in the output voltage which is processed by the rest of the moving system. Thus change in the output voltage is the measure of linear displacement of given system.

From the figure, the current through the potentiometer

$I = \dfrac{V}{R}$, R is total resistance of potentiometer.

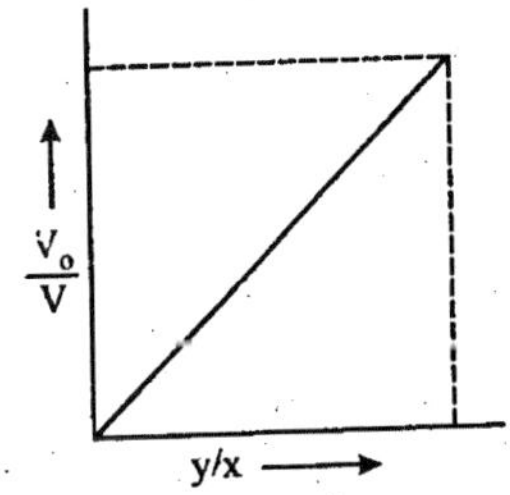

Fig. 10.5.

Output voltage

$$V_0 = \text{Resistance between output terminals} \times \text{current.}$$

$$= R_0 \times I$$

$$= R_0 \times \frac{V}{R} \qquad ...(2)$$

But $$R = \rho \frac{x}{A} \qquad ...(3)$$

where ρ and A have their usual meanings and x is the length of potentiometer.

$$R_0 = \rho \frac{y}{A} \qquad ...(4)$$

where y is the length of potentiometer between output terminals.

From eqn. (3) and (4) , by puting the values of R and R_0 in eqn. (2) we get,

$$V_0 = \frac{\rho y}{A} \times \frac{V}{\frac{\rho x}{A}}$$

$$= \frac{y}{x} V \qquad ...(5)$$

Displacement y of the wiper is the same as displacement of the moving system.

It is, therefore, concluded from eqn. (5) that

$$\frac{v_0}{v} = \frac{y}{x}$$

The graph between linear displacement and the transducer output is shown in Fig. 10.5 which is a straght line. However the graph between $\frac{v_0}{v}$ and $\frac{y}{x}$ is not always linear.

This is due to loading effect. The sensitivity of the potentiometer is given by $\frac{v_0}{y}$ volts per meter. Typical value of the useful sensitivity for linear potentiometer is of range of 0.2 V/mm.

Example 10.2. A linear potentiometer has length of 30 cm. It is used for the measurement of linear displcement. If it is excitied by a dc source of 30 V and output voltage is 5 V what is the displacement of the moving system? Also calculate sensitivity. in V/mm.

Solution. We use relation

$$\frac{v_0}{v} = \frac{y}{x}$$

where $v = 30$ V

$v_0 = 5$ V

$x = 30$ cm

$y = ?$

Hence
$$\frac{5}{30} = \frac{y}{30 \times 10^{-2}}$$

$$y = \frac{5}{30} \times 30 \times 10^{-2} \text{ meter} = 5 \text{ cm}$$

$$\text{Sensitivity } s = \frac{v_0}{y} = \frac{5v}{5\text{cm}} = \frac{5v}{5 \times 10\text{mm}} = 0.1\text{V/mm}$$

Example 10.3. A linear potentiometer has 30 turns of a wire having diameter of 0.2 mm. If it is excited by a source of 20 V and connected to a moving system what will be the displacement of the system if output voltage is 6 V?

Solution. Assuming that wires are in close contact with each other. The length of the potentiometer will be

$$= 0.2 \times 30 \text{ mm} = 6 \text{ mm}$$

Using the formula $\frac{v_0}{v} = \frac{y}{x}$, we get

$$\frac{6}{20} = \frac{y}{6 \times 10^{-3}}$$

or
$$y = \frac{6 \times 6 \times 10^{-3}}{20} = 1.8\text{ mm}$$

2. Measurement of angular displacement : Angular displacement can be measured by rotary potentiometer. The circut diagram is shown in Fig. 10.6. The moving system under test is attached to the spindle of wiper with the help of gear assembly. As the moving part of the system rotates the wiper moves along the pariphery of the potentiometer. As a result the resistance of transducer in the output circuit changes which leads to change in output voltage.

Analysis : Let the periphery of the pot be $l = \alpha a$ where a is the radius of the potentiometer and α is the angle shown. Then total resistance of the pot will be

$$R = \rho \frac{l}{A} = \rho \frac{\alpha a}{A} \qquad ...(6)$$

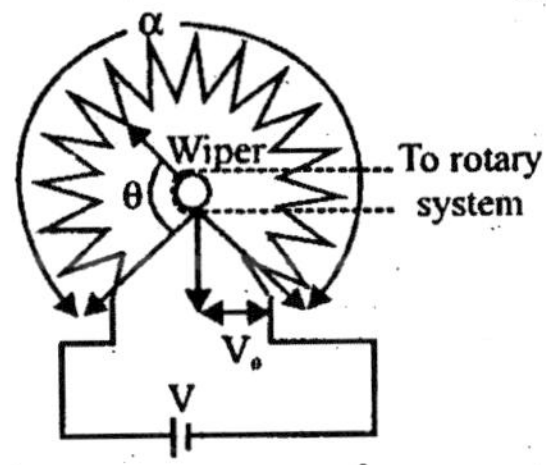

Fig. 10.6. Measurement of angular displacement.

where A is cross sectional area of the wire of the potentiometer. For an angular displacement θ the periphery traversed by the wiper will be

$$S = a \tan \theta, \ \theta \text{ is very small}$$
$$\approx a\theta$$

Resistance of the wire of length S = Resistance per unit length × length S

or
$$R_S = \frac{R}{\alpha a} \times a\theta$$

$$= \frac{\rho}{A}.a\theta \text{ from eqn. (6)}$$

As the wiper moves angular displacement θ, the resistance in the output circuit is reduced by amount of $\frac{\rho}{A} a\theta$. Corresponding decrease in output voltage takes place which is the measure of the angular displacement of the rotary system

Change in the output voltage $\Delta V_0 = V - \frac{R - R_s}{R} V = \frac{R_s}{R} V$

$$= \frac{\frac{\rho}{A} a\theta}{\frac{\rho}{A} \alpha a} V$$

$$\Delta V_0 = \frac{\theta}{\alpha} V$$

$$\frac{\Delta V_0}{V} = \frac{\theta}{\alpha}$$

The plot of $\frac{\Delta V_0}{V}$ with $\frac{\theta}{\alpha}$ is shown in Fig. 10.7. This pot also suffers from the problem of loading. The ratio $\frac{\Delta V_0}{\theta}$ is called sensitivity of the transducer. Typical useful sensitivity is 0.2 Volt per degree rotation.

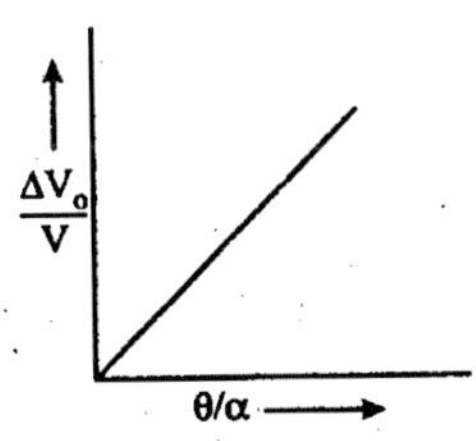

Fig. 10.7.

Example 10.4. A rotary potentiometer has a radius of 0.5 mm. Its wiper coupled at the centre to a rotary system moves along the periphery of the potentiometer. If change in the output voltage is 2.1 volts and a dc battery of 20 volts is used as source what is the total angle of potentometer if wiper moves a displacement of 0.2 mm on the periphery?

Solution : Let total angle of the potentometer is α radians.

$$\text{Angular displacement of the wiper} = \frac{0.2\text{mm}}{0.5\text{mm}} = 0.4 \text{ radian.}$$

We have
$$\frac{\Delta V_0}{V} = \frac{\theta}{\alpha}$$

or
$$\frac{2.1}{20} = \frac{0.4}{\alpha}$$

or
$$\alpha = \frac{0.4 \times 20}{2.1} \text{ radian} = \frac{4}{1.05} \text{ radian}$$

or
$$\alpha = 3.95 \text{ radians. } \textbf{Ans.}$$

3. Measurement of force by potentiometer : Force can not directly be measured by potentiometeric transducer. The force creates displacement which is measured by the transducer. The circut for measurement of force is shown in Fig.. 10.8 *(a)*. It works on torque balance principle. It is an open loop system. Force F applied at the lever in the direction shown displaces the lever which is balanced by spring dash pot system. Under the condition of balance the torque due to spring dash pot system and due to the force are balanced. The wiper attached to the lever moves along the potentiomenter wire. Thus input force is converted into equivalent displacement which is measured by potentiometer. It is therefore a secondary transducer.

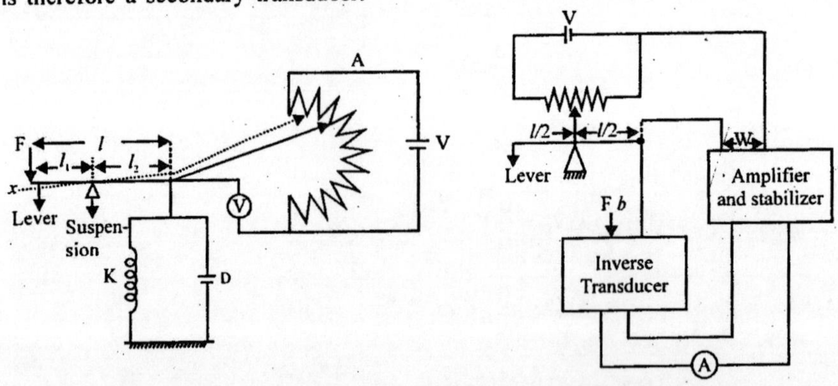

(a) Measurement of force *(b) Closed loop system.*

Fig. 10.8.

Closed loop system for measurement of force is shown in Fig. 10.8 *(b)*. When the force is applied at the lever the output voltage changes which is amplified by amplifier. The output of the amplifier is the input to inverse transducer which is converted into force. F_b . When the torque due to force F and F_b are equal the lever is balanced. The displacement of the wiper along the length of the potentiometer is the measure of force.

The cicuit is complex, bulky and expensive.

10.8. STRAIN GAUGE

A transducer which is used to measure the strain is called strain gauge or strain device. Strain gauge can be used in two cases (1) applications where gauge is used to measure the strain as primary objective such as stress analysis of mechine and structures and (2) where strain is utilized on the transducer as a measure of another parameter such as force, torque, load, acceleration etc. Strain gauge is most widely used in second case. The working principle of strain gauge is as follows:

In an electrical resistance strain gauge, mechanical displacement is converted into a change of resistance. If a metallic conductor is stretched or compressed, change in its length, cross sectional area and resistivity takes place. As a result its resistance changes. The change in resistivity (also called piezoresistance) with mechanical strain indicates the dependence of resistivity on the mechanical strain. Change in resistance of the conductor with strain is utilized for the measurement of strain and hence other quantities.

The resistance R of a conductor of length l, cross sectional area A and resistivity ρ is given by

$$R = \rho\frac{l}{A} \quad ...(7)$$

When the wire is subjected to strain its resistance changes which is found as follows.

Taking natural log on both sides of eqn. (7), we get

$$\ln R = \ln\rho + \ln l - \ln A \quad ...(8)$$

Differentiating eqn. (8) partially, we obtain

$$\frac{dR}{R} = \frac{d\rho}{\rho} + \frac{dl}{l} - \frac{dA}{A} \quad ...(9)$$

If diameter of the wire is D, cross sectional area will be

$$A = \frac{\pi D^2}{4}$$

or

$$\ln A = \ln\pi - \ln 4 + 2\ln D.$$

Differentiating, we get

$$\frac{dA}{A} = 2\frac{dD}{D}$$

Hence from eqn. (9), we obtain

$$\frac{dR}{R} = \frac{d\rho}{\rho} + \frac{dl}{l} - \frac{2dD}{D}$$

or

$$\frac{dR/R}{dl/l} = \frac{d\rho/\rho}{dl/l} + 1 - \frac{2dD/D}{dl/l} \quad ...(10)$$

$\frac{-dD/D}{dl/l}$ is the ratio of lateral strain to longitudinal strain. This is called Poisson's ratio denoted by μ (mu) which has the value between 0 and 0.5 for metals

In other words poisson's ratio $\mu = \frac{\text{lateral strain}}{\text{longitudinal strain}}$

or $$\mu = \frac{-dD/D}{dl/l} = -\frac{dD/D}{\varepsilon}, \varepsilon = \frac{dl}{l} \text{ is longitudinal strain}$$

or $$\frac{dD}{D} = -\mu\varepsilon \qquad ...(11)$$

Minus sign indicates that as the wire stretches in length its cross sectional area decreases

Putting the value of $\frac{dD}{D}$ from eqn. (10), we obtain

$$\frac{dR/R}{dl/l} = 1 + \frac{d\rho/\rho}{\varepsilon} - 2(-\mu)$$

$$= 1 + 2\mu + \frac{d\rho/\rho}{\varepsilon}$$

or $$\frac{dR/R}{\varepsilon} = 1 + 2\mu + \frac{d\rho/\rho}{\varepsilon} \qquad ...(12)$$

The ratio $\frac{dR/R}{\varepsilon}$ is called gauge factor denoted by 'G', of the strain gauge.

Thus $$\frac{dR/R}{\varepsilon} = G = 1 + 2\mu + \frac{d\rho/\rho}{\varepsilon}$$

	1	2μ	$\frac{d\rho/\rho}{\varepsilon}$
	change in resistance due to change in length	change in resistance due to change in area	change in resistance due to change in resistivity (Piezoresistive effect)

Thus strain $\varepsilon = \frac{dl}{l}$ can be measured in terms of gauge factor and change in resistance $\frac{dR/R}{\varepsilon}$. The term $\frac{d\rho/\rho}{\varepsilon}$ can be written as πE where π is called longitudinal piezoresistance coefficient and E is called Young's modulus of elasticity. Eqn. (12) can then be written as

$$G = 1 + 2\mu + \pi E \qquad ...(13)$$

The typical value of gauges factor for most of the metals is 2.0 while for semiconductor strain gauges the gauge factor is about 140.

A strain gauge can give the desired value of the change in resistance if it has minimum resistance value which is dependent upon gauge current and gauge length. Typical values of resistances for strain gauges lie in the range of 100 to 1000 ohms.

Example 10.5. A nickel strain gauge has resistance of 250 Ω and gauge factor of 2.0. It is subjected to a strain of 80 μs. What is the change in its resistivity if it has initial value of resistivity as 70 Ohm meter ? Assume poisson's ratio 0.3.

Solution. We are given

Gauge factor $G = 2.0$

Resistance $R = 250\,\Omega$

Strain $\varepsilon = 80\mu S$, S stands for siemens which is the unit of strain.

Resistivity $\rho = 70\,\Omega$ meter

$\mu = 0.3$

$$G = \frac{dR/R}{\varepsilon}$$

Hence $dR = G\varepsilon R$

$$= 2 \times 80 \times 10^{-6} \times 250 = 4 \times 10^{-2}\,\Omega.$$

$$G = 1 + 2\mu + \frac{d\rho/\rho}{\varepsilon}$$

or
$$2.0 = 1 + 2 \times 0.3 \frac{d\rho/70}{80 \times 10^{-6}}$$

$$0.4 = \frac{d\rho}{70 \times 80 \times 10^{-6}}$$

or $d\rho = 0.4 \times 70 \times 80 \times 10^{-6}$ ohm meter.

$= 22.4$ micro ohm meter **Ans.**

10.9. APPLICATIONS OF STRAIN GAUGE

Strain gauges are widely used in the measurement of displacement, force, torque, pressure, acceleration and related quantities. Measurement of torque and displacement by using strain gauge is described as under.

1. Displacement Measurement by Using strain Gauge : Linear displacement can be measured by using strain gauges. The moving system is coupled to the specimen to which strain gauge is bonded. As the system moves the specimen contracts or expands and hence the wire of strain gauge contracts or expands which leads to a change in its resistance. Consequently the output voltage changes as a result of displacement. The circuit diagram is shown in Fig. 10.9. This is called ballast circuit.

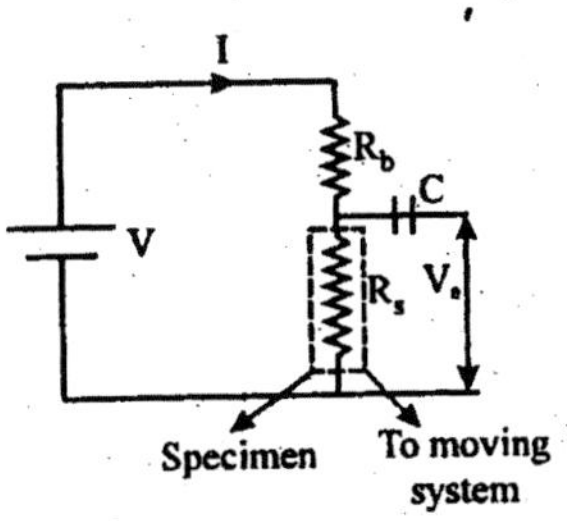

Fig. 10.9. Displacement measurement

Analysis : Current through the strain gauge

$$I = \frac{V}{R_b + R_s}$$

Hence output voltage under the condition of balance

$$V_0 = I \times R_s$$

$$= \frac{R_s}{R_b + R_s} V \qquad ...(14)$$

where R_b is called ballast resistance. If the resistance of strain gauge change to $R_s + \Delta R_s$, the output is changed to $V_0 + \Delta V_0$ and is given by

$$V_0 + \Delta V_0 = \frac{R_s + \Delta R_s}{R_s + \Delta R_s + R_b} V \qquad ...(15)$$

From eqn. (14) and (15), we get

$$V_0 + \Delta V_0 - V_0 = \frac{R_s + \Delta R_s}{R_s + \Delta R_s + R_b} V - \frac{R_s}{R_s + R_b} V$$

or $$\Delta V_0 = \frac{R_s^2 + R_s \Delta R_s + R_b R_s + R_b \Delta R_s - R_s^2 - R_s \Delta R_s - R_b R_s}{(R_s + R_b)(R_s + R_b + \Delta R_s)} V$$

$$= \frac{R_b \Delta R_s}{(R_s + R_b)^2 + \Delta R_s (R_s + R_b)} V \qquad ...(16)$$

If $\Delta R_s << R_s$, then

$$(R_s + R_b)^2 >> \Delta R_s (R_s + R_b)$$

Hence eqn. (16) becomes

$$\Delta V_0 = \frac{R_b \Delta R_s}{(R_s + R_b)^2} V, \text{ neglecting } \Delta R_s (R_s + R_b).$$

$$= \frac{R_b R_s}{(R_b + R_s)^2} \cdot \frac{\Delta R_s}{R_s} V$$

But $\frac{\Delta R_s}{R_s} \Big/ \varepsilon = G$ or $\frac{\Delta R_s}{R_s} = G\varepsilon$, we get.

$$\boxed{\Delta V_0 = \frac{R_b R_s}{(R_b + R_s)^2} G \varepsilon V} \qquad ...(17)$$

From eqn. (17) it is clear that change in output voltage is directly proportional to strain ε which in turn depends upon the displacement *dl* of the moving system. Eqn. (17) can be written as

$$\Delta V_0 = \frac{R_b R_s}{(R_b + R_s)^2} G \frac{dl}{l} V$$

The graph between dl and ΔV_0 is a straight line. The sensitivity of the gauge is given by

$$\frac{\Delta V_0}{\varepsilon} = \frac{R_b R_s}{(R_b + R_s)^2} GV \qquad ...(18)$$

It is clear from eqn. (18) that sensitivity of any strain gauge depends upon the gauge factor. Semiconductor strain gauges are more sensitive than metallic strain gauge. The ballast circuit is not useful to measure static strain. However it can be used for the measurement of dynamic strain. For this purpose a capacitor C is connected as shown in the Fig. 10.9, to pass time variable output through it and to block-dc.

Strain gauge can be connected in the bridge circuit also. The circuit is shown in Fig. 10.10. The change in output voltage as a result of movement of system under test is given by

$$\Delta V_0 = \frac{V}{(1+a)^2}[G\varepsilon]$$

where $a = \frac{R_3}{R_1} = \frac{R_s}{R_2}$

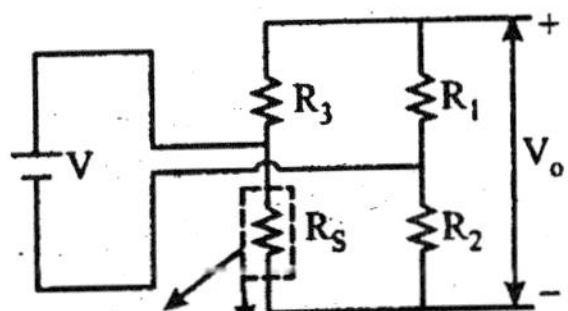

Fig. 10.10. Displacement measurement.

If $R_1 = R_2 = R_3 = R_s = R$ say, then

$$\Delta V_0 = \frac{V}{(1+1)^2} G\varepsilon = \frac{1}{4} VG\varepsilon$$

$$= \frac{1}{4} VG \frac{\Delta l}{l}$$

Hence $\qquad \Delta V_0 \propto \Delta l$

We consider following examples.

Example 10.6. A strain gauge has a resistance of 100 Ω and gauge factor of 2.0. It is connected in series with a ballast resistance of 50 Ω. It is excited by a dc source of 5 V. If output voltage is 20 mV, calculate *(a)* strain, *(b)* change in gauge resistance, *(c)* voltage sensitivity in volt per micro strain.

Solution. *(a)* Output voltage is given by

$$\Delta V_0 = \frac{R_b R_s}{(R_b + R_s)^2} G\varepsilon$$

or $$20 \times 10^{-3} = \frac{50 \times 100}{(50 + 1000)^2} \times 2 \times \varepsilon$$

or $$2 \times 10^{-2} = \frac{10^4}{(150)^2} \varepsilon$$

or $$\varepsilon = \frac{2\times10^{-2}\times(150)^2}{10^4} = \frac{225\times2}{10^4} = \frac{450}{10^4}$$

$$= 4.5\times10^{-2} = 45 \text{ milli siemens}$$

(b) $$\frac{\Delta R}{R} = G\varepsilon$$

or $$\Delta R = \varepsilon GR = 45\times10^{-3}\times2\times100 = 9 \text{ ohms.}$$

(c) Voltage sensitivity

$$\frac{\Delta V_0}{\varepsilon} = \frac{20 \text{ mV}}{45 \text{ milli siemens}} = \frac{20\text{V}}{45 \text{ siemens}}$$

$$= \frac{20\text{V}}{45\times10^6 \mu \text{ siemens}} = 0.44\times10^{-6} \text{ volt / micro strain}$$

$$= 4.4\times10^{-7} \text{ volt/microsiemens} \quad \textbf{Ans.}$$

Example 10.7. A strain gauge is connected in a wheatstone bridge having a resistance of 200 Ω and gauge factor 2.0. All other resistances in the bridge are of 200 Ω each. The bridge is excited by dc source of 30 volts. If output voltage is 4 volt, calculate *(a)* strain *(b)* displacement of the moving system if gauge length is 2 cm.

Solution. We use $\Delta V_0 = \frac{V}{(1+a)^2} G\varepsilon \qquad a = 1$

$$= \frac{1}{4} G\varepsilon V$$

or $$4 = \frac{1}{4}\times2.0\times\varepsilon\times30$$

or $$\varepsilon = \frac{16}{2.0\times30} = \frac{8}{30} \text{ siemens} = 0.027 \text{ siemens}$$

$$= 270 \text{ millisiemens}$$

(b) $$\varepsilon = \frac{dl}{l}$$

or $$dl = \varepsilon l = 27\times10^{-3}\times2\times10^{-2} \text{ meter} = 5.4 \text{ millimeter}$$

2. Measurement of torque using strain gauges : Torque is the torsional twisting movement which tends to twist a rigidly fixed body such as a shaft rotating about its axis of rotation. The circuit arrangement is shown in Fig. 10.11 *(a)*.

Two strain gauges R_1 and R_2 are mounted on the shaft which make an angle of ±45° with the axis of the shaft. When the force F is applied in the direction shown, a torque is developed which will twist the shaft in the direction shown. As the shaft is twisted, strains will be developed in the strain gauges. One strain gauge will develop tensile strain while the other develops compressive strain. As the angle between axes of strain gauge and shaft is 45°, naximum strains in the gauges will be developed along their axes and will lead to a change in the output of the circuit which is the measure of the torque τ working on the shaft.

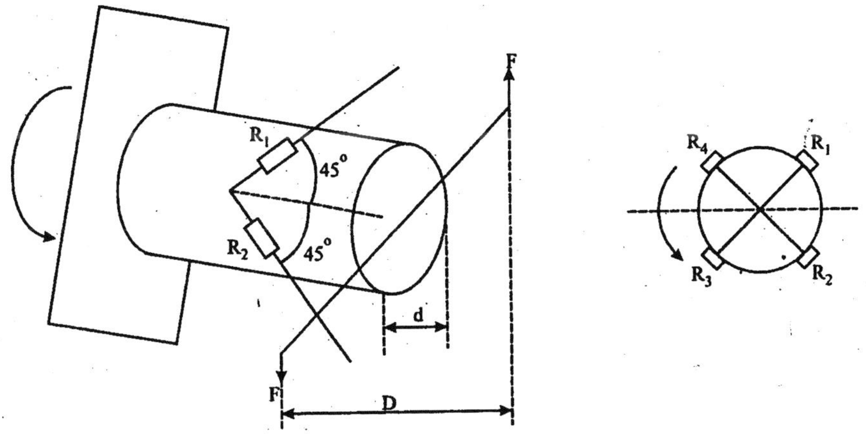

(a) *(b)*

Fig. 10.11. Torque measurement.

Magnitude of shear strain at 45° to the axes of the shaft can be given by the expression

$$\varepsilon_{45^\circ} = \frac{16\tau}{\pi d^3 E}(1+\mu) \qquad ...(19)$$

where τ = torque to be measured on the shaft

d = diameter of the shaft,

E = Modulus of elasticity.

μ = Poisson's ratio.

Higher resistivity compensation against temperature variations and compensation for thrust and bending stress can be achieved by employing two pairs of strain gauge as shown in Fig. 10.11 *(b)*. For the direction of rotation shown strain gauges 1 and 3 undergo compressive strain whole 2 and 4 undergo tensile strain. For measuring the torque the strain gauges are connected as a bridge as shown in Fig. 10.12. It is difficult to mount gauges on the shaft and get output from the circuit. Such a difficulty is overcome by mounting two pairs of slip rings on the shaft-one pair for supplying power to bridge circuit while another pair to collect output from the bridge. The circuit is shown in Fig. 10.13. Carbon brush in contact with the slip rings are used to supply power and collect output signal from the bridge. In place of dc source ac source can be used for excitation of the bridge.

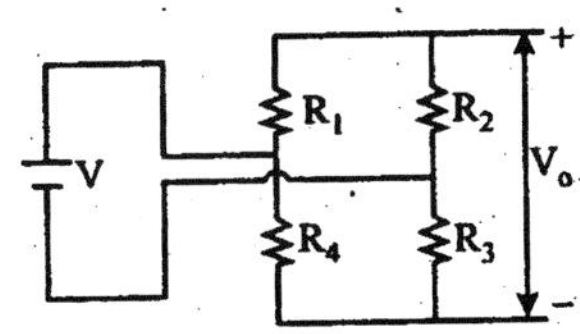

Fig. 10.12. Torque measurement.

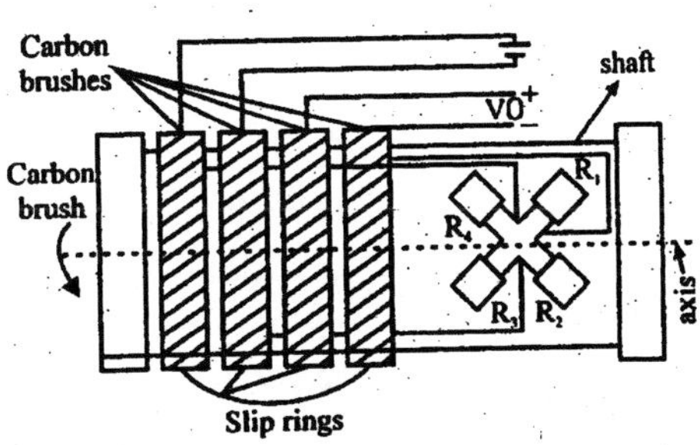

Fig. 10.13. Scheme for torque measurement.

10.10. THERMISTORS

Thermistors are also called thermal resistors as their resistance changes with temperature. These are semiconductor devices having high negative temperature coefficient (usually –0.04 per°C at room temperature) and are used in large temperature range. Their sensitivity is about 100 times that of RTD.

Variation of thermistor resistance with temperature is given by the expression

$$R = ae^{b/T} \quad ...(20)$$

where *a* and *b* are constants depending upon material and manufacturing techniques used.

Differentiating eqn. (20) w.r.t. T we get

$$\frac{dR}{dT} = ae^{b/T}\left(-b/T^2\right) = -\frac{Rb}{T^2}$$

or
$$\frac{dR/dT}{R} = -b/T^2 \quad ...(21)$$

L.H.S. of eqn. (21) is called resistance temperature coefficient of the device which is negative. The resistance temperature characteristics of a thermistor are shown in Fig. 10.14. We observe from the curve that as the temperature rises the resistance of the device falls exponentially. The resistivity of the device is very high.

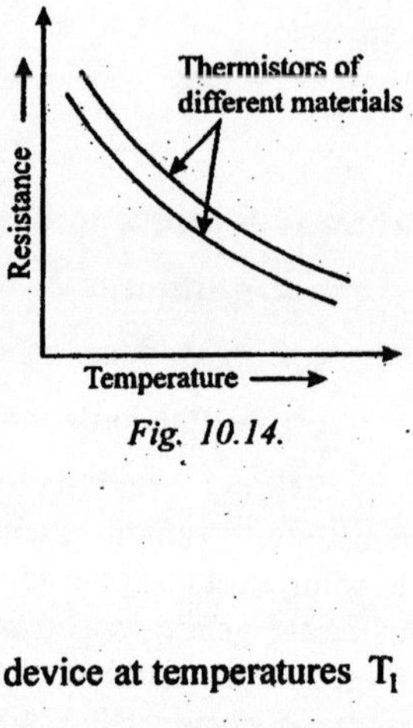

Fig. 10.14.

Again rewriting eqn. (20) for two different temperatures T_1 and T_0, we get

$$R_1 = ae^{b/T_1} \quad ...(22)$$

$$R_0 = ae^{b/T_0} \quad ...(23)$$

Here R_1 and R_0 are respectively, the resistances of the device at temperatures T_1 and T_0.

Dividing eqn. (22) by eqn. (23), we obtain

$$\frac{R_1}{R_0} = \frac{e^{b/T_1}}{e^{b/T_0}} = e^{b\left(\frac{1}{T_1}-\frac{1}{T_0}\right)}$$

or
$$R_1 = R_0 e^{b\left(\frac{1}{T_1}-\frac{1}{T_0}\right)} \quad ...(24)$$

Although the coefficient *b* is assumed constant, its value varies slightly with temperature which is of the order of 4000 Kelvin.

From eqn. (21) temperature coefficient of resistance at 25°C will be

$$a = \frac{dR/dT}{R} = -\frac{4000}{(298)^2} = -0.045/°C$$

The materials used for manufacturing thyristors are oxides of metals such as manganese, nickel, cobalt, copper, iron, zinc, magnesium, titanium etc. These oxides are mixed with silicates, milled and pressed into desired shape. Few of the thermistor shapes are shown in Fig. 10.15.

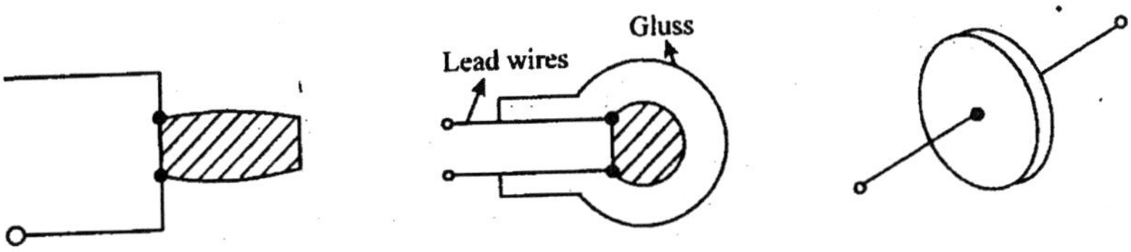

Fig. 10.15. Different shapes of thermistors.

A thermistor has large resistance change per degree change in the temperature. For example for a 2 kΩ thermistor the resistance changes by 78 Ω per degree change in temperature. Hence they can provide good resolution and accuracy in the temperature range of –100 °C to + 300°C. If thermistor is used in series with ammeter to measure temperature, temperature variation as small as 0.1°C can be detected. However it can measure temperature as low as 0.005°C if put in a wheatstone bridge circuit.

10.11. APPLICATIONS OF THERMISTORS

Thermistor due to its properties can be used to control and measure temperature of any system. Thermal conductivity, flow, composition of any gas can be measured with very high degree of accuracy by using this device. Measurement of temperature only will be discussed in this section.

1. Temperature Measurement by Thermistor : The circuit for temperature measurement by using thermistor is shown in Fig. 10.16 *(a)*. In the circuit we have a thermistor connected in series with a battery and microammeter calibrated to read the temperature directly. As the temperature changes the thermistor resistance changes which in turn changes the current through the microammeter used to measure the temperature directly. A temperature change as low as 0.1°C can be measured by the circuit. To measure further lower temperature the device is used in bridge circuit as shown in Fig. 10.16 *(b)*. The resolution of this circuit is 0.005°C

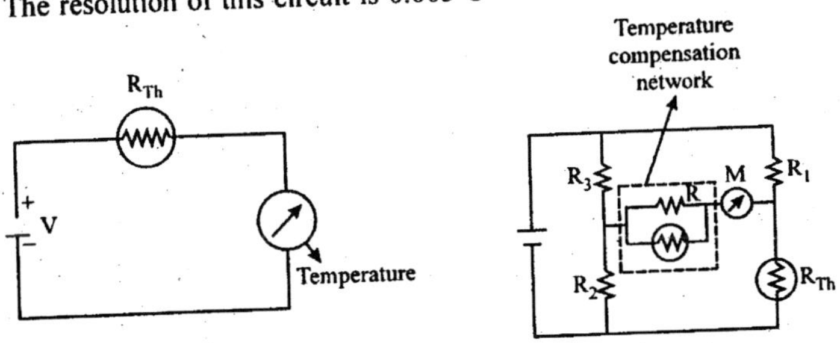

(a) *(b)*

Fig. 10.16. Scheme for temp measurement using thermistor.

Initially let the bridge be balanced. Under this condition $\frac{R_1}{R_{th}} = \frac{R_3}{R_2}$. We take the values of resistances such that $R_1 = R_3$ and $R_2 = R_{th}$. As the temperature changes, the thermistor resistance changes which produces output voltage measured by meter M. A temperature compensation network is used in series with the meter to compensate for the change in resistance of meter coil as a result of change in temperature. Resistance R is connected in parallel with thermistor to make its characteristics linear. The meter is calibrated to read the temperature directly.

2. Measurement of temperature difference by thermistor : For the measurement of difference of two temperatures, the thermistors are connected in adjacent arms of wheatstone bridge as shown in Fig. 10.17. The thermistors are connected to two systems or to two different parts of a system whose temperature difference is to be measured. Initially the bridge is balanced to give zero output voltage. As the temperatures of the systems change the resistance of the thermistors changes according to temperature.

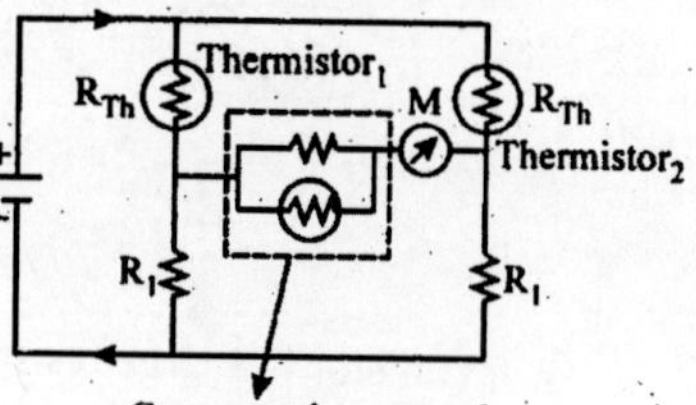

Fig. 10.17. Measurement of differential temperature.

The difference of the temperature produces non zero output which is measured by the meter directly. The compensation network is used to compensate the change in meter resistance with temperature.

10.12. RESISTANCE HYDROMETER

Humidity is the measure of water content in dry air or other gas. It is measured as absolute humidity and relative humidity which are defined as below.

Absolute humidity : It is the actual amount of water present in air and is usually expressed as gram per kg. or parts per million.

Relative humidity : It is the ratio of water vapour actually present in the air to the maximum amount of water that the air can hold at the same temperature. It is expressed in percentage.

The working of resistance hydrometer is based on the principle that with increase in humidity the resistance of a substrate decreases. Dunmore elements and Pope cell are used for percent relative humidity transducers.

Dunmore sensor consists of bifilar -wound inert wire grid on an insulative substrate coated with lithium chloride salt solution of given concentration. The nature of this salt is hygroscopic and it absorbs water from the surrounding. As relative humidity of surrounding increases Lithium chloride absorbs more and more water and its resistance decreases correspondingly. The resistance which acts as one arm of a bridge excited by ac power gives changed output as a result of change in relative humidity. The related circuit is shown in Fig. 10.18. Dunmore transducers can be used to measure relative humidity from 40 to 60 % RH with resolution of about 0.15%

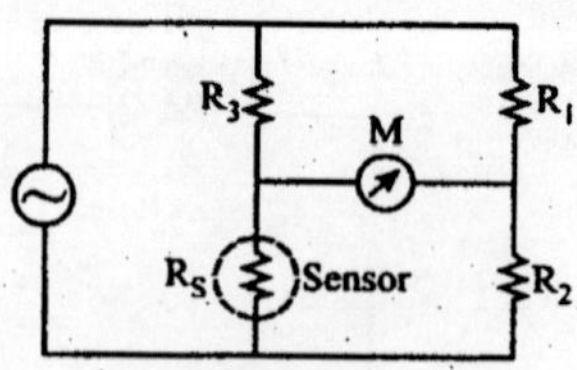

Fig. 10.18. Measurement of humidity.

relative humidity and an inaccuracy of 1.5% relative humidity. The working temperature range of this transducer is –40°C to 65 °C.

In Pope cell transducer also same type of conductive grid is used. However the substrate is made from polystyrene treated with sulphuric acid. This transducer exhibits nonlinear resistance humidity characteristics and covers from 15% to 99%. RH.

Example 10.8. Resistance of a thermistor at room temperature is 0.5 KΩ. What will be the resistance at 50°C. Assume b = 4000 kelvin.

Solution. We use relation

$$R_1 = R_0 e^{b\left(\frac{1}{T_1} - \frac{1}{T_0}\right)R}$$

where R_1 = ?

R_0 = 0.5 kΩ

T_0 = 298 K

T_1 = 273 + 50 = 323 K

b = 4000

Hence $$R_1 = 500\, e^{4000\left(\frac{1}{323} - \frac{1}{298}\right)} \text{ ohm} = 1.39 \text{ K}\,\Omega \quad \textbf{Ans.}$$

10.13. CAPACITIVE TRANSDUCER

In order to explain as to how a capacitor can be used as a transducer we write the expression for capacitance of a parallel plate capacitor which is

$$C = \frac{\varepsilon_0 \varepsilon_r A}{d}$$

where ε_0 = Permittivity of free space. It has value 8.854×10^{-12} F/m

A = Area of overlapping of the capacitor plates, m^2

d = Separation between the plates in meter

ε_r = Relative permittivity of the medium between the plates.

We observe that the capacitance can be varied in three ways - one by changing the area of overlapping between the plates, two by changing the distance between the plates and three by changing the dielectric constant of the medium between the plates. Above changes can be caused by physical variables such as displacement, force or pressure. The change in permittivity can be caused in the case of measurement of levels of liquids or gases and thickness. The method for the measurement of displacement, pressure liquid level and thickness will be discussed as below :

1. Displacement measurement by capacitive transducer : For the measurement of displacement of a moving system by capacitive transducer movable part of the system is coupled to the moving plate of a parallel plate capacitor as shown in Fig. 10.19. The capacitance of the capacitor is

$$C = \frac{\varepsilon_0 \varepsilon_r A}{l} \qquad ...(25)$$

Differentiating eqn. (25) w.r.t. l, we get

$$dC = -\frac{\varepsilon_0 \varepsilon_r A}{l^2} dl$$

or $$\frac{dC}{dl} = -\frac{\varepsilon_0 \varepsilon_r A}{l^2} \qquad ...(26)$$

$\frac{dC}{dl}$ is called sensitivity of the transducer

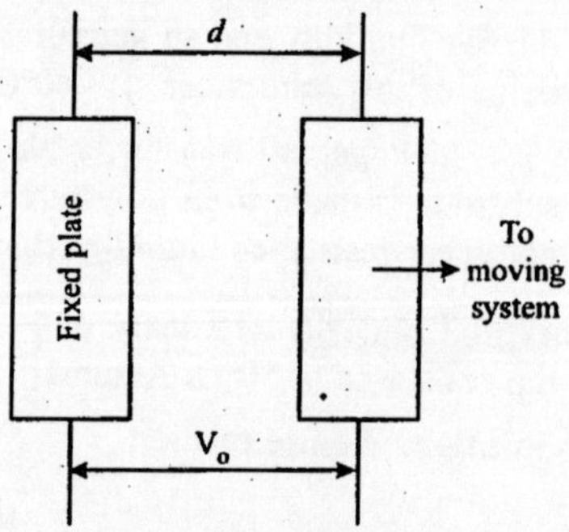

Fig. 10.19. Displacement measurement.

which varies as $\frac{1}{l^2}$. It means the variation of capacitance with displacement is non-linear. This is the limitation of this transducer. For very small displacement the relationship between capacitance and displacement can be assumed to be linear. The curve between capacitance variation and displacement change is shown in Fig. 10.20. The circuit arrangement is shown in Fig. 10.21. In the Figure the current between inverting and non-inverting ends is zero. The current I_f and I are same. At point A, let potential be V_A then

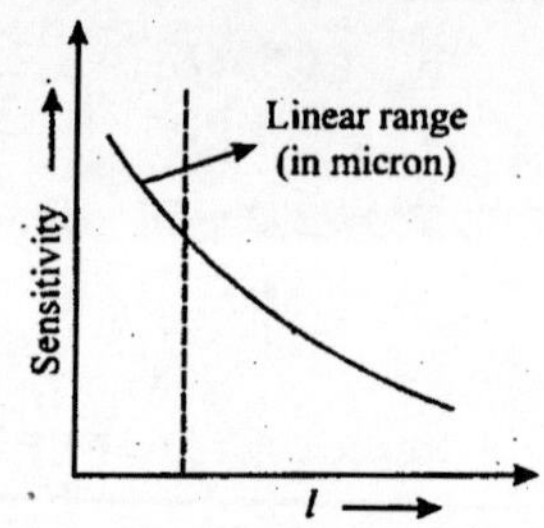

Fig. 10.20.

$$V_I - V_A = IR$$

and $$V_o - V_A = I_f / SC$$

But $$I_f = -I, V_A = 0 = V_B$$

Hence $$V_i = IR = -I_f R = -V_o R SC$$

or $$\frac{V_o}{V_i} = -\frac{1}{RSC} = -\frac{d}{SR\varepsilon_0\varepsilon_r A}$$

i.e., the transfer function of the system is

$$\frac{V_o(s)}{V_i(s)} = -\frac{d}{SR\varepsilon_0\varepsilon_r A}$$

or $$V_o(s) \propto d$$

or $$V_o(s) \propto \Delta d$$

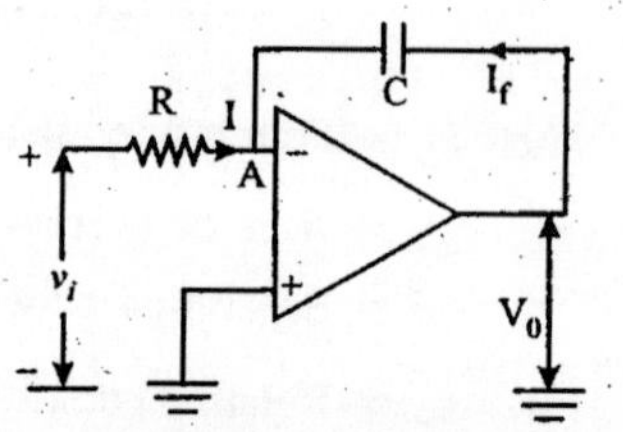

Fig. 10.21. Circuit arrangement using capacitor transducer.

As the displacement changes the output voltage changes. The amplifier chosen should be such as to give linear output for wide frequency range.

Example 10.9. Separation between the plates of a capacitve transducer is 2 mm and the dielectric between them has relative permittivity of 10. Its one plate is coupled to a moving system. If change in the capacitance is 0.1 pF, what is the displacement of moving system ? Area of the plate is 6 cm^2.

Solution : The change in the capacitance is given by

$$dc = -\frac{\epsilon_o \epsilon_r A}{l^2} dl$$

Here $\epsilon_o = 8.854 \times 10^{-12}$ F/m

$\epsilon_r = 10$

$A = 6 \text{ cm}^2 = 6 \times 10^{-4} \text{ m}^2$

$l = 2 \text{ mm} = 2 \times 10^{-3}$ m

$dl = ?$

$dc = 0.1 \ \mu\text{F} = 1 \times 10^{-13}$ F

Putting these values in the above eqn., we get

$$0.1 \times 10^{-12} = \frac{8.854 \times 10^{-12} \times 10 \times 6 \times 10^{-4}}{(2 \times 10^{-3})^2} dl$$

or

$$dl = \frac{4 \times 10^{-6} \times 10^{-13}}{8.854 \times 10^{-15} \times 6}$$

$$= \frac{2 \times 10^{-4}}{3 \times 8.854} \text{ metter}$$

$$= 7.5 \times 10^{-6} = 7.5 \text{ micron}$$ **Ans.**

2. Pressure measurement by capacitive transducer : Although the pressure cannot be measured directly by capacitor, yet it can be converted into equivalent displacement which can be measured by capacitive transducer. It is, thus, a secondary transducer. The arrangement for the measurement of pressure is shown in Fig. 10.22 *(a)*. Movable plate

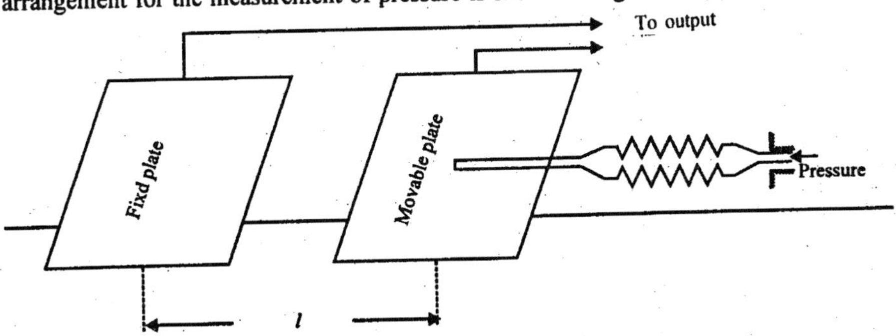

Fig. 10.22 (a) Pressure measurement.

of the capacitor is coupled to a pressure sensing element (Bellows in this case). As the pressure is applied the length of the bellows changes as it expands. As a result the displacement between capacitor plates changes which leads to a change in output corresponding to change in displacement and hence the pressure. Small range of pressures can be measured by this method. The analysis of the circuit is the same as in the displacement measurement.

In another method for pressure measurement a tank circuit can be employed as shown in Fig. 10.22 *(b)*. A pressure sensor (diaphragm in this case) is coupled to the plate of capacitor of tank circuit through mechanical linkage. When the pressure is applied the sensor expands which changes the separation between capacitor plates and hence the frequency of the tank circuit which is given by $f = \frac{1}{2\pi\sqrt{LC}}$, changes. As a result the output of the oscillator circuit is frequency modulated which is demodulated and calibrated in terms of applied pressure or force. The drawback of this method is low accuracy.

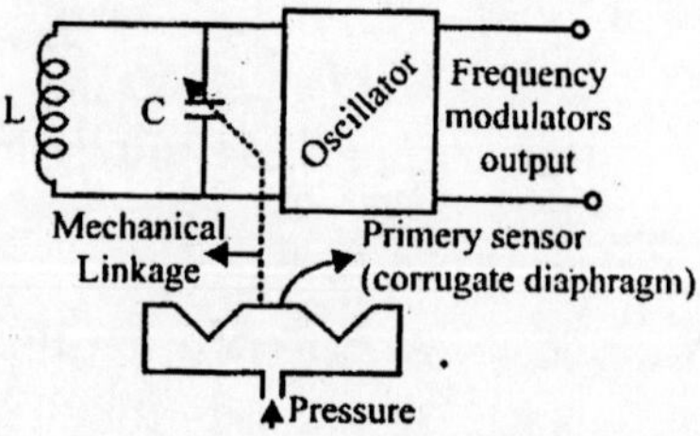

Fig. 10.22 (b) Pressure measurement.

3. Liquid level measurement by capacitive transducer : The measurement of liquid level is required for meeasuring and controlling the liquid content in a tank, height of a liquid in any column, etc. For liquid level measurement various schemes can be used depending upon the nature of the liquid. Capacitve transducer can be very effectively used for this purpose. Some of the schemes for measurement of liquid level using capacitive transducers are described as below :

(a) A cylindrical capacitor in the form of concentric electric probe is inserted in the liquid whose level is to be measured. As the liquid rises between the plates of cylinder the dielectric constant and hence the capacitance of the transducer changes, which is the measure of the liquid level and measured by using capacitor as one of the arms of an ac bridge. The scheme is shown in Fig. 10.23. The cpacitance of cylindrical capacitor is given by

$$C = 2\pi \in_o \frac{\in_v h_v + \in_g h_g}{\ln r_2 / r_1}$$

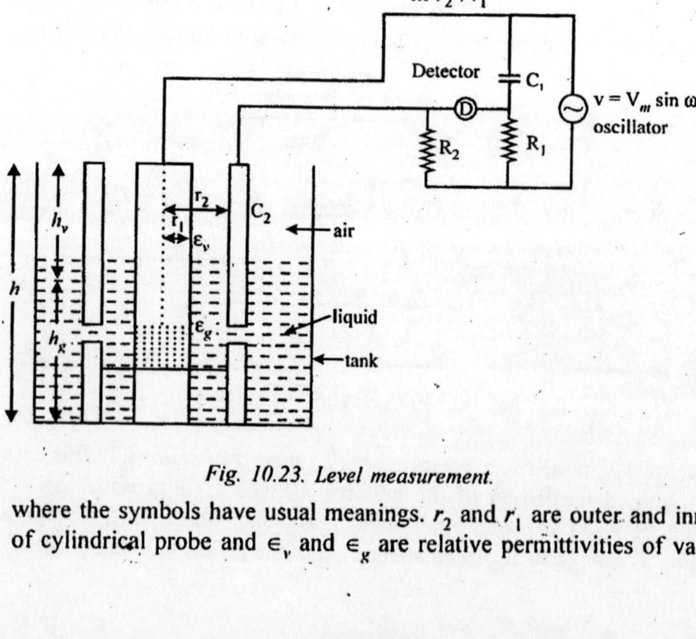

Fig. 10.23. Level measurement.

where the symbols have usual meanings. r_2 and r_1 are outer and inner radius of cylindrical probe and $\in_v$ and $\in_g$ are relative permittivities of vaccum and medium between capacitor plates changes which gives rise to a change in capacitance and hence the output of the ac bridge changes which directly indicates the liquid level. This method is used for measuring the level of non conducting liquid (having conductivity less than 10^{-7} per ohm cm).

(b) In Fig. 10.24, schemes for measurement of level of conducting liquids is shown. As the liquid is highly conducting its top surface acts as a plate of a parallel plate capacitor. In the Fig. two capacitors connected in series are shown — one formed by plates P_1 and P_2 is designated C_1 and the second formed by plate P_1 and conducting liquid surface is designated C_2. The capacitance C_2 varies as the level of the liquid changes. This change in capacitance C_2 causes a change in output voltage which is the direct indication of liquid level.

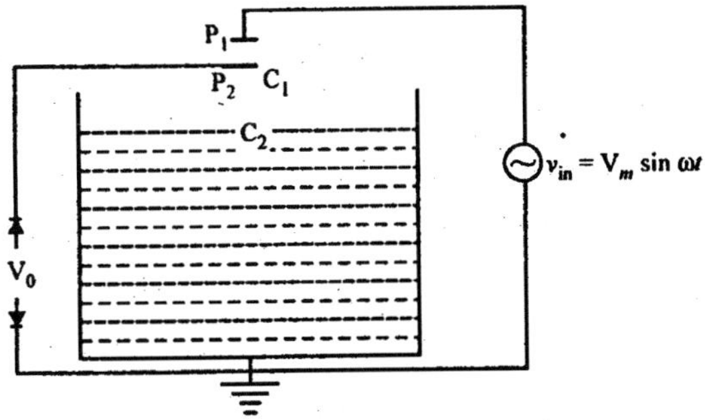

Fig. 10.24. Level measurement.

Analysis of the circuit : The capacitive reactance of capacitor C_1 is $X_{C_1} = \dfrac{1}{SC_1}$,

and that of C_2 will be $X_{C_2} = \dfrac{1}{SC_2}$. As the capacitors are connected in series total

capacitance
$$C = \frac{C_1C_2}{C_1+C_2}$$

and total impedance
$$= \frac{1}{SC} = \frac{1}{SC_1} + \frac{1}{SC_2} = \frac{C_1+C_2}{SC_1C_2}$$

Output voltage will be given by

$$V_o = \frac{X_{C_2}}{X_{C_1}+X_{C_2}} \times V_{in}$$

$$= \frac{1/SC_2}{\dfrac{C_1+C_2}{SC_1C_2}} = \frac{C_1}{C_1+C_2}V_{in}$$

or
$$V_o = \frac{C_1}{C_1+C_2}V_{in} \qquad ...(27)$$

Thus changes in capacitance C_2 changes the output voltage which indicates the level of the liquid.

Example 10.10. A cylindrical capacitive transducer is used to meaure the level of oil in a tank. The height, inner radius and outer radius of the cylinder are respectively one meter, 1 mm and 3 mm. The relative permittivity of the oil is 16. What is the change in the capacitance for 2 cm rise in liquid level ? Initial value of liquid level is 20 cm.

Solution : The capacitance of cylindrical capacitor is given by

$$C = 2\pi \epsilon_o \frac{\epsilon_v h_v + \epsilon_g h_g}{\ln r_2 / r_1}$$

Differentiating,

$$dC = 2\pi \epsilon_o \frac{\epsilon_v (-dh_v) + \epsilon_g dh_g}{\ln r_2 / r_1}$$

$$= 2\pi \epsilon_o \frac{1 \times (-2 \times 10^{-2}) + 16 \times 10^{-2} \times 2}{\ln 3/1} \text{ Farad}$$

$$= 2 \times 3.14 \times 8.854 \times 10^{-12} \frac{32 \times 10^{-2} - 2 \times 10^{-2}}{\ln 3}$$

$$= \frac{6.28 \times 8.854 \times 10^{-12} \times 30 \times 10^{-2}}{\ln 3} = 0.13 \mu F \textbf{ Ans.}$$

Example 10.11. A differential capacitive transducer is used to measure the level of a conducting liquid. The capacitance of a fixed capacitor is 0.47 μF and that of one formed by liquid is 0.10 μF. A power source of 5 volt ac is used to excite the circuit. What is the output voltage of the circuit ?

Solution : Using the formula

$$V_o = \frac{C_1}{C_1 + C_2} V_{in}, \text{ we obtain}$$

$$V_o = \frac{0.47 \times 10^{-6}}{0.47 \times 10^{-6} + 0.1 \times 10^{-6}} \times 5 \text{ volt}$$

$$= \frac{0.47}{0.57} \times 5 \text{ volt}$$

$$= 3.6 \text{ volts}$$

4. Thickness measurement by using capacitive transducers : Various methods such as inductive, resistive, capacitive, nuclear radiation etc. are used for the measurement of thickness of any material. The utility of any method for the measurement of thickness of any material is governed by the nature of the material. In this section capacitive transducer to measure the thickness of any material will be discussed.

Capacitive transducer is used to measure the thickness of insulating materials. The material to be tested is placed between the plates of parallel plate capacitor. If the dielectric constant of the material is known, its thickness is computed from the measurement of value of its capacitance. The arrangement is shown in Fig. 10.25. The capacitance C_1 is given by

$$C_1 = \frac{\epsilon_0 \epsilon_r A}{t} \quad ...(28)$$

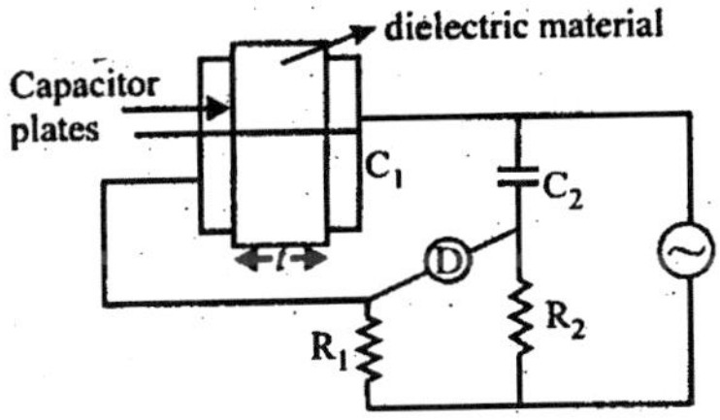

Fig. 10.25. Thickness measurement.

Initially the bridge is balanced for known value of C_1. Now the insulating material of known dielectric constant is inserted between its plates. As a result the capacitance C_1 is changed and unbalanced bridge output is measured by meter D. The bridge is balanced again to calculate the value of capacitance. Now from eqn. (28) the thickness t of the dielectric material is computed.

From the bridge,
$$C_1 = C_2 \frac{R_2}{R_1} \quad ...(29)$$

From eqn. (28) and (29), we obtain

$$\frac{\epsilon_0 \epsilon_r A}{t} = C_2 \frac{R_2}{R_1}$$

or
$$t = \frac{\epsilon_0 \epsilon_r A R_1}{C_2 R_2} \quad ...(30)$$

Eqn. (30) can be used to compute value of t when all the quantities on R.H.S. are known.

Example 10.12. A dielectric sheet is introduced between two parallel plates of a capacitor. Area of each plate is 3 cm². The dielectric constant of the sheet is 30. The capacitor is connected as one arm of an ac bridge. In the condition of balance $C_2 = 0.3$ μF, $R_2 = 470\ \Omega$, $R_1 = 300\ \Omega$, compute the thickness of the material.

Solution : Using eqn. (30), the thickness is given by

$$t = \frac{\epsilon_0 \epsilon_r A R_1}{C_2 R_2}$$

Here
$$\epsilon_0 = 8.854 \times 10^{-12} \text{ F/m}$$
$$\epsilon_r = 30$$
$$A = 3 \text{ cm}^2 = 3 \times 10^{-4} \text{ m}^2$$
$$R_1 = 300\ \Omega$$
$$R_2 = 470\ \Omega$$
$$C_2 = 0.3\ \mu\text{F} = 3 \times 10^{-7} \text{ Farad}$$

Putting the values, we get

$$t = \frac{8.854 \times 10^{-12} \times 30 \times 3 \times 10^{-4} \times 300}{3 \times 10^{-7} \times 470} \text{ meter}$$
$$= 0.18 \text{ micron.}$$

10.14. VARIABLE INDUCTANCE TYPE TRANSDUCER

This is a passive transducer which can work on the principle of variation of following three quantities.

1. By variation in self inductance
2. By variation in mutual inductance

3. By variation in eddy current

1. Inductive transducers using variation in self inductance : The self inductance L of an inductor is given by $L = \frac{N^2}{R}$, where N is the no of turns in the coil and R is the reluctance which is given by $R = \frac{l}{\mu A}$. Thus the expression for self inductance is L $= \frac{N^2}{l/\mu A} = N^2 \frac{\mu A}{l} = N^2 \mu G$, where $G = \frac{A}{l}$ is called geometric form factor. These transducers are usually used for measurement of linear and angular displacements. The displacement under measurement causes change in self inductance of the coil by varying any of the three variables N, μ and G as described.

(a) **Change in Self inductance by variation in number of turns :** Such transducers are used for measurement of linear and angular displacements as shown in Fig. 10.26 *(a)* and *(b)* respectively.

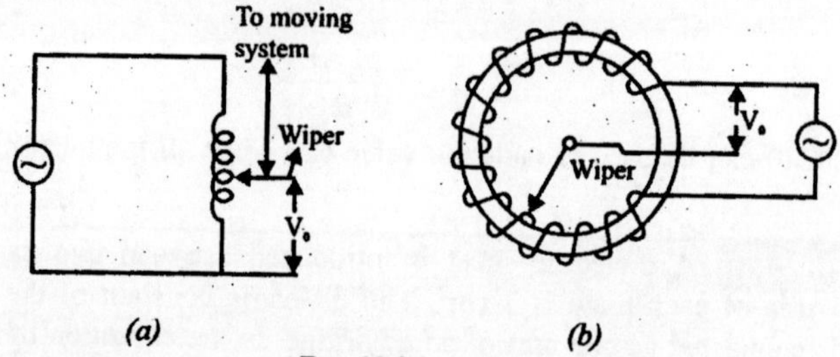

Fig. 10.26.

(b) **By Variation in permeability of the medium :** Variation in permeability to change inductance is shown in Fig. 10.27. A coil is wound on a former (non conducting) and an iron core attached to movable part of the system moves inside the coil. The reluctance of air gap is very high while that of iron is negligible compared to air gap reluctance. When core is completely inside the coil inductance will be maximum. As the core moves out the inductance and hence the output voltage decreases.

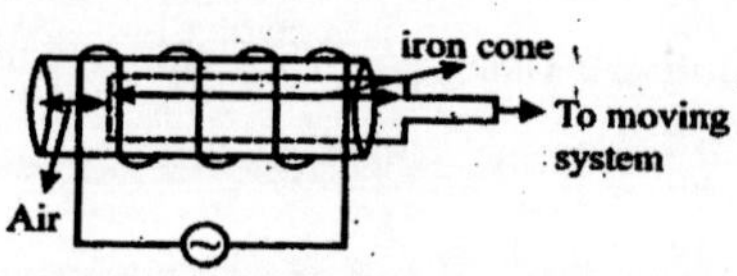

Fig. 10.27.

2. Transducer based on the principle of mutual induction : The mutual inductance between two coils depends upon self inductance of the coils and the coefficient of coupling between the two as shown in Fig. 10.28. The mutual inductance between the coils is given as $M = K\sqrt{L_1 L_2}$ where K is coupling

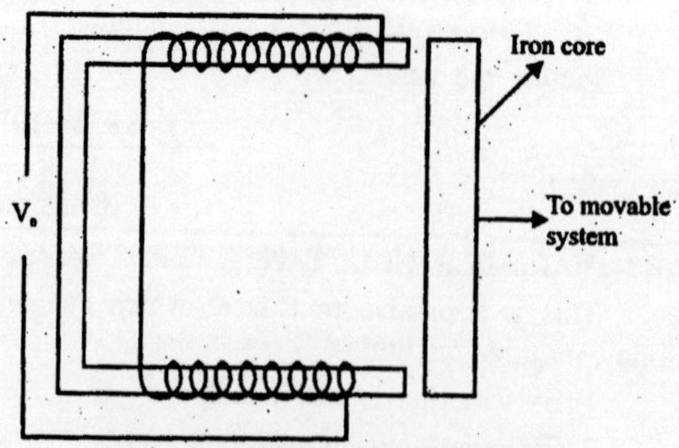

Fig. 10.28. Displacement measurement.

coefficient. In the Fig. two coils are connected in series hence self inductance of such an arrangement varies between $L_1 + L_2 - 2M$ and $L_1 + L_2 + 2M$. An iron core is coupled to a moving system. As the core is moved towards the coils the flux increases which gives an increase in self inductance. When the core is moved away from the coils the flux and hence self inductance decreases which is measured in the output.

3. Transducers based on the principle of production of eddy currents : When a conducting plate is placed near coil carrying alternating current, eddy currents are induced in the plate which produce their own field in a direction opposite to the main field created by the alternating current. As a result net flux linking the coil and hence its self inductance decreases. When the plate is moved away from the coil the field produced by eddy current is reduced. Hence the net flux linking the coil increases, which in turn increases the self inductance of the coil. In this way the displacement of any moving system is measured by linking its moving part to the conducting plate. As the plate moves the self inductance of the coil and hence the output changes which is the measure of displacement of the system under test.

10.15. LINEAR VARIABLE DIFFERENTIAL TRANSFORMER (LVDT)

LVDT is an inductive transducer which is used in measurement of various physical quantities such as displacement, force, pressure, position etc. The construction of LVDT is shown in Fig. 10.29. It has a primary winding P and two secondary windings S_1 and S_2 wrapped around a hollow bobbin of non magnetic material. A soft iron core which moves inside the bobbin is attached to the moving part of the system whose displacement

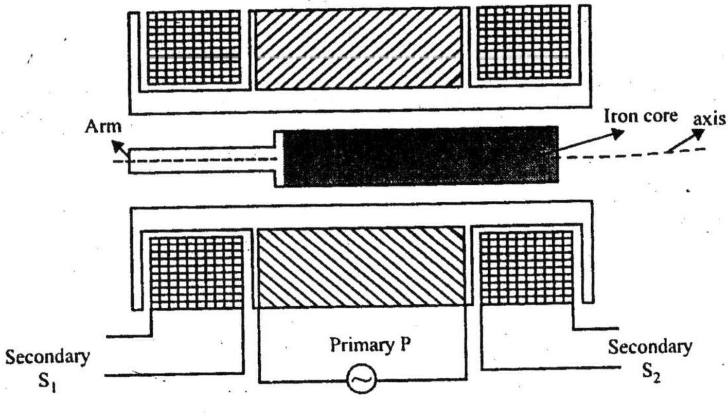

Fig. 10.29. LVDT Structure.

is to be measured. The primary is connected to oscillator having frequency range of 50 Hz to 20 kHz and voltage of 5 V to 25 V. When the iron core is in the middle of the tube, the flux coupled to the secondary by the primary is equal. This is considered reference position. Now the core is moved to left or right. As a result of its movement the flux coupled to the secondary near the core is increased while that coupled to the secondary far from the core is reduced. Hence the induced e.m.f. in both the secondaries are different and their difference (differential output) is obtained across the output as shown in Fig. 10.30 *(a)* where the two secondaries are connected in series opposition. The output voltage varies linearly with displacement for small displacement while for large displacements the variation between displacement and output is parabolic as shown in Fig. 10.30 *(b)*.

It is clear from the graph that output voltage is non-zero at reference position. This is due to presence of stray magnetic and electric fields and harmonics present in the input to primary. For commercially available LVDT the range of displacement varies between ± 0.01 mm to ± 25 mm. Following are few applications of LVDT.

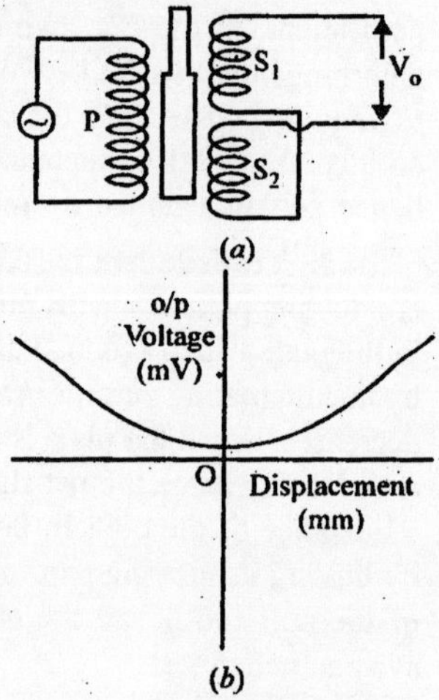

Fig. 10.30.

1. Displacement measurement by using LVDT : The circuit used for displacement measurement is shown in Fig. 10.30 *(a)* and relationship between output and linear displacement is shown in Fig. 10.30 *(b)*.

2. Angular displacement measurement using LVDT : LVDT used to measure angular displacement is also called RVDT. The scheme is shown in Fig. 10.31. In the circuit input shaft fastened to the core is mounted on the centre of the coil former on which primary and secondary are mounted symmetrically. The core is so chosen that it gives linear output over a specified angle of rotation. As angular displacement of, the system takes place the core coupled to it moves correspondingly.

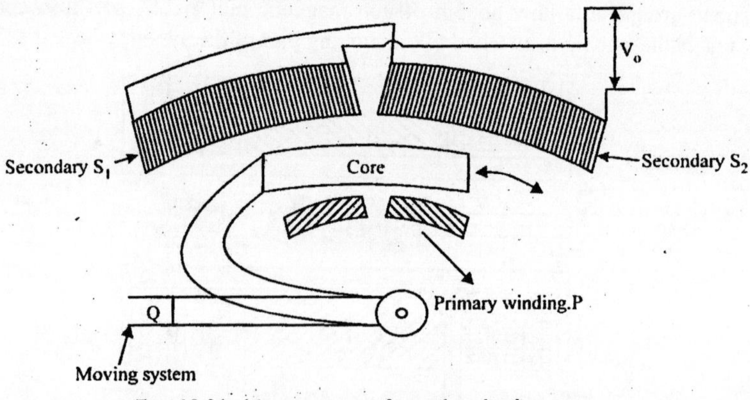

Fig. 10.31. Measurement of angular displacement.

In null position of the core, *i.e.,* central position output voltage is zero as induced e.m.f. in both secondaries S_1 and S_2 are equal and opposite. As the core is displaced from its null position, induced e.m.f. in coil near the core is more than the one far from the core. The difference of the two e.m.fs. appears across the output which is the measure of angular displacement.

These devices give linearity about ±1% for travel of ± 40 with sensitivity of the order of 10 mV per degree.

3. Measurement of pressure by LVDT : Measurement of pressure is very important in process industries where it is measured in the unit of force per unit area. It is in SI unit expressed as pascal. Low pressure is expressed as torr where 1 torr = 1 mm column of mercury. For the measurement of pressure various methods employed for measurement of force, displacement, temperature etc. can be used. Some of the methods for pressure measurement using LVDT are described as under :

(a) **Using bourdon tube :** The circuit is shown in Fig. 10.32. The pressure is applied to a primary sensor or summing device which is bourdon tube in this case. The change in dimension of bourdon tube is accompanied by a displacement of the LVDT core. This LVDT acts as secondary transducer here. The change in the position of the core results in a change in the output as the induced e.m.f.s in the secondaries are opposite and unequal. The schematic is shown in Fig. 10.32. *(a)*.

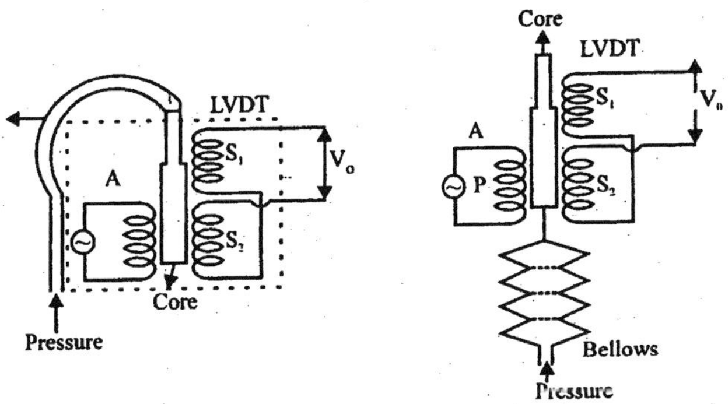

(a) using bourdon tube. *(b) using bellows.*

Fig. 10.32. Pressure measurement.

(b) **Using bellows :** This scheme is shown in Fig. 10.32 *(b)*. When the pressure is applied to the sensor (bellows) it expands in length. The core attached to this sensor moves which is obtained across the output.

3. Measurement of force by LVDT : Force and pressure are related by the expression Force = pressure × area in the direction of force. The measurement of force is, therefore, similar to the measurement of pressure. We apply a pressure on a given area and measure it. The force is thus computed by the relation given above.

4. Measurement of position : There is not much difference between position and displacement. The position of any part of the system is measured in terms of the displacement of the LVDT core. For zero position (origin of the system) of the part of the system the output of the LVDT is zero. As the position of the system changes, corresponding displacement of the LVDT occurs which is measured across output.

The circuit diagram is shown in Fig. 10.30.

10.16. OTHER TYPES OF TRANSDUCERS

In addition to the transducers discussed above various other transducers are also used to meassure different quantities. Some of the transducers will be discussed in this section.

1. Measurement of temperature by using solid state sensor : Silicon diodes and transistors are commonly used as solid state temperature sensing device. As the junction temperature of a p-n junction changes its forward bias voltage decreases for constant forward current. In a transistor the collector current I_C is given by

$$-I_C = \alpha I_E - I_{CO} \quad ...(31)$$

where α is common base current gain and I_{CO} is leakage current. The emitter current with emitter base voltage varies as

$$I_E = I_{EO}\left(e^{\frac{qV_{BE}}{KT}} - 1\right) \quad ...(32)$$

where K is Boltzman's constant*, T is absoute temperature of the system under test.

From eqn. (32)

$$\frac{I_E}{I_{EO}} + 1 = e^{\frac{qV_{BE}}{KT}}$$

or
$$\frac{qV_{BE}}{KT} = \ln\left(\frac{I_E + I_{EO}}{I_{EO}}\right)$$

or
$$V_{BE} = \frac{KT}{q}\ln\left(\frac{I_E + I_{EO}}{I_{EO}}\right) \quad ...(33)$$

where q is charge of the charge carrier.

From the above eqn. (33) as the emitter current is constant, the base to emitter voltage V_{BE} is directly proportional to temperature. The temperature of any system can be measured by using eqn. (33). The emitter base voltage is measured by a meter calibrated to read the temperature directly. The accuracy of this sensor is very high. For dynamic temperature measurement this device is generally used. The relationship between V_{BE} and T is not linear below 170 K due to incomplete ionization of donor or acceptor atoms. Higher temperature can be measured accurately by this sensor.

Example 10.13. An LM 335 solid state sensor is placed in a medium having a temperature of 45°C. The emitter current is at this temperature, equal to 8 mA. What is the base emitter voltage ?

Solution : Using eqn. (33), we get

$$V_{BE} = \frac{1.38\times10^{-23}\times(273+45)}{1.6\times10^{-19}}\ln 8\times10^{-3}$$

Neglecting I_{EO},

$$= \frac{1.38\times10^{-23}\times(318)}{1.6\times10^{-19}}\ln 8\times10^{-3}$$

$$= -0.23 \text{ volts} \quad \textbf{Ans.}$$

2. Measurement of temperature using thermocouple : If two dissimilar metals or alloys are joined at their free ends and the two junctions are placed at different temperature a potential difference is developed between the junctions. This effect is called Seebeck effect. The structure thus formed is called thermocouple whose working is due to Seebeck effect. The current flow takes place from lower conductivity material to higher conductivity materials. For example copper-constantan pair can be used as a thermocouple, platinum, iridium and various other metal pairs can be used as thermocouple. Construction of a thermocouple is shown in Fig. 10.33 *(a)*. In the Fig. two metals A and B are used to form thermocouple. The junctions J_1 and J_2 are kept respectively, at temperatures T_1 and T_2.

* K = 1.38 × 10^{-23} Joules/°K

system while a measuring instruments is connected between the ends of conductors forming junction J_2 as shown in Fig. 10.33 *(b)*. As the temperature of the system changes the current through the thermocouple circuit changes which results in change in the output voltage which is directly used to read the temperature of the system. The relationship between temperature and output voltage of the thermocouple is given by

$$e = a\,\Delta T + b\,(\Delta T)^2 \qquad ...(34)$$

where e is called thermoelectric e.m.f.

At lower temperature the e.m.f. varies linearly with the temperature while at higher temperature the term containing "t^2" becomes predominant.

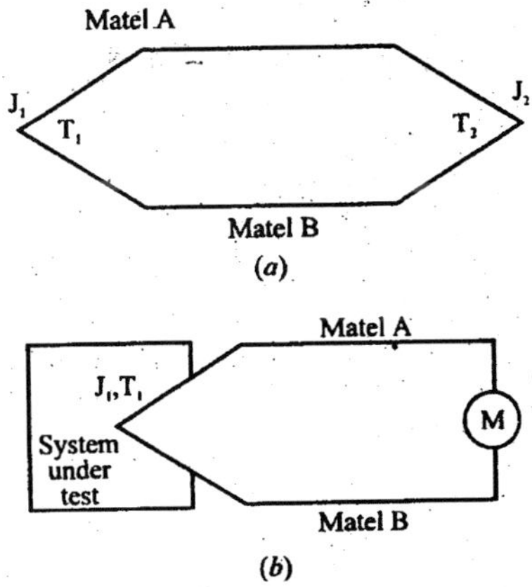

Fig. 10.33. Temperature measurement.

Laws governing thermocouples : Thermocouple circuit obeys the following law.

(a) **Law of Homogeneous metals :** A thermoelectric current cannot be retained in a circuit of a single homogeneous material by the application of heat even if it is of varying cross-section. Thus two dissimilar metals are essential.

(b) **Laws of Intermediate metals :** This law states that a third homogeneous material can always be added to the circuit with no effect in the e.m.f. as long as its extreme ends are at the same temperature. This law enables us to extend the length of thermocouple.

(c) **Law of Intermediate temperature :** We consider two dissimilar metals forming a thermocouple. Let its extreme ends are at temperature T_1 and T_2 and induced e.m.f. is E_1. Now the temperature of extreme ends is changed to T_2 and T_3 and the generated e.m.f. is E_2. According to this law if the junctions are at temperature T_1 and T_3 the induced e.m.f. will have the value equal to $E_1 + E_2$. The law is depicted in Fig. 10.34.

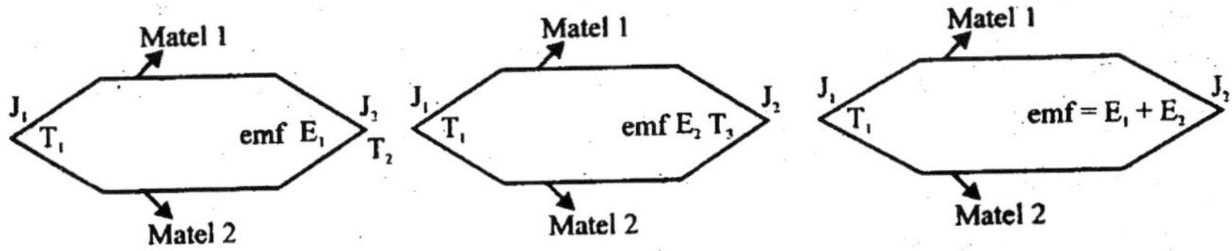

Fig. 10.34.

Thus it is clear that the induced e.m.f. by a thermocouple depends only on the temperature difference between the extreme ends.

Measurement of thermocouple output : Various methods are used to measure the output of a thermocouple. As the thermocouple output depends upon the temperature difference of two junctions, the temperature can be measured accurately by maintaning one of the junction (called reference junction) at constant temperature and the other junction is kept in touch with the system whose temperature is required to be measured. The circuit is shown in Fig. 10.35. It is used in laboratories. The temperature of reference junction should remain same as it is during calibration of the thermocouple. If reference junction temperature is not maintained constant the accuracy in output at low and medium temperature will be very poor. For this purpose the circuit shown in Fg. 10.35 is replaced by automatically balanced thermocouple circuit as shown in Fig. 10.36.

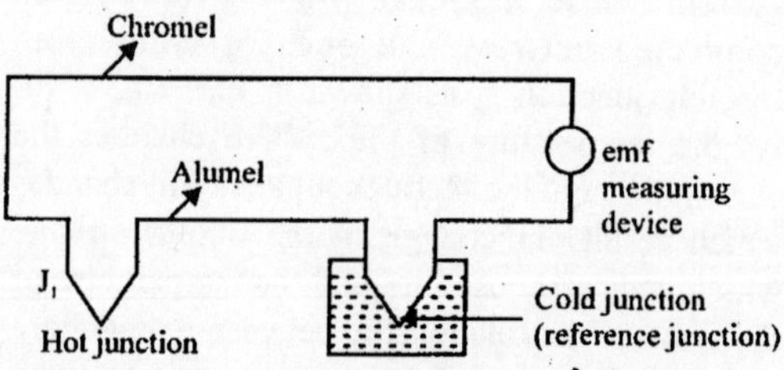

Fig. 10.35. Temperature measurement.

The e.m.f. measuring device may be a

1. sensitive permanent magnet moving coil millivoltmeter
2. a dc potentiometer or
3. a digital voltmeter.

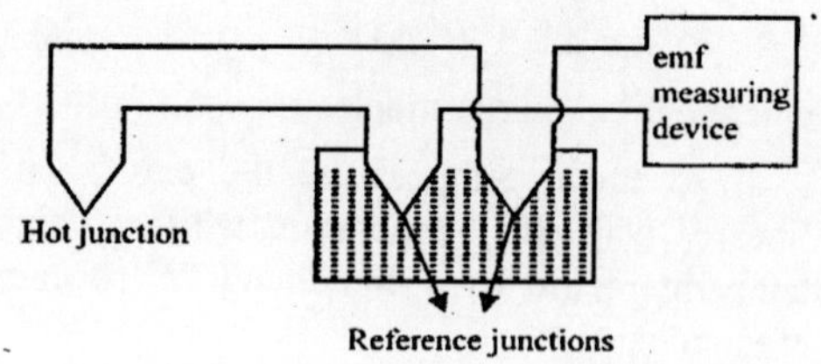

Fig. 10.36.

The thermocouple can be used to measure the temperature upto 1400° C accurately.

Lead compensation in thermocouples: For the measurement of temperature the metering device is kept at a distance from the hot junction. As a result whole length of the thermocouple wire is not at same temperature throughout, which may cause errors. These errors are eliminated by connecting compensation leads from the thermocouple circuit to the meter as shown in Fig. 10.37.

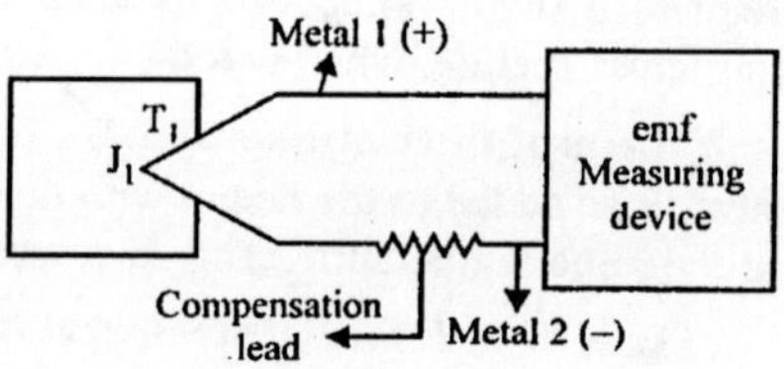

Fig. 10.37. Lead compensation in thermocouple

Example 10.14. A thermocouple is made up of iron and constantan. The thermoelectric e.m.fs. of iron and constantan w.r.t. platinum are 16 and – 34 μV/°C respectively. Calculate the e.m.f. developed per °C.

Solution : e.m.f. of iron w.r.t. constantan

= e.m.f. of iron w.r.t. platinum
– e.m.f. of platinum w.r.t. constantan
= 16 – (–34) μV/°C (by law of intermediate metals)
= 50 μV/°C **Ans.**

Example 10.15. A copper constantan thermocouple has $a = 37.5$ μV/°C and $b = 0.0045$ μV/°C. Detemine the e.m.f. developed when hot junction of thermocouple is kept at temperature of 200°C and cold junction is at temperature of ice.

Solution : Thermo e.m.f. is given by

$$e = a\,\Delta T + b\,\Delta T^2 + ... = a\,(T - T_0) + b\,(T - T_0)^2$$

Here T = 200 + 273 = 473 K

$T_0 = 0 + 273 = 273$ K

$$e = 375\ \mu V/°C\ (473 - 273) + 0.0045\ \mu V/°C\ (473 - 273)^2$$
$$= 8.17\ mV$$ **Ans.**

3. Measurement of Temperature by electrical resistance thermometers: The resistance of metals and alloys varies with temperature. This property of such materials is used for the measurement of temperature. The variation of resistance with variation in temperature is governed by the equation.

$$R_t = R_0 (1 + \alpha t) \text{ in linear range}$$

In the above expression;

R_t is the resistance of material at temperature t, R_0 is resistance at 0° temperature and α is temperature coefficient of metal. Its unit is Ω/Ω° C. Platinum and Nickel are generally used metals for temperature measurement. Once the values of α and R_0 are known, R_t can be standardized to read the temperature directly. A very important drawback, however, of this thermometer is non linearity in the output. Some schemes are employed to overcomes non linearity in the output.

For the measurement of temperature, the thermometer is used, in one arm of a whetstone bridge, in the form of a coil as shown in Fig. 10.37 *(a)*. The measurement coil in placed in contact with the system whose temperature is to be measured. R_1 and R_2 are fixed resistors while resistor R_3 is variable to obtain balanced output by balancing the bridge. A compensating lead is connected as shown in the circuit to compensate for the variation in coil resistance due to disturbance in external temperature. As the coil L comes in contact with the system, its changes. As a consequence, the circuit generates the output voltage which is detected by the detector D. A graph between R_1 and temperature is shown in Fig. 10.37 (*b*). It is obvious from the graph that P_t is more suitable device for large range of temperature variations.

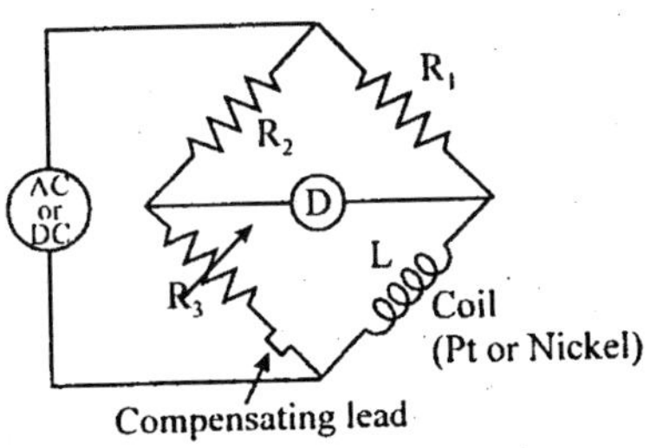

(a) Temperature measurement using electrical resistance thermometer.

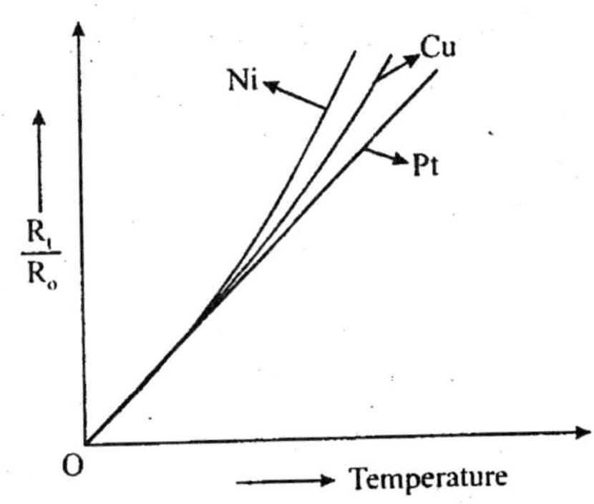

(b) Resistance variation with temperature for different materials.

Fig. 10.37

Example 10.16. Temperature coefficient of resistance for platinum is 0.00392 Ω/Ω°C. If its resistance at °C is 6Ω what is its value at 50°C.

Solution. Using the relations

$$R_t = R_0 (1 + \alpha t) \text{ we get}$$

$$R_t = 6 (1 + 6.00392 \times 50)$$

$$= 6.12 \ \Omega \text{ Ans.}$$

4. Measurement of force by piezoelectric device : Some materials have the property of developing potential difference across opposite faces as a result of dimensional change when a mechanical force is applied to it, between these faces. This effect is called piezo-electric effect and the materials are called piezoelectric materials. Piezoelectric materials are inverse transducer. When potential difference is applied across opposite faces, dimensional change occurs in the crystal. Generally, the crystals are used as mechanical input and elecrtrical output transducers, for the measurement of dynamic pressure, force, vibratory motion, shock etc. Piezoelectric effect is possible only in the crystal having asymmetrical charge distribution. In such case lattice deformation occurs due to relative displacement of positive and negative charges. This displacement of charge produces equal and opposite charge on opposite faces which is measured as potential difference between the faces. A piezoelectric transducer is shown in Fig. 10.38. In the Fig. a force F is applied to the crystal in the direction shown. Initially the thickness of the crystal is t and it is changed by Δt as force is applied. The potential difference developed as a result of dimensional change is fed to signal conditioning unit to measure the force applied.

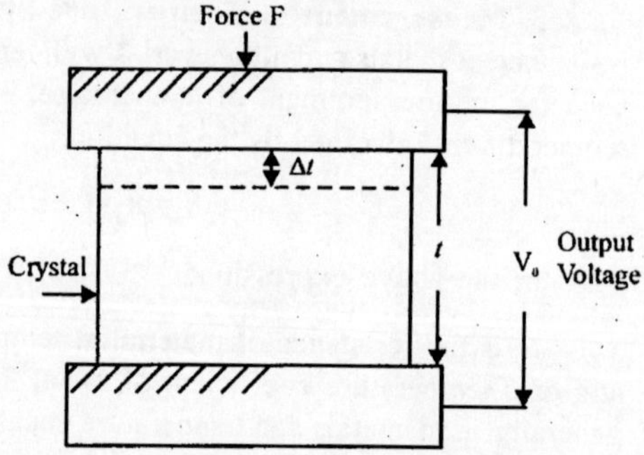

Fig. 10.38. Force measuring device.

Various piezoelectric crystals such as quartz, rochelle salt, lithium sulphate, barium titanate etc. are used as transducer. We consider quartz crystal (SiO_2) structure as shown in Fig. 10.39. In equilibrium it is represented by equisided two triangles having charges such that centre of positive and negative charges coincide so that net charge is zero as shown in Fig. 10.39 *(a)*. When the force F is applied to the crystal the centres of positive and negative charges are displaced w.r.t. each other. The potential difference appears across the output terminals as shown in Fig. 10.39 *(b)*.

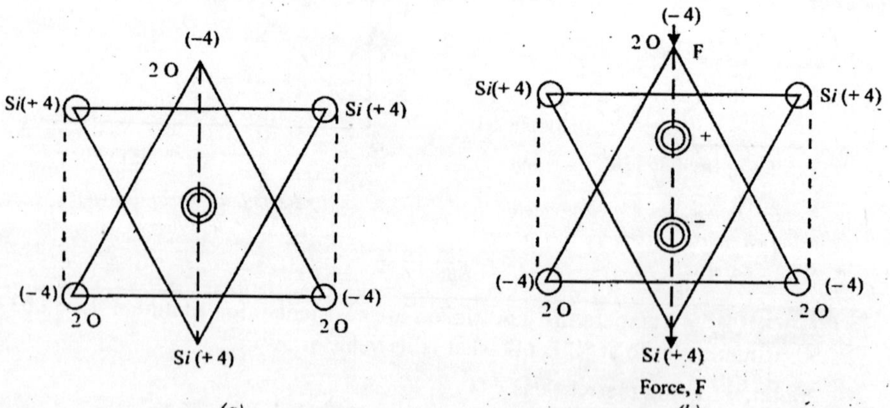

Fig. 10.39. Configuration of SiO_2 crystal.

Analysis : The magnitude and polarity of the induced surface charge depends on the direction and magnitude of the applied force, *i.e.*,

$$Q = dF \quad ...(35)$$

where Q = Induced surface charge in coulomb.

d = crystal charge sensitivity in coulomb/newton.

F = force applied in newton.

As the force is applied the dimensional change in crystal takes place. If change in crystal thickness is Δt, the stress in the crystal

$$\sigma = \text{Strain} \times \text{Young's modulus}$$

$$= \frac{\Delta t}{t} \times E$$

$$= \frac{\text{Force applied}}{\text{area of crystal}}, \text{ definition of } \sigma$$

$$= \frac{F}{A}$$

or

$$F = \sigma A$$

$$= \frac{AE\Delta t}{t} \text{ Newton} \quad ...(36)$$

The charge at the electrodes gives rise to voltage which is given by

$$V_o = \frac{Q}{C_C} \quad ...(37)$$

where C_C is capacitance of the capacitor formed by the crystal faces. From eqn. (35) and (36), we get

$$Q = d \times \frac{AE\Delta t}{t} \quad ...(38)$$

From eqn. (37) and (38), one obtains

$$V_o = d \times \frac{AE\Delta t}{t \times C_C}$$

where $C_C = \frac{\epsilon_r \epsilon_o A}{t}$. Here ϵ_r is relative permittivity of the crystal medium. Hence the expression for output voltage V_o becomes

$$V_o = d \times \frac{AE\Delta t}{t \times} \frac{\times t}{\epsilon_r \epsilon_o A}$$

$$= \frac{dE\Delta t}{\epsilon_r \epsilon_o}$$

$$= gE\Delta t \quad ...(39)$$

where $g = \frac{d}{\epsilon_r \epsilon_o}$ is called voltage sensitivity of the crystal.

Rewriting eqn. (37), we get

$$V_o = \frac{Q}{C_C} = \frac{dF}{C_C} = \frac{dF \times t}{\epsilon_o \epsilon_r A}$$

$$V_o = g\frac{F}{A}t \qquad \text{...(40)}$$

Eqn. (40) gives the relationship between applied force and the output of a piezoelectric crystal.

Example 10.16. A piezoelectric crystal has a thickness of 2 mm and voltage sensitivity of 0.05 mV/N. Determine the output voltage when it is subjected to a pressure of 1.0×10^6 N/m^2.

Solution : Output voltage is given by

$$V_o = g\frac{F}{A}t = gpt$$

$$= 0.05 \times 1.0 \times 10^6 \times 2 \times 10^{-3} \text{ volt}$$

$$= 10^2 \text{ volts}$$

$$= 100 \text{ volts } \textbf{Ans.}$$

Example 10.17. A piezoelectric pick up has dimension 3 mm × 2 mm × 1.5 mm. It is subjected to a pressure of 1.5×10^6 N/m^2 and if the voltage output of the crystal is 100 V calculate its voltage sensitivity and the force applied.

Solution : Output voltage of a piezoelectric transducer is given by

$$V_o = g\frac{F}{A}t = gpt$$

Here V_o = 100 V

g = ?

$p = 1.5 \times 10^6$ N/m^2

t = 1.5 mm

$$100 = g \times 1.5 \times 10^6 \times 1.5 \times 10^{-3}$$

or

$$g = \frac{100}{1.5 \times 10^6 \times 1.5 \times 10^{-3}}$$

$$= 0.044 \text{ V}_m\text{/N } \textbf{Ans.}$$

Force applied $F = PA$

where $A = 3 \times 10^{-3} \times 2 \times 10^{-3}$ m^2

$= 6 \times 10^{-6}$ m^2

$F = 1.5 \times 10^6 \times 6 \times 10^{-6}$ Newton = 9 Newton

5. Photoelectric devices : These devices are light operated ones. As a result of light intensity impinging on such devices their parameter varies in accordance with the change in intensity of light. The variation in any parameter is used to measure the intensity of light. Various photoelectric devices used to measure intensity of light are described below:

(a) **Photoemissive cells or phototubes :** These devices are of two types namely vacuum phototubes and gas filled phototubes. The device is shown in Fig. 10.40 *(a)* and circuit arrangement to measure intensity of light is shown in Fig. 10.40 *(b)*. As the intensity of light changes the current and hence the voltage across the load changes. The graph between intensity of light and anode current is shown in Fig. 10.40 *(c)*.

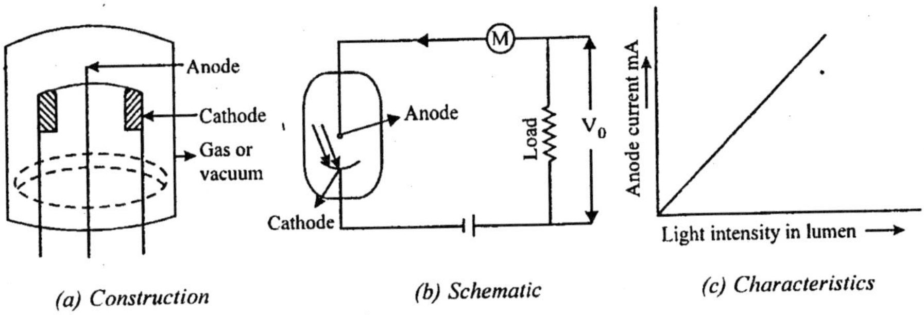

Fig. 10.40. Phototube.

However, if the luminous intensity is very low, very sensitive device is required for its detection. Photomultiplier tube serves this purpose. This device is shown in Fig. 10.41. The electrodes used in the device are called dynodes denoted by D_1 to D_4 in the figure. A ray impinging on the dynode D_1 to lead the generation of two rays which when impinge on dynode D_2 generates four rays. Thus the intensity of light can be increased.

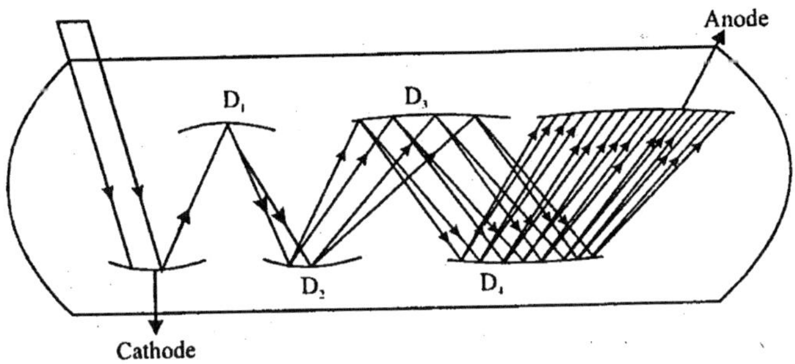

Fig. 10.41. Photomultiplier tube.

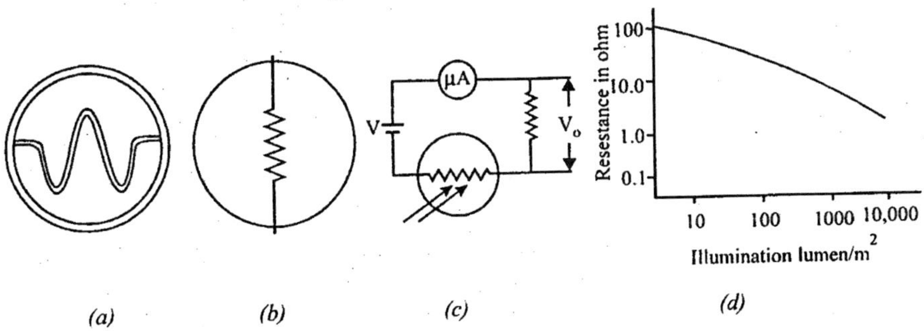

Fig. 10.42. Photoconductive cell.

(b) **Photoconductive Cells :** In some materials, the flow of the current varies as the intensity of light falling on them changes. As a result of light energy falling on such

materials the no of charge carriers changes which means a change in the resistance of the material with intensity of incident radiation. Such materials are called photo conductive cells or photoresistors or sometimes light dependent resistors (LDR). The cunstruction and symbol of the device are shown in Fig. 10.42 *(a)* and *(b)* respectively. The circuit using phototube for measurement of light intensity is shown in Fig. 10.42 *(c)* As the luminous intensity changes the resistance of the device and hence the output voltage changes. The graph between luminous flux and resistance is shown in Fig. 10.42 *(d)*. When the cell is not illuminated its resistance is of the order of MΩ. This resistance is called dark resistance and the current flowing through the device is called dark current. One major drawback of this device is that its resistance changes with temperature even if the light intensity is constant. Temperature compensation is required to overcome this difficulty.

(c) **Photovoltaic Cells :** These are self generating devices which convert radiation energy into electrical energy. Thes are called solar cells or photovoltaic cells for their voltage generating capability. The symbol for the device is shown in Fig. 10.43 *(a)*. Because of their voltage generating capability they are used as source of electrical energy. When the cell is open circuited the voltage output of the deivce is maximum, the current is maximum under short circuit condition. The voltage and current characteristics of the device are shown in Fig. 10.43 *(b)* and *(c)*.

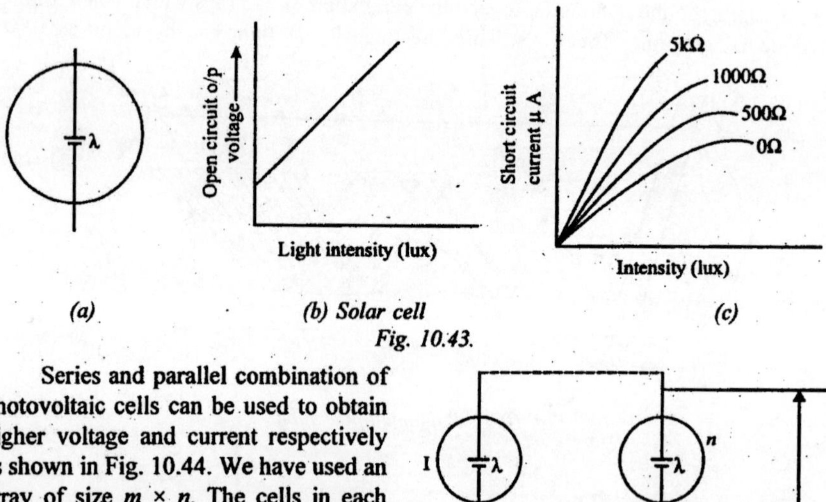

(a) *(b) Solar cell* *(c)*

Fig. 10.43.

Series and parallel combination of photovoltaic cells can be used to obtain higher voltage and current respectively as shown in Fig. 10.44. We have used an array of size $m \times n$. The cells in each column are connected in series which are used to give higher voltage. We have n columns connected in parallel. The total current delivered by the circuit will be n I where I is the current output of one column. Thus voltage and current output can be enhanced by using suitable combination of the cells.

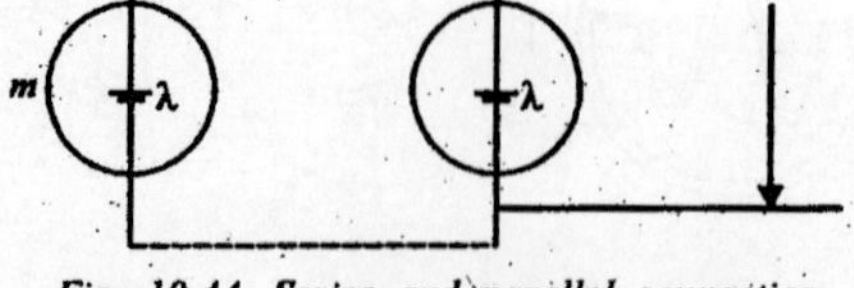

Fig. 10.44. Series and parallel connection of solar cells.

5. Proximity probes for measurement of r.p.m. : Proximity probes are used for measurement of r.p.m.

(a) **Photoelectric pickup to measure r.p.m. :** The scheme is shown in Fig. 10.45. A disc with holes on its periphery is mounted on a shaft of machine whose r.p.m. is to be measured. A light source and a photocell are placed in such a way that the light falls on the photocell through the holes on the disc which is counted by electronic counter.

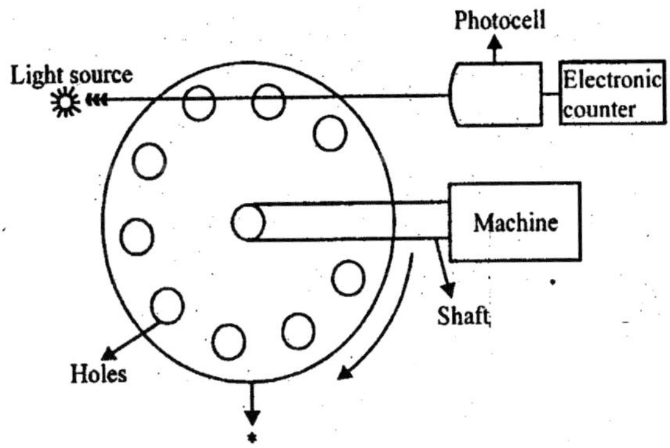

Fig. 10.45. Photoelectric pick up for r.p.m. measurement.

Let the speed of disc be n r.p.m. and no. of holes on the disc = p

$$\text{No. of rotations per second} = \frac{n}{60}$$

Frequency of output pulses is given by

$$f = \text{speed in rps} \times \text{total no. of holes}$$

$$= \frac{n}{60} \times p$$

or $$n = \frac{60f}{p}$$

Thus speed in r.p.s. $$n = \frac{60f}{p} \qquad ...(41)$$

As the light falls on photocell it is turned ON and sends a pulse to the counter. In case of opaque portion of the disc between light source and photocell the cell remains in off condition. No count occurs in the counter. Total no. of counts per minute are displayed as r.p.m. of the disc.

Example 10.18. A photoelectric pick up is used to measure rotational speed. The no. of holes on the disc are 30 and counting frequency is 100 cycles/sec. What is the speed of the disc ?

Solution : Speed of the disc is given by

$$n = \frac{60f}{p} \text{ r.p.m.}$$

$$= \frac{60 \times 100}{30} = 200 \text{ r.p.m.}$$

(b) **Magnetic pick up for r.p.m. measurement :** Basic scheme for r.p.m. measurement using magnetic pickup is shown in Fig. 10.46. A toothed rotor rotates between the poles of a magnet with a coil wound over it. When air gap faces the magnet, the induced e.m.f. in coil is small as reluctance of air is very high. When tooth is under magnet the induced e.m.f. is high which sends a pulse to an electronic counter. Thus the speed of the rotor can be measureed.

Let the no. of teeth on the rotor be $= t$

and speed of the rotor $= n$ r.p.m.

$$\text{no. of rotations per sec.} = \frac{n}{60} \text{ r.p.s.}$$

$$\text{Hence frequency of rotation} = \frac{n}{60}$$

$$\text{Frequency of counter} = \frac{n}{60} \times \text{total no. of teeth}$$

$$f = \frac{n}{60} \times t$$

or

$$n = \frac{60f}{t} \text{ r.p.s.} \qquad ...(42)$$

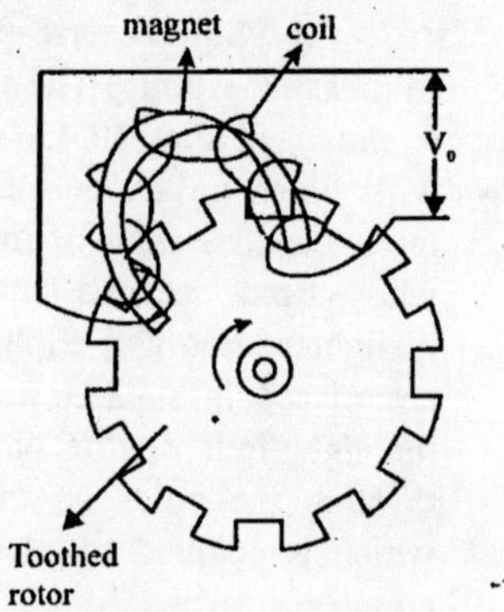

Fig. 10.46. Magnetic pickup to measure r.p.m.

Example 10.19. A rotor has 10 teeth. It is rotating under a magnetic field when the counting freequency is 100 counts/sec. If radius of the rotor is 6 mm what is its linear velocity ?

Solution : Speed of the rotor is given by

$$n = \frac{60f}{t} \text{ r.p.m.}$$

$$= \frac{f}{t} \text{ r.p.s} = \frac{100}{10} = 10 \text{ r.p.s.}$$

$$\text{angular velocity, } \omega = 2\pi n$$

$$= 2\pi \times 10 = 20\pi$$

$$\text{Linear velocity, } u = r \times \omega$$

$$= 6 \times 10^{-3} \times 20\pi = 0.12\pi \text{ m/sec}$$

$$= 0.3768 \text{ m/sec}$$

$$\approx 0.38 \text{ m/sec } \textbf{Ans.}$$

6. Digital displacement transducer : In digital displacement transducers for displacement, the transducer output is in the form of pulse counts, sinusoids or binary numbers 1 or 0. The pulse count, e.g., scintillation counter in radioactivity, frequency of sinusoid, sequence of 1's and 0's are not necessarily dependent on the amplitude of the signals. These signals can, therefore, be transmitted over long distances without any appreciable distortion due to phase shift and amplitude variation.

In most of the cases the transducer converts input quantity to analog form which is digitized by analog to digital converter. For example, a change in resistance or tension in the wire can be converted into frequency or pulse train through an electronic circuitry. These transducers are threfore, called encoder and are used for linear and rotary displacements.

Digital transducer are classified into three categories viz. tachometer, incremental and absolute digital transducer.

Out of the above mentioned types last two are used for displacement measurement.

The incremental encoder uses at least two signal generating elements. The two tracks used are shifted mechanically by 1/4 cycle relative to each other as shown in Fig. 10.47. In the Fig. two tracks 1 and 2 are shifted by 1/4 cycle. A third track called reference track produces signal pulse per revolution. By using incremental encoder tracks can be rotated through as many revolutions as desired. However the pulse signal appears in the output due to noise present in the input.

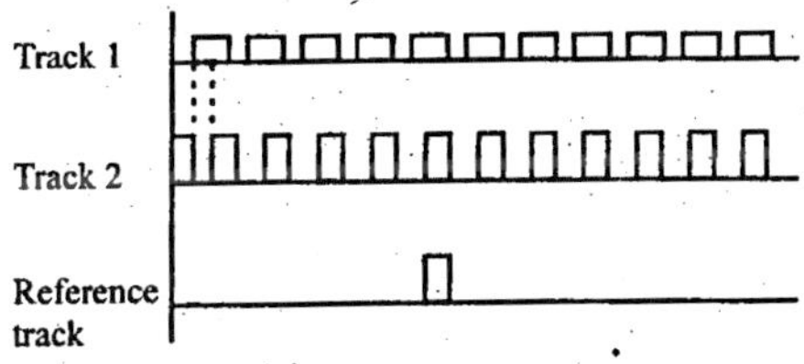

Fig. 10.47.

Construction of encoders : Encoders can be constructed in two ways viz. contact type and non contact type. Contact type rotary encoder is shown in Fig. 10.48 *(a)*. It consists of a cylindrical disc, the coding pattern arranged in concentric rings on one side of the disc. On the rings the black areas represent the conducting material while white areas are used as non conducting materials and are made by depositing thin layer of insulating materials on conducting materials. For measurement of linear displacement linear encoder is used with sector along its length. Linear encoder is shown in Fig. 10.48 *(b)*.

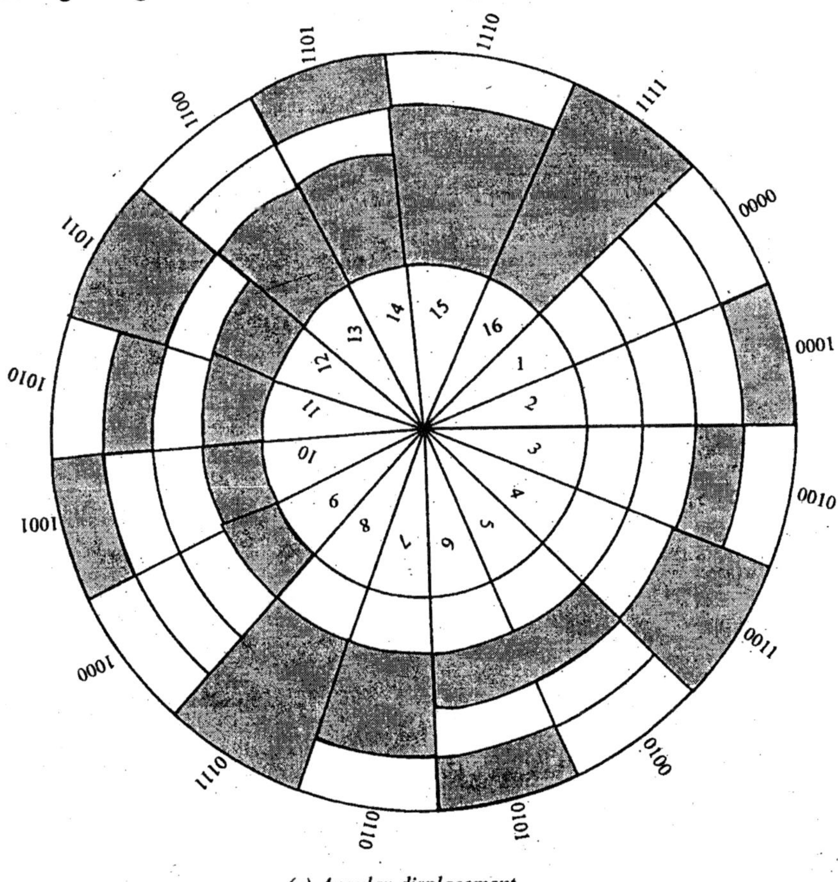

(a) Angular displacement.

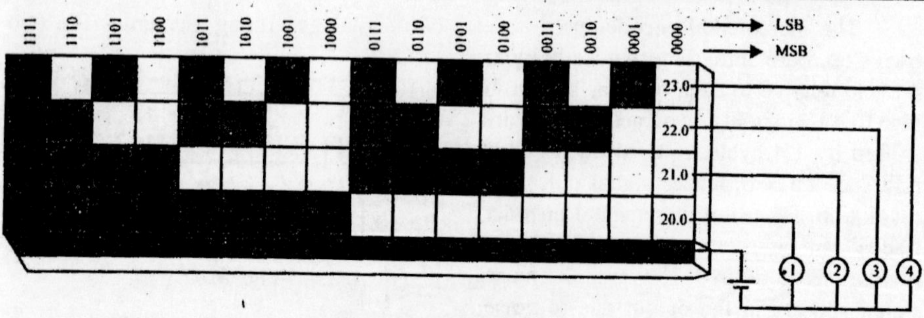

(b) Linear displacement.

Fig. 10.48. Digital displacement transducer.

For measurment of displacement the sectors in the layers move corresponding to the input. For example if the displacement is 6 mm the code will be 0110 and the lamp 2 and 3 will glow. The number will be displayed by the display device.

EXERCISE

1. What is a transducer ? What are its requirements ?
2. Distinguish between active and passive transducers with example.
3. What is an inverse transducer ? Describe an inverse tranducer in detail.
4. Describe working principle of resistive transducer. How can force be measured by a potentiometer ?
5. Describe how torque can be measured by strain gauge.
6. What is a thermistor ? Describe scheme for measurement of temperature by using thermistor.
7. Describe scheme for humidity measurement.
8. Explain how small displacement can be measured by variable seperation capacitive transducer?
9. How can liquid level be measured by using differential capacitive transducer ?
10. By using dielectric gauge explain how thickness of an insulator can be measured ?
11. Explain the construction and working of an LVDT.
12. How can LVDT be used to measure force and displacement.
13. What is temperature range of solid state devices ? Explain its working.
14. Explain the working principle of a thermocouple. Also describe the thermocouple laws.
15. What is piezoelectric effect? How can force be measured by a piezoelectric transducer? Derive expression between force and piezoelectric transducer output ?
16. What is meant by photoelectric device ? Describe photovoltaic cell.
17. Describe r.p.m. measurement by using proximity probes.
18. Describe digital displacement transducer.

NUMERICAL PROBLEMS

1. Length of a potentiometer is 10 cm and it is excited by a battery of 30 volts. If the length of potentiometer between wiper and ground point is 7 cm what is output voltage ?
2. Diameter of a wire wound potentiometer wire is 0.8 mm and the potentiometer is 30 cm long. If resistivity of the wire is 8×10^{-8} ohm-meter compute its resistance. Periphery of a turn = 9 cm.
3. A wheatstone bridge consists of three fixed resistor of 120 Ω each and a strain gauge also

of 120 Ω and its gauge factor is 2.1. If the bridge is excited by 10 volt D.C. source and bridge output is 1.2 volt what is the strain in the gauge ?

4. Resistance of a thermistor is given by $R = ae^{b/T}$, where $a = -0.045$ and $b = 3985$ Kelvin. What is the thermistor resistance at *(a)* 40°C, *(b)* 60°C ?

5. Separation between the plates of a parallel plate capacitor is 2 mm and area of plate is also 2 mm^2. The medium between the plates is air. If plate separation is changed by 0.1 mm what will be the change in the capacitance ?

6. In the above problem a resistance of 20 Ω is connected in series with the capacitor and the network is excited by a source of 5 V dc. What is the change in output voltage ?

7. In problem 5, if an insulator of dielectric constant 16 is introduced between the capacitor plates what will be the change in the capacitance ?

8. An LVDT transducer output is 1.2 mV for a displacement of 2 mm what will be the output if displacement is increased in the same direction by 0.08 mm in opposite direction ?

9. A solid state transistor has emitter current of 5 mA at 40°C. What is emitter base voltage?

10. A thermocouple output is given by $e = a\,\Delta T + b\,\Delta T^2$. If its output at 25°C is 1.2 mV and 2.6 mV at 45°C what is the value of a and b ?

11. A piezoelectric transducer is used for measurement of pressure. If voltage sensitivity is 3×10^{-6} mV/N and pressure applied is 2×10^6 N/m^2 what will be the output of the transducer for thickness of 2.1 mm ?

12. In the above problem if relative permittivity of the crystal material is 25 what is its charge sensitivity ?

13. A photoelectric pickup is used to measure angular speed of a motor. The number of holes in the disc is 28. If the counting frequency is 115 counts/sec, calculate angular speed in r.p.s.

14. The no. of teeth in a rotor is 8 and it rotates under the influence of magnetic field. If the speed of rotor is 700 r.p.m. what is counting frequency ?

OBJECTIVE TYPE QUESTIONS

1. Transducer
   (a) converts electrical energy into mechanical energy
   (b) converts one form of energy into another
   (c) converts any energy into electrical energy
   (d) is a passive device.

2. Active transducer
   (a) needs no power source *(b)* needs power source
   (c) is called self generating *(d)* both *(a)* and *(c)*

3. Example of a passive transducer is
   (a) solar cell *(b)* LVDT
   (c) Thermcouple *(d)* Tachometer

4. Example of an inverse transducer is
   (a) solar cell *(b)* piezoelectric crystal
   (c) LVDT *(d)* potentiometer

5. Resistivie transducer is

(a) an active transducer (b) a passive transducer
(c) used for measurement of light intensity
(d) active as well as passive transducer

6. Unit of strain is
(a) Newton/m^2 (b) unit less
(c) siemens (d) (b) and (c)

7. Typical value of gauge factor for metallic strain gauge is about
(a) 1.0 (b) 0.3
(c) 1.2 (d) 50

8. Value of gauge factor for semiconductor strain gauge is
(a) 100 (b) 140
(c) 130 (d) 148

9. Thermistor has
(a) negative temperature coefficient (b) positive temperature coefficient
(c) zero temperature coefficient (d) no resistance

10. Thermistor is
(a) an active transducer (b) a passive transducer
(c) inverse transducer (d) not a transducer

11. Capacitive transducer
(a) is an active transducer (b) is a passive transducer
(c) is not a transducer (d) is inverse transducer

12. Capacitve transducer is not used for measurement of
(a) light intensity (b) displacement
(c) thickness (d) pressure

13. LVDT has
(a) one secondary (b) two secondaries
(c) sometimes three sceondaries (d) no core

14. The LVDT core
(a) is made of soft magnet (b) is made of hard magnet
(c) is made of iron (d) is made of brass

15. In displacement measurement LVDT
(a) acts as primary sensor (b) acts as secondary transducer
(c) is a passive transducer (d) both (a) and (c)

16. Thermocouple works on
(a) seebeck effect (b) peltier effect
(c) thomson effect (d) none of the above

17. Thermocouple is
(a) an active transducer (b) a passive transducer
(c) inverse transducer (d) is not a transducer

18. Piezoelectric transducer is
(a) an active transducer (b) a passive transducer

(*c*) an inverse transducer (*d*) none of the above

19. Piezoelectric transducer can be used to measure
    (*a*) displacement (*b*) pressure
    (*c*) temperature (*d*) (*a*) and (*b*)

20. Phototube is
    (*a*) an active transducer (*b*) a passive transducer
    (*c*) inverse transducer (*d*) none of the above.

21. Solar cell is
    (*a*) not a transducer (*b*) an active transducer
    (*c*) a passive transducer (*d*) none of the above.

22. In an LVDT
    (*a*) The minimum output is zero (*b*) The minimum output is non-zero
    (*c*) The harmonics in the primary make the output non-zero
    (*d*) (*b*) and (*c*).

23. Reluctance of an inductor
    (*a*) is given by $R = \frac{N^2}{L}$
    (*b*) is more of a magnetic material than that of vacuum
    (*c*) is a vector quantity
    (*d*) both (*a*) and (*b*).

24. Bourdon tube is
    (*a*) a primary sensor (*b*) a transducer
    (*c*) senses the pressure (*d*) (*a*) and (*b*)
    (*e*) (*a*) and (*c*).

ANSWERS

1. (*b*)	2. (*d*)	3. (*b*)	4. (*b*)	5. (*b*)	6. (*d*)
7. (*a*)	8. (*b*)	9. (*a*)	10. (*b*)	11. (*b*)	12. (*a*)
13. (*b*)	14. (*a*)	15. (*d*)	16. (*a*)	17. (*a*)	18. (*c*)
19. (*d*)	20. (*b*)	21. (*b*)	22. (*d*)	23. (*d*)	24. (*e*)

11
Signal Conditioners

11.0. INTRODUCTION

As we have discussed in earlier chapter about the trasnducer as a device which converts one form of energy into another. At this stage by transducer is meant a device which converts any form of energy into electrical form. The output of the transducer is to be modified before it becomes usable and satisfactory to drive the data presentation stage. The intermediate stage between transducer stage and data presentation stage is called signal conditioner. The process carried by signal conditioner is called signal conditioning. The block diagram of signal conditioning units are shown in Fig. 9.2 and 9.3.

Without being familiar with opeartiol amplifier our discission on signal condisioning will be incomplete as operational amplifier is the most important unit of signal conditioner. In this chapter we shall have discussion on operational amplifiers and other elements.

11.1. OPERATIONAL AMPLIFIERS

Operational amplifier abbreviated as OP AMP is such an amplifier which can perform mathematical operation such as additon, subtraction, mulitiplication as well as logical operation such as comparsion, inversion conversion etc. It is maufactured with integrated circuit technology and contains transistors, diodes, resistors and capacitors. The construction of on OP AMP is shown in Fig. 11.1. This mode is called differatial mode. Two signals have been fed one each to the bases of transistors Q_1 and Q_2. The difference of outputs of the transistors

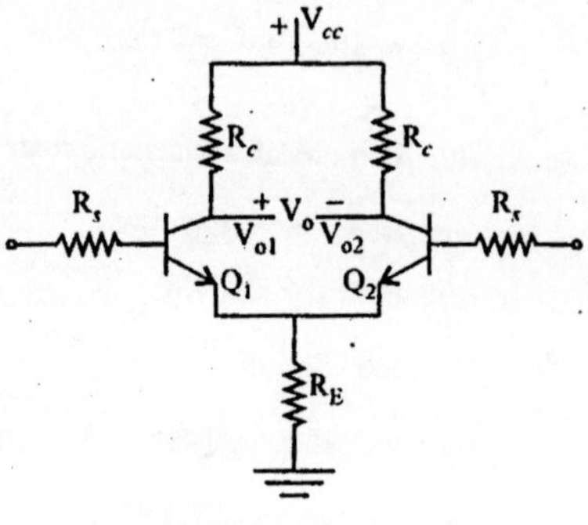

Fig. 11.1. Operational amplifier.

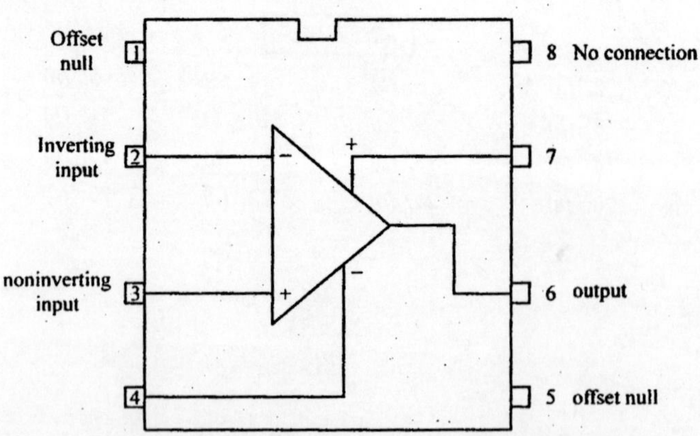

Fig. 11.2. OPAMP IC 741 pin configuration.

appears as the output of the OP AMP. The pin configuration of operational amplifier IC 741 is shown in Fig. 11.2.

11.2. IDEAL OPERATIONAL AMPLIFIER

An ideal operational amplifier has the following properties.

1. It should have infinite input impeduce, *i.e.*, $Z_i = \infty$
2. It should have infinite open loop gain, *i.e.*, $A_{OL} = \infty$
3. Its CMRR should be infinite, *i.e.*, CMRR = ∞
4. Its output impedance should be zero, *i.e.*, $Z_{out} = 0$
5. It should have flat frequency response over wide range of frequency, *i.e.*, its bandwidth = ∞.

However above properties are not satisfied by commercially available operational amplifiers. For practical amplifiers

1. Input impedance is of the order of several megaohms.
2. Open loop gain is 100000.
3. Output impedance is not zero, the typical value being about 100 Ω.
4. Bandwidth is of the order of megahertz.

Operatinal amplifier is used in the following modes.

(a) **Non inverting mode :** The circuit of the OPAMP in this mode is shown in Fig. 11.3. Let the current I_f flow in the direction shown, then we can write

$$I_f = \frac{V_0}{R_2 + R_1} \qquad ...(1)$$

$$V_- = I_f R_1 \qquad ...(2)$$

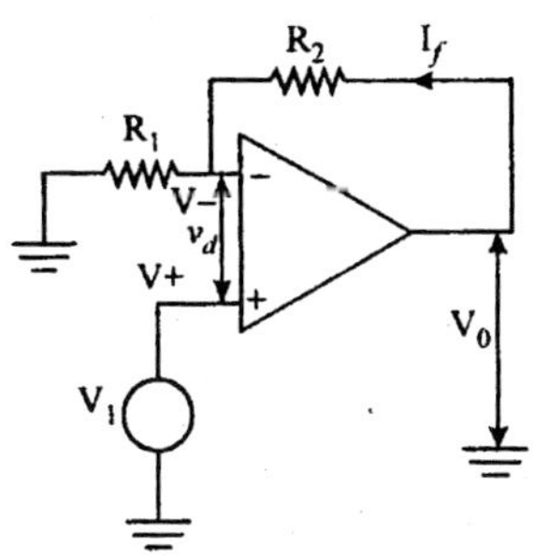

Fig. 11.3. Non inverting amplifier.

Since the op amp is assumed to be ideal, no current flows through the device. If open loop gain is A_{vol}, we can write

$$\begin{aligned} V_o &= A_{vol} \times V_d \\ &= A_{vol}(V_+ - V_-) \\ &= A_{vol}(V_+ - I_f R_1) \end{aligned}$$

$$V_o = A_{vol}\left(V_+ - \frac{V_o}{R_1 + R_2} R_1\right)$$

or
$$\frac{V_o}{A_{vol}} = V_1 - \frac{R_1 V_o}{R_1 + R_2} \qquad ...(3)$$

In eqn. (3) A_{vol} on left hand side is ∞ for ideal op amp rendering L.H.S zero. Hence, we can write

$$0 = V_1 - \frac{R_1}{R_1 + R_s} V_o$$

or
$$\frac{V_o}{V_I} = \frac{R_1 + R_2}{R_1} = \left(1 + \frac{R_2}{R_1}\right)$$

$$\frac{V_o}{V_I} = \left(1 + \frac{R_2}{R_1}\right) \quad ...(4)$$

Eqn. (4) is the relationship between input and output voltage of an amplifier in non-inverting mode.

Example 11.1. Feedback resistance R_2 in Fig. 11.3 is 100 kΩ and output voltage of 3 volts. If R_1 = 1 k what is the value of V_I ?

Solution : By using relation

$$Vo = \left(1 + \frac{R_2}{R_1}\right) V_I \text{, We obtain}$$

$$3 = \left(1 + \frac{100 \times 10^3}{1 \times 10^3}\right) V_I$$

or
$$\frac{3}{101} = V_I \approx 0.029 \text{ Volts.}$$

(b) **Isolation (Voltage follower) mode :** The gain of the amplifier in this made is unity. This is used as buffer stage between high and low impedance circuits. The circuit is shown in Fig. 11.4. The analysis of the circuit is as follows :

Analaysis : Let the current I_f flow in the direction shown. Clearly

$$V_o = V_- = I_f R_1$$

If open loop gain of the device be A_{vol} then

$$V_o = A_{vol} V_d$$
$$= A_{vol} (V_+ - V_-)$$
$$= A_{vol} (V_I - V_o)$$

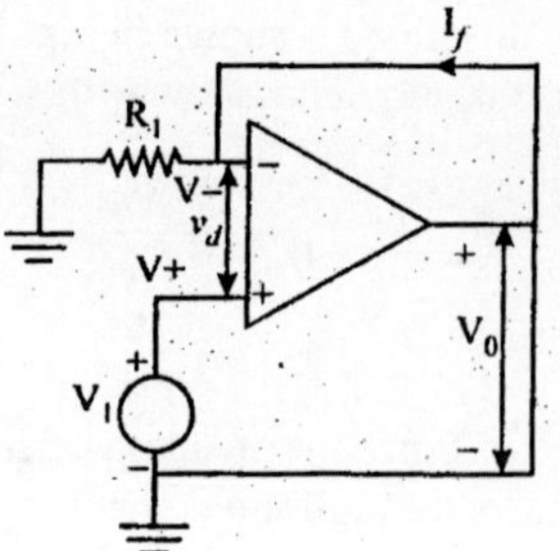

Fig. 11.4. Voltage follower.

or
$$\frac{V_o}{A_{vol}} = V_I - V_o$$

or
$$V_o\left(1 + \frac{A_{vol}}{A_{vol}}\right) = V_I$$

or
$$\frac{V_o}{V_I} = \frac{A_{vol}}{1 + A_{vol}} \approx 1 \quad ...(5)$$

The feedback when output is connected to inverting terminal is positive. If output voltage is fedback to non-inverting terminal the feedback is negative. By analysing equivalent circuit of OPAMP it can be proved that its input impedance is $A_{vol}\ R_i$ where R_i is input impedance without feedback. The input impedance and A_{vol} for a practical amplifier is 1 to 2 MΩ and about 10^5 respectively. As a result the input impedance of voltage follower increases to 10^5 times its nominal value. Similarly the output impedance of voltage follower is $\frac{R_o}{A_{vol}}$ where R_o is the nominal value of output impedance of an amplifier which is about 100 Ω. Hence for voltage follower the output impedance is $\frac{100}{10^5}$ $= 10^{-3}$ Ω. Thus a voltage follower is characterized by high input and low output impedance. For this property it acts as a very good buffer (Isolator).

Example 11.2. Gain of a voltage follower is 0.99999, its output impedance is 1.3×10^{-3} ohms. Calculate output impedance of the amplifier without feedback.

Solution. Gain of voltage follower

$$\frac{A_{vol}}{1+A_{vol}} = 0.99999$$

or $$A_{vol} - 0.99999\ A_{vol} = 0.98$$

or $$0.00001\ A_{vol} = 0.998$$

$$A_{vol} = \frac{0.99999}{0.00001}$$

Impedance of voltage follower $= \frac{R_o}{A_{vol}} = 1.3 \times 10^{-3}$

or $$R_O = A_{vol} \times 1.3 \times 10^{-3}$$
$$= 99999 \times 1.3 \times 10^{-3}$$
$$= 129998.7 \times 10^{-3}$$
$$= 129.99 \text{ ohm}$$

(c) **Inverting mode :** In this mode the output and input signals have 180° phase difference between them. The circuit of an inverting amplifier is shown Fig. 11.5. Non inverting terminal has been tied to ground hence $V_+ = 0$. The output voltage of the amplifier is given by:

$$V_o = A_{vol}\ V_d$$
$$= A_{vol}\ (V_- - V_+)$$
$$= A_{vol}\ V_- \text{ since } V_+ = 0 \qquad ...(6)$$

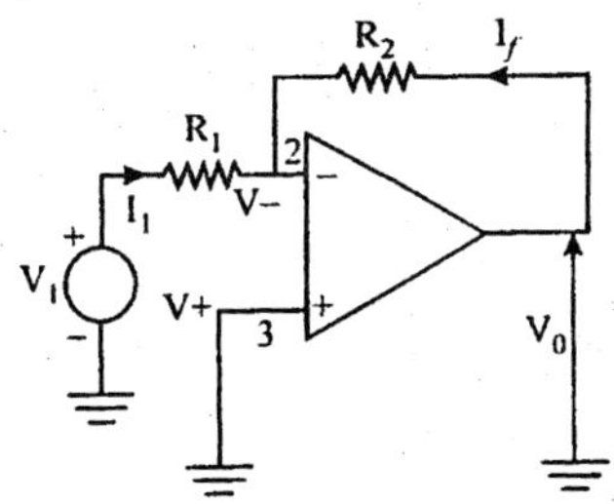

Fig. 11.5. Inverting OPAMP

Applying Kirchoff's current law at node 2, we get

$$I_1 + I_f = 0$$

or $$\frac{V_1 - V_-}{R_1} + \frac{V_o - V_-}{R_2} = 0$$

or $$\frac{V_1}{R_1}+\frac{V_o}{R_2}-\left(\frac{1}{R_1}+\frac{1}{R_1}\right)\frac{V_o}{A_{vol}}=0 \quad \text{from eqn. (6)}$$

Since for ideal amplifier $A_{vol} = \infty$, hence third term in above eqn. will be equal to zero, reducing the eqn. to

$$\frac{V_1}{R_1}+\frac{V_o}{R_2}=0$$

or $$\frac{V_0}{V_1}=-\frac{R_2}{R_1} \quad ...(7)$$

Eqn. (7) is the relationship between input and output of an inverting amplifier. The output impeduce of the circuit = 0

The input impedance of the circuit is defined by the ratio

$$Z_1=\frac{V_1}{I_1} \quad ...(8)$$

but $$I_1 = \frac{V_1 - V_d}{R_1}$$

for one ideal amplifier V_d=0 hence

$$I_1 = \frac{V_1}{R_1}$$

or $$R_1 = \frac{V_1}{I_1} \quad ...(9)$$

From eqn. (8) and (9), we get

$$Z_1 = R_1$$

The input impedance in inverting mode is less than that in non inverting mode of operation.

Example 11.3. An inverting amplifier shown in Fig. 11.4 is an ideal one. The values of resistors R_1 and R_2 respectively are 5 KΩ and 1 MΩ. If input voltage is 3 mV, calculate the value of feedback current and output voltage.

Solution. Current $$I_1=\frac{V_1}{R_1}=\frac{3\times10^{-3}}{5\times10^{3}}=0.6\times10^{-6} \text{ Amp.}$$

Feedback current $I_f = -I_1 = -0.6 \times 10^{-6}$ Amp.

Output voltage $$V_o = -\frac{R_2}{R_1}\times V_1$$

$$= -\frac{9\times10^{6}}{5\times10^{3}}\times3\times10^{-3} \text{ Volt}$$

$$= -0.2 \times 3$$

$$= -0.6 \text{ volts}$$

(–) sign indicates phase reversal between input and output

Example 11.4. In the circuit shown in Fig. 11.3 the OPAMP shown is to have open loop gain 10^5 and resistors have values $R_1 = 1\ k\Omega$ and $R_2 = 50\ k\Omega$.

(a) Find actual value of Voltage gain

(b) Find ideal value of gain.

(c) Calculate percentage error between ideal and actual voltage gain.

Solution : *(a)* From the Fig potential of –ve terminal is given by:

$$V_- = \frac{R_1}{R_1 + R_2} V_o \quad \text{...(1)}$$

$$V_d = V_+ - V_- = V_1 - \frac{R_1}{R_1 + R_2} V_o$$

$$\frac{V_o}{A_{vol}} = V_1 - \frac{R_1}{R_1 + R_2} V_o$$

or
$$V_o \left[\frac{1}{A_{vol}} + \frac{R_1}{R_1 + R_2} \right] = V_1$$

or
$$\frac{V_o}{V_1} = \frac{(R_2 + R_1) A_{vol}}{A_{vol} R_1 + R_1 + R_2}$$

$$= \frac{(50+1) \times 10^3 \times 10^5}{(50+1) \times 10^3 + 10^3 \times 10^5} = 50.97$$

(b) Ideal value of voltage gain $= 1 + \frac{R_2}{R_1}$

$$1 + \frac{50}{1} = 51$$

11.3. OPERATIONAL AMPLIFIER CHARACTERISTICS

There are certain characteristics of operational amplifier which must be taken into account for design and applications.

When an OPAMP is used as d.c amplifer its d.c output voltage should be zero for zero dc input voltage.However it has non zero o/p for zero input which is called dc offset voltage.The offset values should be reduced to zero. Various parameters of OPAMP are described below :

1. Input offset current : We refer to Fig. 11.6 The two transistors forming OPAMP may not be perfectly matched. Hence input bias current for both the transistors is different and the difference of these currents is called input offset current, *i.e.*,

$$I_{os} = I_{B1} - I_{B2}$$

2. Input bias current : For an operational amplifier to work properly it is necessary to supply some dc bias current to the transistors. The average of these two currents shown in Fig. 11.6 is called input bias current, *i.e.*,

$$\text{Input bias current } I_B = \frac{I_{B1}+I_{B2}}{2}$$

This current is of the order of pA to nA.

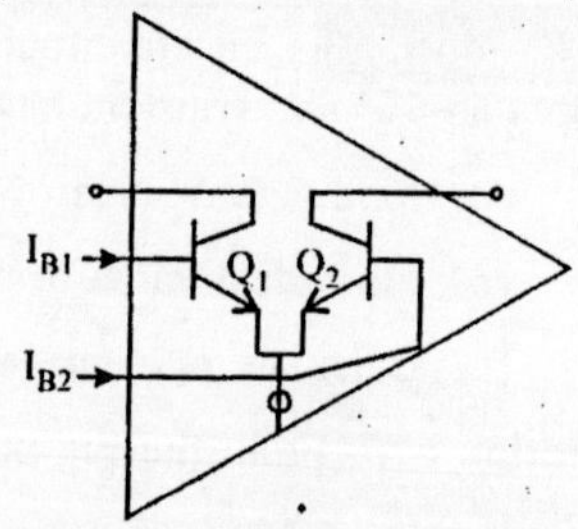

Fig. 11.6. Input bias currents.

3. Input offset voltage : Due to mismatch in the base to emitter voltage of transistors forming OPAMP, the output voltage is non zero even when input voltage is zero. Input offest voltage is defined as input voltage required to OPAMP to redcue output offset voltage to zero.

4. Output dc offset voltage : It is defined as the voltage across output in the absence of any input signals to the amplifier. This occurs due to mismatch in the characteristics of the transistors forming the OPAMP. It is illustrated in Fig. 11.7.

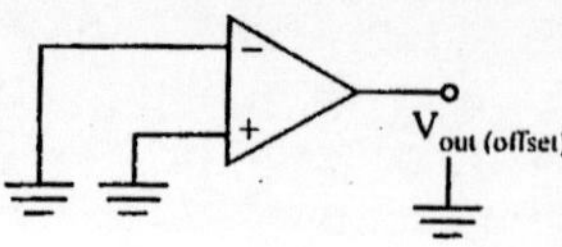

Fig. 11.7. Illustration of output offset voltage.

5. Input impedance : The ratio of input voltage to input current through the OPAMP is called input impedance of the amplifier, *i.e.*,

$$Z_i = \frac{V_1}{I_1}$$

where Z_i = Input impedance

V_1 = Input voltage

I_1 = Input current

6. Output Impedance : It is defined as the ratio of output voltage to output current shown in Fig. 11.8.

$$Z_{out} = R_{out} = \frac{V_0}{I_0}$$

Ideally the output impedance of OPAMP is zero. For a practical amplifier its value lies between 75 and 100 ohms.

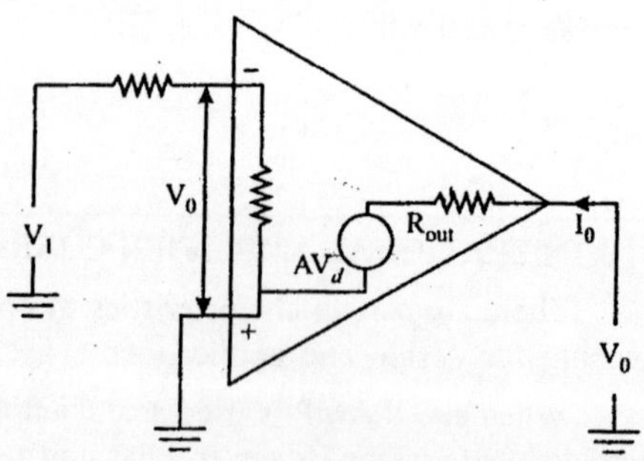

Fig. 11.8. Output impedance.

7. Drift : Rate of change of output of amplifier with temperature is called its drift expressed in mV/°C. If for a change ΔT in temperature the change in output voltage is ΔV_0, the drift is given by

$$\frac{\Delta V_0}{\Delta T} \text{ mV/°C}$$

8. Slew rate : It is defined as rate of change of output voltage. This is the most important parameter of OPAMP. It places a severe limit on the large signal operation of the amplifier. Typical value of slew rate given by $\frac{\Delta V_o}{\Delta t}$ lies between 0.04 V/μs to 70 V/μs.

9. Common mode rejection ratio (CMRR) : It defines the ability of the amplifiers to reject certain types of undesired voltage signals. These undesired voltage signals are called noise which is to be rejected by the amplifier.

Noise in this case are not signals which are to be amplified. These are the signals which appear equally at both inputs. The ability of OPAMP to reject these signals at both inputs called common mode signal is the measure of its CMRR. Common mode rejection ratio is numerical value of the ability of amplifer to reject common mode signal. We can thus write

$$\text{CMRR} = \frac{A_d}{A_c} \quad \text{...(10)}$$

where A_d in the gain of differential amplifier and A_c is the common mode gain. For a perfect differential amplifier A_c is zero hence CMRR is infinite. The output of differential amplifier is

$$V_{out} = A_d(V_1 - V_2) \quad \text{...(11)}$$

The differential signal is given by

$$V_d = V_1 - V_2 \quad \text{...(12)}$$

and common mode signal as

$$V_c = \frac{1}{2}(V_1 + V_2) \quad \text{...(13)}$$

From Fig. 11.9, let the output V_{out} be expressed as linear combination of two input voltages as

$$V_{out} = A_1V_1 + A_2V_2 \quad \text{...(14)}$$

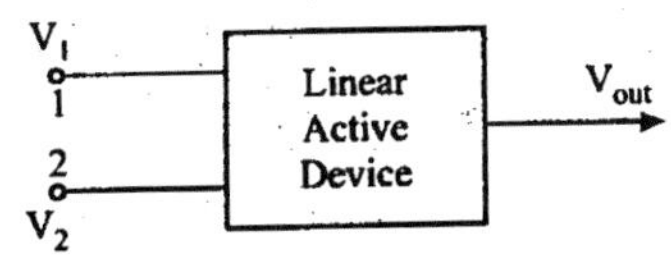

Fig. 11.9. Two port network.

But from eqn. (12) and (13), we obtain

$$V_1 = V_c + \frac{1}{2}V_d$$

and

$$V_2 = V_c - \frac{1}{2}V_d$$

Putting the values of V_1 and V_2 in eqn. (14), we get

$$V_{out} = A_1\left(V_c + \frac{1}{2}V_d\right) + A_2\left(V_c - \frac{1}{2}V_d\right)$$

$$= (A_1 + A_2)V_c + \frac{1}{2}V_d(A_1 - A_2)$$

$$V_{out} = A_cV_c + A_dV_d \quad \text{...(15)}$$

where $A_c = A_1 + A_2$ and $A_d = \frac{1}{2}(A_1 - A_2)$

For measurement of A_d, we select $V_1 = -V_2 = 0.5$ volt so that $V_d = 1$ volt and $V_c = 0$. Thus from eqn. (15)

$$V_{out} = 0 + A_d \times 1 = A_d$$

i.e., output voltage is equal to A_d

Similarly for measuring A_c, put $V_1 = V_2 = 1$ volt. Then we get

$$V_c = 1 \text{ volt and } V_d = 0 \text{ and}$$

$$V_{out} = A_c \times 1 + 0$$

In this way A_c can be measured. From above two values of A_d and A_c we can calculte CMRR by using relation.

$$\text{CMRR} = \frac{A_d}{A_c}$$

Example 11.5. Input voltage $V_1 = 1$ Volt and $V_2 = 200$ mV are applied to a differential amplifier. The output of the amplifier is 4.2 volts. Calculate CMRR.

Solution. We use relations

$$V_{out} = A_d (V_1 - V_2) \qquad ...(i)$$

and

$$V_{out} = A_c V_c + A_d V_d \qquad ...(ii)$$

Hence

$$V_d = 1 \text{ Volt} - 200 \text{ mV}$$

$$= 800 \text{ mV}$$

$$4.2 = 800 \times 10^{-3} (A_d) \text{ from eqn. } (i)$$

or

$$A_d = \frac{4.2 \times 10^3}{0.800} = 5.25$$

$$V_c = \frac{V_1 + V_2}{2} = \frac{1200}{2} \text{ mv} = 600 \text{ mv}$$

From eqn. *(ii)*

$$4.2 = A_c \times 0.600 + 5.25 \times 0.800$$

$$= 4200$$

or

$$A_c = 0$$

$$\text{CMRR} = \frac{A_d}{A_c} = \frac{5.25}{0} = \infty \text{ Ans.}$$

10. Gain of OPAMP : Gain of OPAMP in inverting and non inverting mode have been discussed earlier. The gain of differential amplifier is given by expression.

$$V_{out} = A_{vol} V_d$$

$$A_{vol} = \frac{V_{out}}{V_d}$$

A_{vol} is the open loop gain of the amplifier which is about 200000 and maximum output voltages is equal to supply voltage, *i.e.,* 10 V. Hence the differentical input will be

$$V_d = \frac{10}{200000} = 5 \times 10^{-5} = 50\,\mu V$$

which is very small and can be assumed zero for all practical purposes. The gain of a buffer amplifier is unity which is obtained when total output voltage is fedback to the amplifier. By intorducing a variable resistance between output and input terminal the gain of the amplifier can be varied to the desired value.

11. Noise in OPAMP : Noise is unwanted signal that is not related to input. When an input is fed to the device, the noise in the device is superimposed on the signal. In case of amplifiers the noise signal gets amplified with input and appear with output which is undesirable. Various phenomena of noise which occur in amplifier are as follows:

(a) **Thermal or Johnson Noise :** The motion of free electrons drifting around in the material constitute tiny current which give rise to voltage. The amplitude of the voltage rises with increase in temperature. For this reason it is called thermal noise.

The thermal noise is generated in resistors. When the resistors are at input stage of amplifier, the noise will appear amplified at the output. The value of thermal resistance noise voltage V_n in frequency range $f_2 - f_1$ is given by :

$$V_n = 2\sqrt{KTBR}$$

where K = Boltzmann's constant.

T = Absolute temperature

B = Circuit bandwidth

R = Input resistance.

The noise present in output of any device is expressed quantitatively as noise, figure which is defined as

$$NF = 10 \log \frac{\text{Total noise Power Output}}{\text{Noise Power output due to source resistance}}$$

$$= 10 \log \frac{N_{po}}{\dfrac{S_{po}}{S_{pi}} \times N_{pi}}$$

$$= 10 \log \frac{N_{po}/N_{pi}}{S_{po}/S_{pi}}$$

Here S_{po} = Signal power output

S_{pi} = Signal power input

N_{pi} = Noise power input

N_{po} = Noise power output

manipulating, we get $$NF = 10 \log \frac{S_{pi}/N_{pi}}{S_{po}/N_{po}} \quad ...(16)$$

12. Frequency response of operational amplifier : Frequency response is defined as the curve drawn between voltage gain and frequency of signal applied to amplifier. The gain of the amplifier varies with fequency due to capacitive effects. The gain of the amplifier remains constant for large range of frequency. This is called mid frequency gain. At lower and higher frequencies than the mid frequency the gain falls below mid frequency gains. The frequencdis f_1 and f_2 in Fig. 11.10 at which the gain falls by 3 db (0.707 of mid frequency gain) are called respectively, as lower and upper cut off frequencies. The frequency range $f_2 - f_1$ is called bandwidth of the amplifier. The gain of the amplifier falls at low frequency due to coupling and bypass capacitances. At high frequencies the capacitive reactance given by

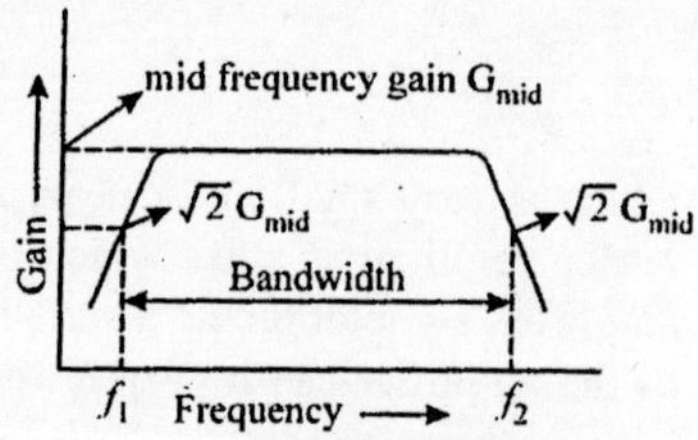

Fig. 11.10. Frequency response of an OPAMP

$X_c = \dfrac{1}{2\pi f_c}$ is very small and therefore all coupling and bypass capacitors have short circuits. At low frequencies capacitive reactance has large value and, therefore, some of the signal voltage is lost across the capacitors. Consequently the output voltage and hence the gain falls.

11.4. INSTRUMENTATION AMPLIFIER

It is a very high adjustable gain amplifier. The small signals cannot be amplified to measurable level by operational amplifier. This drawback is overcome by instrumentation amplifiers which are available in the maket in the form of ICs. Simply by varying a resistance its gain can be adjusted. Instrumentation amplifier possesses the following characteristics.

1. High adjustable gain
2. High (CMRR) = ∞
3. Low output impedance = 0
4. High input impedance = ∞
5. High gain bandwidth product.

Construction of an intrumentation amplifier is shown in Fig. 11.11. Three amplifiers are configured to form the instrumentation amplifier. Resistance R_g is variable which can be adjusted to obtain the desired output.

$R_3 = R_4 = R_5 = R_6 = R$ (say)

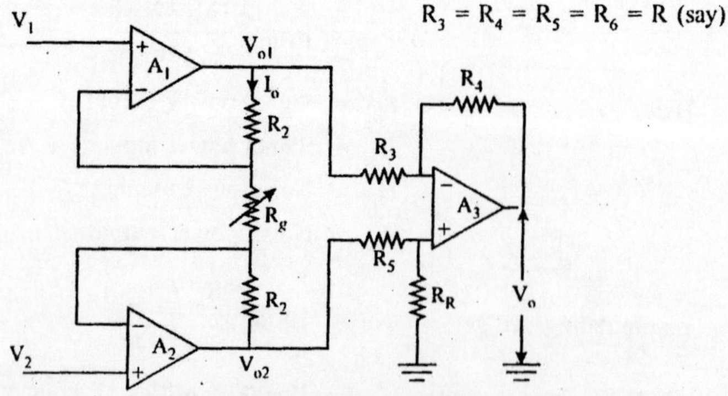

Fig. 11.11. An instrumentation amplifier.

From the circuit, we get

$$\frac{V_{o1} - V_1}{R_e} = \frac{V_1 - V_2}{R_g} = \frac{V_2 - V_{o2}}{R_2} = I_o \qquad \text{...(17)}$$

Solving eqn. (17) we obtain

$$V_{o1} - V_{o2} = -\left(1 + \frac{2R_2}{R_g}\right)(V_2 - V_1)$$

and

$$V_o = -(V_{o1} - V_{o2})$$

$$V_o = \left(1 + \frac{2R_2}{R_g}\right)(V_2 - V_1) \qquad \text{...(18)}$$

In eqn. (18) only resistance R_g can be varied to adjust the output.

The characteristics of operational or instrumentation amplifier discussed above are very important parameters and need to be taken into consideration while using the amplifier for amplifying the signal generated by any transducer. For example capacitive, inductive, resistive and all active transducer generate output at different frequencies. The frequency response of the amplifier should match the frequency of the signal generated by a particular transducer. In another example if we consider the impedance of the source and the load, loading effect should be taken into account. To avoid loading effect buffer amplifier is used. Thus while using an amplifer for a particular transducer above said parameters should be considered.

Example 11.6. Output of a displacement transducer is 6 mV. It is to be amplified to 3.5 volt by an instrumentation amplfier to measure displacement accurately. If resistance R_2 =10 kΩ calculate value of R_g.

Solution : Grounding one of the input terminals, we get $V_1 = 0$. Using expression

$$V_o = \left(1 + \frac{2R_2}{R_g}\right)(V_2 - V_1), \text{ we get}$$

$$3.5 = \left(1 + \frac{2 \times 10 \times 10^3}{R_g}\right)(6 \times 10^{-3})$$

$$(3500 - 6) \times 10^{-3} = 6 \times 10^{-3} \frac{\times 2 \times 10^4}{R_g}$$

or

$$R_g = \frac{12 \times 10^4}{0.3494} \Omega$$

$$R_g = 34\ \Omega \textbf{ Ans.}$$

11.5. MODULATION

Modulation is a term which means "to modify". The need of modulation of signal arises for various advantages of modulated signals. Modulation consists basically of two parts.

(1)**Carrier.** This part carries the information.

(2)**Modulating signal.** It represents the information to be carried out.

The modulation is of various kinds. Some of the kinds are described as under.

Amplitude modulation. In amplitude modulation the amplitude of the carrier wave changes in proportion to the amplitude of the modulating (information) signal. The modulating signal may be of any form like sinusoidal wave, square wave, any periodic function or transient phenomenon. The carrier wave also may be sinusoidal, square wave etc. How the amplitude modulation takes place can be explained by using Fig. 11.12. In the figure the two are taken of sinusoidal nature. It can be seen that the output of the process has amplitude different from the amplitude of the inputs.

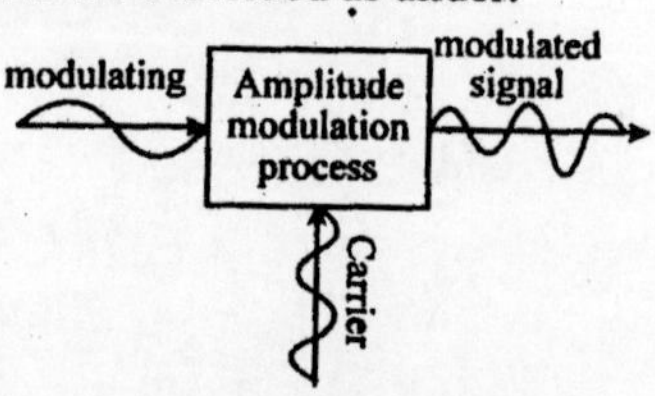

Fig. 11.12. Schematic of amplitude modulation

Analysis. Let the information (modulating signal) and carrier waves be represented as:

$$v_m = V_m \sin \omega_m t \quad \text{...(19), modulating signal}$$

$$v_c = V_c \sin \omega_c t \quad \text{...(20), carrier signal}$$

Then the peak amplitude of the modulated signal can be written as

$$v_{cm} = V_c + V_m \sin \omega_m t \quad \text{...(21)}$$

The instantaneous value of modulated signal will then be

$$\begin{aligned} v_{cm} &= V_{cm} \sin \omega_c t \\ &= (V_c + V_m \sin \omega_m t) \sin \omega_c t \\ &= V_c \sin \omega_c t + V_m \sin \omega_m t \sin \omega_c t \\ &= V_c \sin \omega_c t + \frac{V_m}{2} \times 2 \sin \omega_m t \sin \omega_c t \end{aligned} \quad \text{...(22)}$$

From eqn. (21)

$$\begin{aligned} v_{cm} &= V_c\left(1 + \frac{V_m}{V_c} \sin \omega_m t\right) \\ &= V_c (1 + m \sin \omega_m t) \end{aligned} \quad \text{...(23)}$$

where $m = \frac{V_m}{V_c}$ is called modulation index, its value lying between 0 and 1. The maximum value of modulation index is 1. From the definition of m we can put in eqn. (22).

$$V_m = mV_c, \text{giving}$$

$$V_{cm} = V_c \sin \omega_c t + \frac{mV_c}{2} \cdot 2 \sin \omega_c t \sin \omega_m t$$

or. $v_{cm} = V_c \sin \omega_c t + \frac{mV_c}{2}[\cos(\omega_c - \omega_m)t - \cos(\omega_c + \omega_m)t]$

$$= V_c \sin \omega_c t + \frac{mV_c}{2} \cos(\omega_c - \omega_n)t - \frac{mV_c}{2} \cos(\omega_c + \omega_m)t \quad \text{...(24)}$$

From above eqn. it is clear that the modulated signal has, in addition to carrier component, two components having frequency $\omega_c + \omega_m$ and $\omega_c - \omega_m$. The frequency $\frac{\omega_c - \omega_m}{2\pi}$ is less than $\frac{\omega_c}{2\pi}$ and $\frac{\omega_c + \omega_m}{2\pi}$ has value higher than $\frac{\omega_c}{2\pi}$ has. These are called, respectively, as lower side and upper side frequencies. The modulated signal and its frequency spectrum are shown, respectively, in Fig. 11.13 *(a)* and *(b)*.

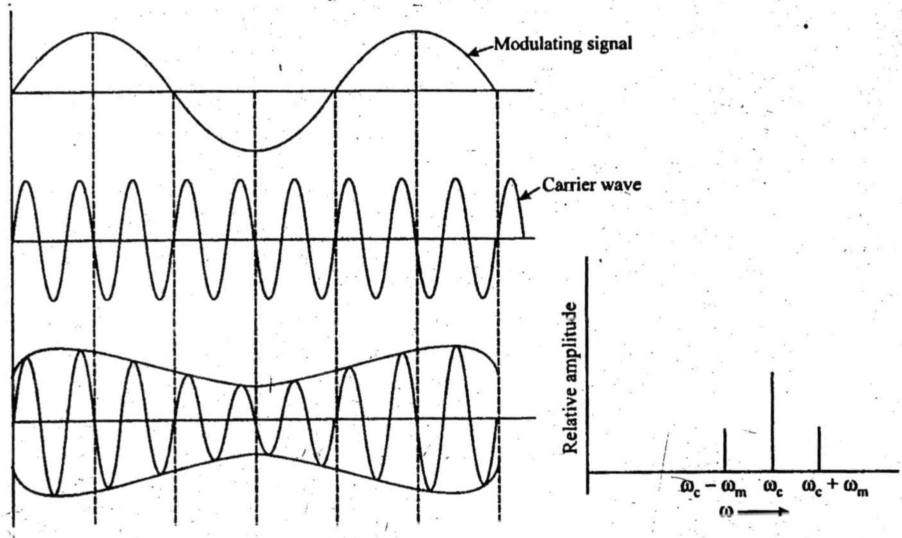

(a) Modulated wave *(b) Frequency spectrum.*

Fig. 11.13.

As an example we consider measurement of temperature by use of thermistor, as shown in the circuit shown in Fig. 11.14. The temperature measured may be static or dynamic. The thermistor acts as one arm of the wheatstone bridge which is excited by a source of carrier waves.

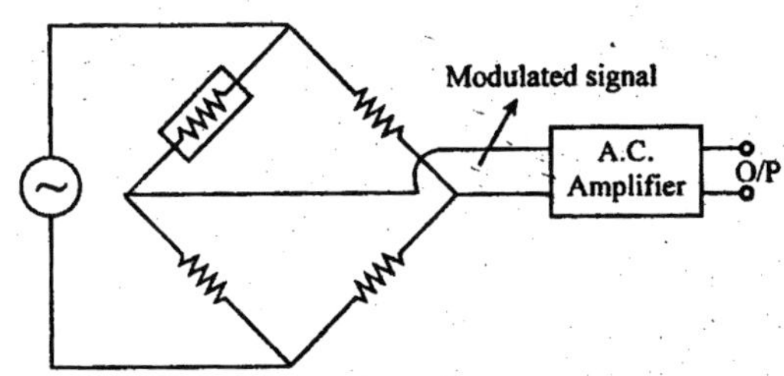

Fig. 11.14. An example of amplitude modulation.

As the temperature changes the bridge output changes which is a.c. having the frequency of carrier (excitation) signal for static temperature and modulated one for dynamic temperature. For instance let the frequency of excitation signal be 1000 Hz. The output will have the frequency of 1 KHz for static temperature. However for dynamic temperature the output will be amplitude modulated having the frequency between $\omega_c - \omega_m$ and $\omega_c + \omega_m$. For ω_m = 100Hz the frequency of modulated output will lie in the range of 900 Hz to 1100 Hz.

The output of the wheatstone bridge is amplified by ac amplifier. By using ac amplifier the output shown are more reliable and accurate as these do not suffer from the problem of drift which is the limitation of dc amplifier.

From above example it is clear that in some cases the modulated signals are essential in signal conditioning.

Amplitude Demodulation. The modulated signal can not be applied to the read out system directly. For this purpose the transducer output mixed with carrier signal needs to be separated before it is applied to the read out device. The process of recovery of the transducer output in its original form is called demodulation or detection. It is shown in Fig. 11.15.

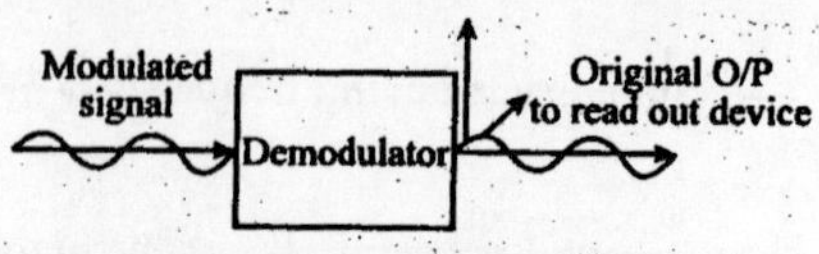

Fig. 11.15. (a) Demodulation.

In addition to amplitude modulation other techniques such as frequency modulation, Pulse code modulation, pulse width modulation, pulse amplitude modulation etc. are used.

Frequency modulation : In frequency modulation the frequency of the carrier wave varies in accordance with the amplitude of modulating (intelligence) signal. Thus generated modulated signal has unaltered amplitude. The change in the modulated frequency over the carrier frequency is called the frequency deviation. The process of frequency modulation is shown in Fig. 11.15 *(b)*.

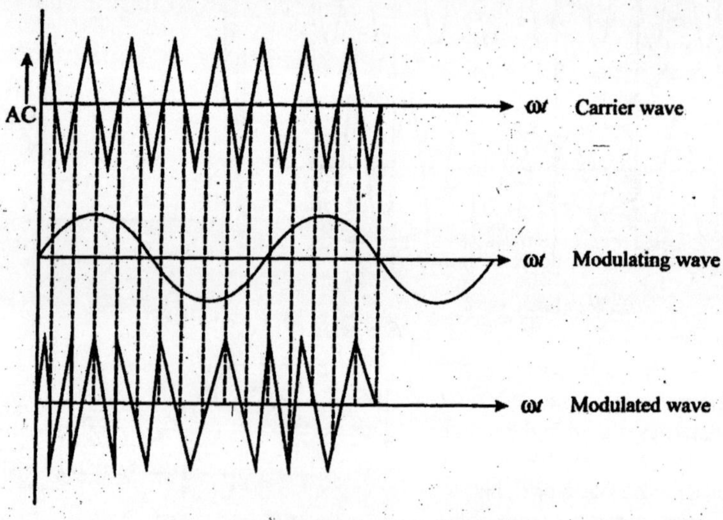

Fig. 11.15. (b) Frequency modulation.

The transimission of signal using frequency modulation (FM) has some advantages over the amplitude modulated (AM) signal. As the amplitude of the FM does not change, the noise signals available in the media of propagation of the wave do not mix with it which happens in AM. The distance travelled by FM is larger than AM does. The modulation index in FM may take any value where it has maximum value of unity in AM. Small power is required to produce modulation.

However, the major drawback of FM is that it is limited to small area of reception. For this reason a wider channel is required to transmit FM.

Analysis : Let the carrier and modulating waves be represented by

$$v_c = V_c \sin (\omega_c t + \phi) \quad ...(25)$$

$$v_m = V_m \sin \omega_m t \quad ...(26)$$

Let us write

$$\omega_c t + \phi = \phi_c$$

or
$$\omega_c = \frac{d\phi_c}{dt} \qquad \text{...(27)}$$

The frequency of the carrier wave changes with the instantaneous amplitude of the modulating wave. At any instant the increasing frequency of the modulated signal can be written as

$$\omega = \omega_c + K\, v_m$$

where K is proportionality constant.

$$\omega = \omega_c \pm KV_m \sin \omega_m t \qquad \text{...(28)}$$

The phase of modulated signal is then

$$\phi = \int \omega dt$$

$$= \int (\omega_c \pm KV_m \sin \omega_m t) dt$$

$$= \frac{1}{\omega} \int (\omega_c + KV_m \sin \omega_m t) d(\omega_m t)$$

$$= \frac{\omega_c \omega_m t + KV_m (\mp \cos \omega_m t)}{\omega_m}$$

$$= \omega_c t \mp \frac{KV_m}{\omega_m} \cos \omega_m t$$

The equation of FM wave is then

$$V_{FM} = V_c \sin \phi(t)$$

$$= V_c \sin\left[\omega_c t \mp \frac{KV_m}{\omega_m} \cos \omega_m t\right] \qquad \text{...(29)}$$

The frequency of the modulated signal can be obtained from eqn. (28). From this equation

$$f = \frac{\omega}{2\pi} = \frac{\omega_c}{2\pi} = \frac{\omega_c}{2\pi} \pm \frac{KV_m}{2\pi} \sin \omega_m t$$

Maximum frequency $f_{max} = \dfrac{\omega_c}{2\pi} + \dfrac{KV_m}{2\pi}$

$$= f_c + \frac{KV_m}{2\pi}$$

Minimum frequency $f_{min} = f_c - \dfrac{KV_m}{2\pi}$

Hence frequency deviation

$$f_d = f_{max} - f_c = \frac{KV_m}{2\pi} = f_c - f_{min} \qquad \text{...(30)}$$

and the carrier swing $= f_{max} - f_{min}$

$$= \frac{KV_m}{\pi} \qquad ...(31)$$

The frequency modulation index is given by

$$m_f = \frac{f_d}{f_m} = \frac{KV_m/2\pi}{\omega_m/2\pi}$$

$$m_f = \frac{KV_m}{\omega_m} \qquad ...(32)$$

FM finds wide applications for signal conditioning in an instrumentation system. The data to be processed is frequency modulated before it is transmitted. At the display or receiving end it is reproduced through frequency demodulation. The block diagram representing FM is shown in Fig. 11.15 *(c)*.

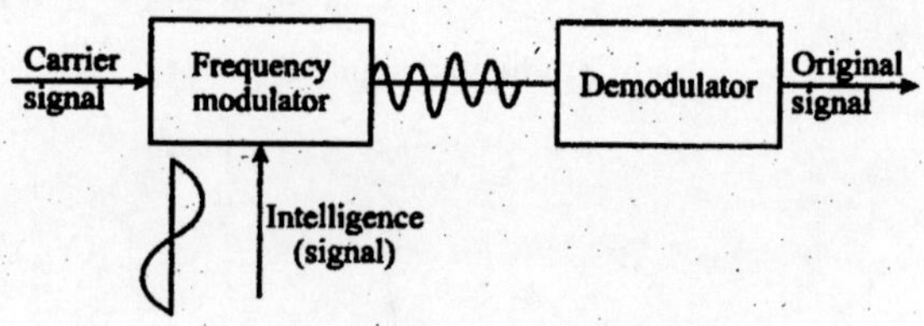

Fig. 11.15. (c) Illustration of frequency modulation.

11.6. FILTERS

The transducer output is required in original form for obtaining the desired output. The variation in its output due to the presence of spurious signals causes difficulty in this process. Such spurious or unwanted signals are removed by use of what is called filter. The filters are categorized as follows:

1. Passive Filter. The passive filters use only passive elements like resistance, capacitor and inductors.

2. Active Filters. Active filters use active elements like operational amplifiers in addition to the passive elements described above.

Classification of Filters. The passive and active filters are classified as:

(a) Low pass filter

(b) High pass filter

(c) Band pass filter

(d) Band elimination filter.

The description of the filters is given below:

***(a)* Low Pass Filters.** The low pass filter is shown in Fig. 11.16 *(a)* with its characteristics shown in Fig. 11.16 *(b)*.

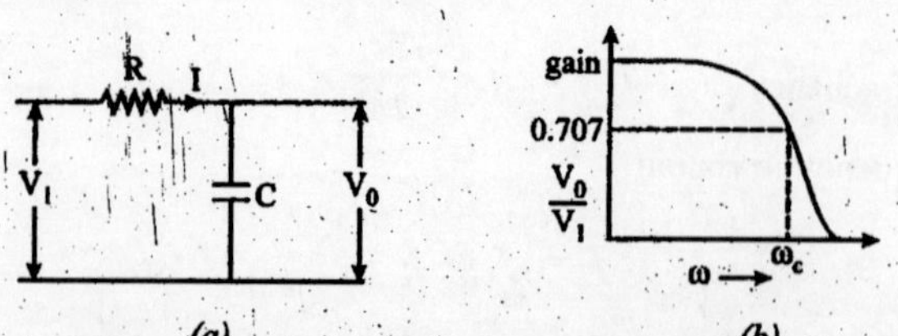

(a) *(b)*

Fig. 11.16. (a) Lowpass filter, (b) characteristics.

Let the current I flows through the network. Then

$$I = \frac{V_i}{R + \frac{1}{SC}} \quad ...(33)$$

$$V_o = I \times \frac{1}{SC}$$

$$= \frac{V_i}{R + \frac{1}{SC}} \times \frac{1}{SC}$$

$$= \frac{V_i \times SC}{(1 + RSC)\, SC}$$

or

$$\frac{V_o}{V_i} = \frac{1}{1 + RSC} \quad ...(34)$$

At low frequency the denominator in eqn. (26) has lower value which increases with frequency. As a result, the gain of the filter falls as the frequency increases. It is clear from Fig. 11.16 *(b)*.

***(b)* High Pass Filter.** A high pass filter allows high frequency signal to pass through it. The circuit is given in Fig. 11.17 *(a)*. From the figure, we can write.

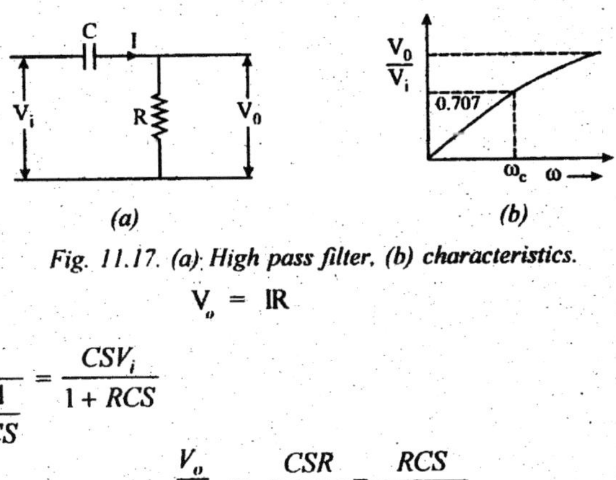

Fig. 11.17. (a) High pass filter, (b) characteristics.

$$V_o = IR$$

where $I = \frac{V_i}{R + \frac{1}{CS}} = \frac{CSV_i}{1 + RCS}$

or

$$\frac{V_o}{V_i} = \frac{CSR}{1 + RS} = \frac{RCS}{1 + RCS}$$

$$= \frac{1}{1 + \frac{1}{RCS}} \quad ...(35)$$

From above eqn. it is clear that with an increase in frequency the gain of the network increases as shown by the characteristics shown in Fig. 11.17 *(b)*.

Band Pass Filters. A band pass filter is the combination of a low pass and a high pass filter as shown in Fig. 11.18 *(a)*. The transfer function of band pass filter is given by

$$\frac{V_0}{V_c} = \frac{1}{1 + R_1C_1S} \times \frac{R_2C_2S}{(1 + R_2C_2S)} \quad ...(36)$$

Using above eqn. the characteristics of bandpass filter is shown in Fig. 11.18 *(b)*.

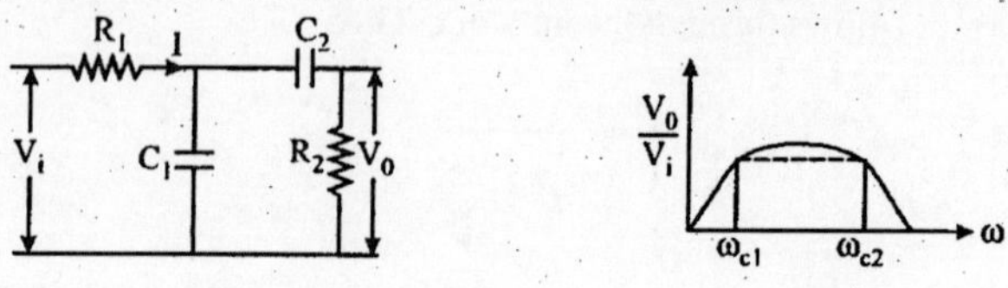

(a) (b)

Fig. 11.18. Bandpass filter, (a) circuit, (b) characteristics.

(d) **Band Reject Filter.** Band reject filter stops the signal having frequency in a band $\omega_2 - \omega_1$ where ω_2 and ω_1 are upper and lower frequencies of the band. Example of such a filter are Wein bridge and twin T network. Its output in the stop band is zero at a particular frequency. This frequency is called notch frequency and the filter is called a notch filter. The circuit of a band reject filter is shown in Fig. 11.19 *(a)*. The characteristics shown in Fig. 11.19 *(b)* shows the notch frequency which is given as

$$f_0 = \frac{1}{2\pi RC} \quad ...(37)$$

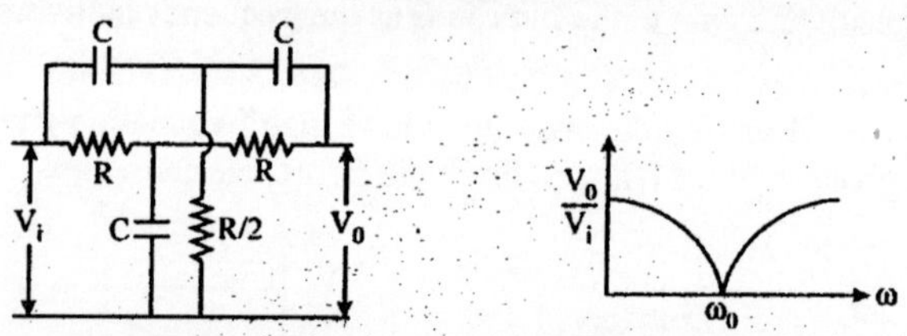

(a) (b)

Fig. 11.19. Band reject (a) Circuit, (b) Characteristics.

LC Filters. The combination of inductors and capacitors also can be used to remove ripples in the signal. Some of the configurations of LC filters are:

(a) L-Section, *(b)* π-section and *(c)* T-section

***(a)* L-section.** An L-section is shown in Fig. 11.20 *(a)*. It is a low pass filter. The output is given as

$$v_0 = I \times \frac{1}{SC}$$

where

$$I = \frac{V_i}{SL + \frac{1}{SC}}, \text{ giving}$$

$$\frac{v_0}{v_1} = \frac{1}{SC} \times \frac{1}{\left(SL + \frac{1}{SC}\right)}$$

$$= \frac{1}{1 + S^2 LC} \quad ...(38)$$

Above eqn. shows that with an increase in frequency the gain of the network decreases. At lower frequencies the gain has higher value.

***(b)* π-section.** π-section is shown in Fig. 11.20 *(b)*. At high frequencies the capacitor is short circuited while inductor offers high resistance. The characteristic impedance of the circuit is given by $Z_0 = \sqrt{L/C}$. The cut off frequency of the circuit is $f_0 = \frac{1}{2\pi\sqrt{LC}}$

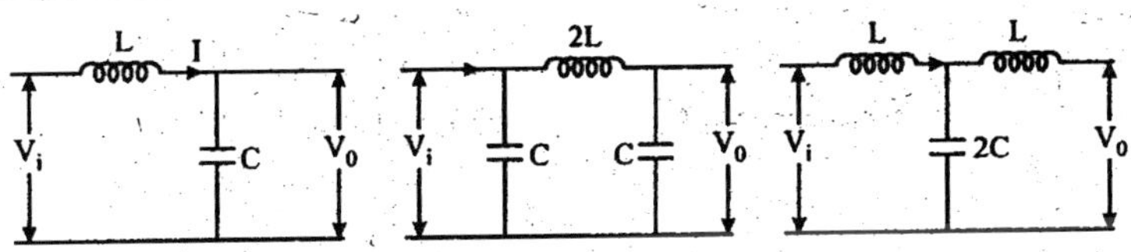

(a) (b) (c)

Fig. 11.20. Low pass LC filters, (a) L-section, (b) π-section, (c) T-section.

***(c)* T-section.** The circuit is shown in Fig. 11.20 *(c)*. The capacitor and inductors form a T-type network.

In high pass LC filters the position of inductor and capacitor are interchanged as shown in Fig. 11.21.

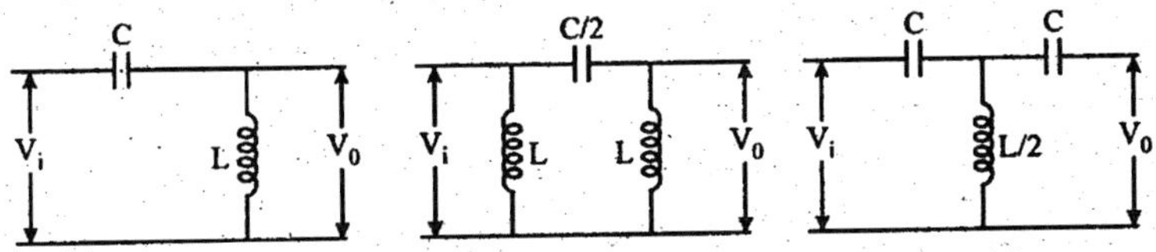

(a) (b) (c)

Fig. 11.21. High pass LC filters (a) L-section, (b) π-section, (c) T-section.

Example. 11.11. A basic T section low pass filter has inductance L = 200 mH and C = 10 μF. Calculate its characteristic impedance and cut off frequency.

Solution. Characteristic impedance is given by

$$Z_0 = \sqrt{L/C} = \sqrt{\frac{200 \times 10^{-3}}{10 \times 10^{-6}}}\ \Omega$$

$$= \sqrt{20 \times 10^{+3}} = 1.41 \times 10^2\ \Omega$$

$$= 141\ \Omega$$

$$\text{Cut off frequency } f_0 = \frac{1}{2\pi\sqrt{LC}} = \frac{1}{2\pi\sqrt{200 \times 10^{-3} \times 10 \times 10^{-6}}}\ \text{Hz}$$

$$= \frac{1}{2\pi\sqrt{2 \times 10^{-3}}} = \frac{1 \times 10^2}{2\pi \times 4.4}\ \text{cycles/sec}$$

$$\approx 3.5\ \text{cycles/sec.}\quad \textbf{Ans.}$$

11.7. A/D AND D/A CONVERTERS

The output of a transducer remains generally in analog form. However for some applications it is required in digitized form. For this purpose A/D converters are used. A D/A converter does the reverse of an A/D converter. It is required in a signal conditioner as the control of the system is performed by using analog signals. Some A/D and D/A converters for these purposes are described in what follows.

***(a)* A/D Converter.** The block diagram of a voltage to time (ramp type) A/D converter is shown in Fig. 11.22. In this counter when ramp voltage rises to the level of input voltage the input comparator sends the stop pulse to the main gate. The output of the gate is counted and displayed.

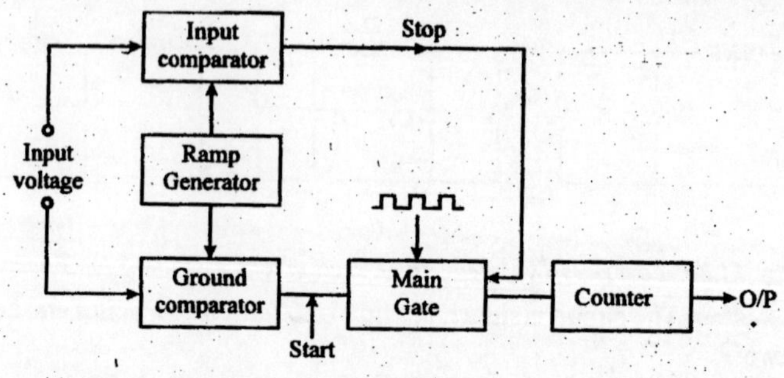

Fig. 11.22.

***(b)* D/A converter.** Basic scheme of digital to analog converter known as DAC is shown in Fig. 11.23. In normal state the switches have connections with ground. When any of the digital input goes HIGH, the switch gets connected to reference voltage for that input. For example if the input is 10000000 all switches except the lowest will remain connected to ground. The input A which goes HIGH will send the input equal to reference voltage V_{ref} at the inverting terminal of the OP Amp. The output voltage in this case has the value equal to reference voltage. In general

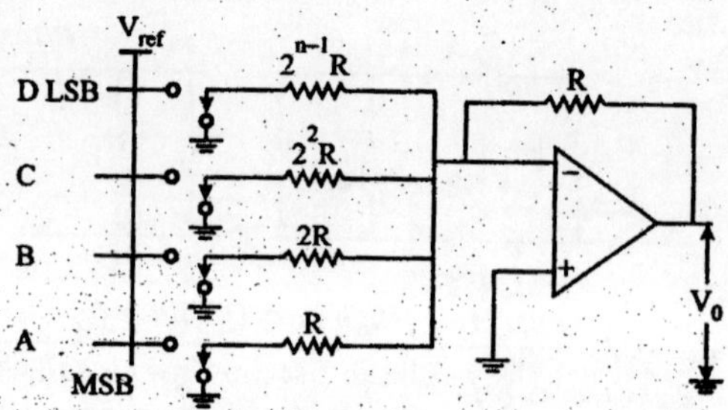

Fig. 11.23. D/A Convertor.

$$V_o = [V_A + 2^{-1} V_B + 2^{-2} V_C + 2^{-(n-1)} V_N]$$

11.8. MULTIPLEXERS AND DEMULTIPLEXERS

In a multiplexer more than one inputs are switched sequentially. Thus a multiplexer converts many inputs into one output. Two types of multiplexing techniques namely time division multiplexing (TDM) and frequency division multiplexing (FDM) are used. The function of demultiplexer is opposite to that of a multiplexer, *i.e.*, for one input it gives many outputs. These circuits are described as under.

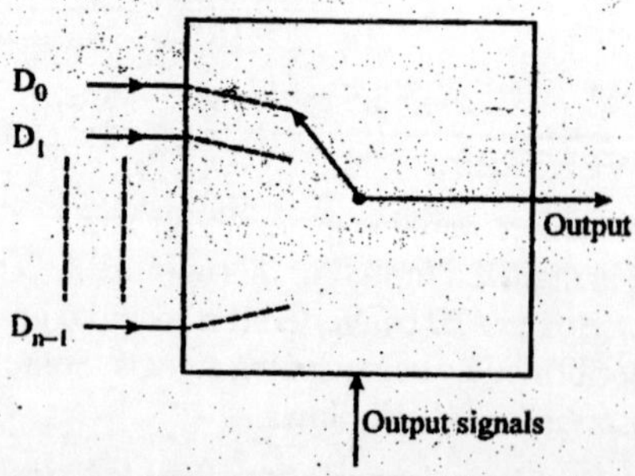

Fig. 11.24. Block diagram of a multiplexer.

Multiplexer. Block diagram of a multiplexer is shown in Fig. 11.24 where *n* inputs D_0 to D_{n-1} are fed to the multiplexer (acting as a switch) which selects one of the inputs and sends it to the output. This is known as *n* to 1 multiplexer. A 4 to 1 multiplexer is given in Fig. 11.25.

From the figure it is clear that the input data that is sent to the output depends upon the status of control signals A and B. Suppose A is HIGH and B is LOW, then for only AND gate A_3 control signals are HIGH. The input data D_2 (LOW or HIGH) can therefore be

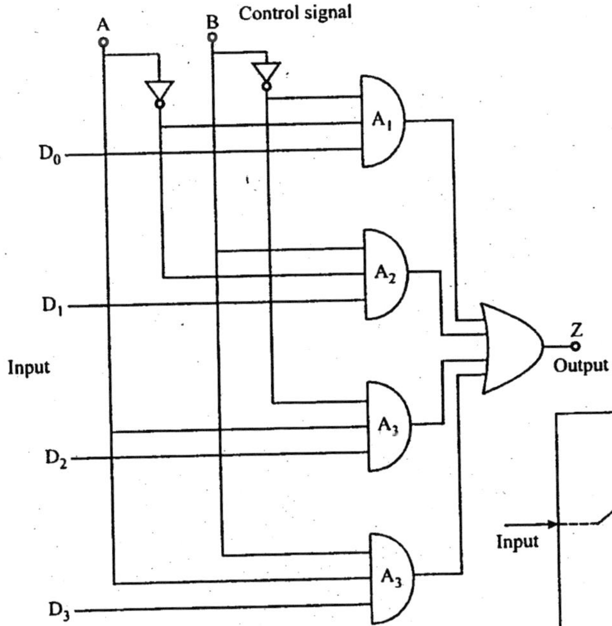

Fig. 11.25. 4 to 1 multiplexer.

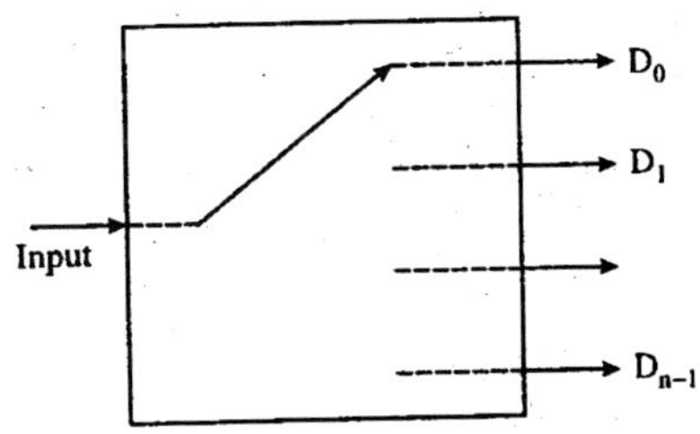

Fig. 11.26. Block diagram of a demultiplexer

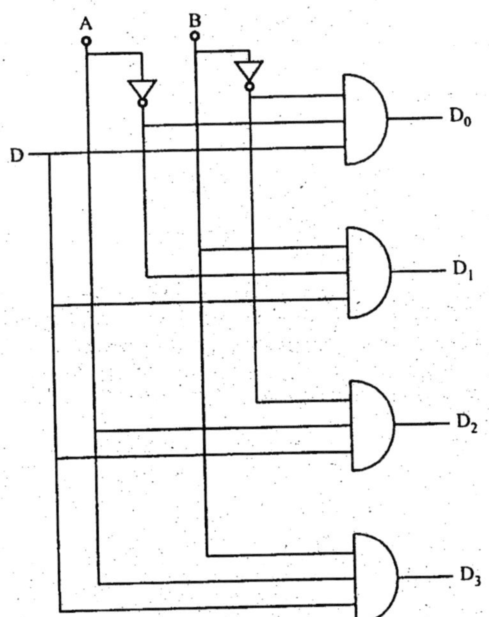

Fig. 11.27. A 1 to 4 demultiplexer

A	B	Z
0	0	A_0
0	1	A_1
1	9	A_2
1	1	A_3

Table 1: Truth table for multiplexer shown in Fig. 11.25

A	B	O/P
0	0	D_0
0	1	D_1
1	0	D_2
1	1	D_3

Table 2: Truth table for demultiplexer

switched to OR gate whose output depends upon the status of the input signal. Similarly other inputs can be passed as output for different combinations of control signals A and B The truth table is shown by table 1.

Demultiplexers. A demultiplexer which performs operation reverse to that of a multiplexer can be shown by the block diagram shown in Fig. 11.26. It is used to distribute a given input to a number of output channels. A one to four demultiplexer is shown in Fig. 11.27. The working of this circuit also can be explained in the same manner as that of a multiplexer. For example, when A and B both are HIGH, the input will appear as output D_3 etc. The truth table is shown by table 2.

SUMMARY

Signal conditioner is an important stage of measurement system. Operational amplifier is an important unit of signal conditioner. In this chapter emphasis has been laid on operational amplifier. Also use of the amplifier in different case has been discussed. In addition other elements of signal conditioning system such as filters, multiplexers, demultiplexers, A to D and D to A converters etc. have been described in the present unit.

EXERCISE

1. Derive expression for the gain of OPAMP is non inverting mode.
2. In inverting mode derive expression for the gain of OPAMP.
3. What is a voltage follower ? where is it used ?
4. What are the characteristics of an ideal OPAMP?
5. Explain the following
   (a) Input bias current
   (b) Input offset current
   (c) Input offset voltage
   (d) Output offset voltage
   (e) Slew rate
   (f) Drift
6. What is the significance of CMRR ? Derive expression for CMRR.
7. What is main source of noise in operational amplifier ? Explain.
8. What do you understand by frequency response ? Explain the fall in gain at low and high frequencies.
9. What is an instrumentation amplifier ? Derive expression for output voltage of instrumentation amplifier.
10. What are the characteristics of instrumentation amplifier ?
11. What is the meaning of modulation? Describe amplitude modulation.
12. Write a note on low pass passive filter.
13. Draw the circuit of band pass filter. Explain its working.
14. Explain the working of high pass passive Filter.
15. Describe band reject filter. Why is it called notch filter.
16. Describe low pass and high pass LC filters.
17. Write a short note on multiplexing.
18. Describe 1 to 4 demultiplexer.

NUMERCIAL PROBLEMS

1. An input signal of 10 mV is applied at the inverting end of the amplifier. If input resistance is 50 Ω and feedback resistance is 10 kΩ calculate o/p voltage.
2. In the above problem if non-inverting amplifier is used, what is the output voltage?
3. Open loop voltage gain of an OPAMP is 99985. If input signal of 12 mV is applied to a voltage follower calculate output voltage.
4. Voltage of 10 mV and 1 volts are applied to two terminals of an OPAMP. Calculate common mode and differential voltage.
5. Open loop voltage gain and common mode gain of an OPAMP are respectively 10^5 and 1.5. Calculate CMRR.
6. An instrumentation amplifier has a adjustable resistance of 100 Ω and fixed resistor 100 kΩ. If signal applied at both terminals are 10 mV and 9.2 mV, what is output voltage ? **(Ans.** 1.6. Volt)
7. In a DAC technique input data is 1001. If reference voltage used is 8 volts, determine output voltage.
8. In the above problem determine maximum output voltage.

OBJECTIVE TYPE QUESTIONS

1. Maximum value of input impedance of ideal instrumentation amplifier is

   (a) 100 kΩ   *(b)* 1 MΩ
   (c) ∞   *(d)* none of the above

2. Output impedance of ideal OPAMP is

   (a) 50 Ω   *(b)* 0 Ω
   (c) 75 Ω   *(d)* 130 Ω

3. CMRR of an ideal instrumentation amplifier is

   (a) 100000   *(b)* ∞
   (c) 0   *(d)* none of the above

4. CMRR is defined as

   (a) $\frac{A_c}{A_d}$ *(b)* Ac × Ad
   (c) $\frac{Ad}{Ac}$ *(d)* Ad

5. CMRR of two amplifiers are 10^5 and 10^6 then

   (a) First amplifier is better than second
   (b) There is no comparison between the amplfiers
   (c) Second amplifier is better than first
   (d) None of the above.

6. Gain of operational amplifier in inverting mode is given by

   (a) $\frac{R_f}{R_1}$ *(b)* $\frac{-R_f}{R_1}$

(c) $\left(1+\frac{R_f}{R_1}\right)$ (d) either (a) or (b)

7. Input impedance of practical operational amplifier is

(a) ∞ (b) about 2 MΩ

(c) very high (d) very small

8. Gain of OPMAP in noninverting mode is given by

(a) $\frac{R_f}{R_1}$ (b) $1+\frac{R_f}{R_1}$

(c) $\frac{R_1}{R_f}$ (d) 1.5

9. Gain of buffer amplifier is given as

(a) 1 (b) $\frac{A_{VOL}}{1+A_{VOL}}$

(c) $\frac{1+A_{VOL}}{A_{VOL}}$ (d) (a) or (c)

10. Output of instrumentation amplifier is given by

(a) $V_o = \left(1+\frac{2R_1}{R_g}\right)(V_2+V_1)$ (b) $V_o = \left(1+\frac{2R_2}{R_g}\right)(V_2-V_1)$

(c) $V_o = \left(1+\frac{R_2}{R_g}\right)(V_2-V_1)$ (d) none of the above

11. Output impedance of a practical amplifier is

(a) about 100 Ω (b) 1 kΩ

(c) 500 Ω (d) zero.

12. In amplitude modulation.

(a) amplitude of signal changes

(b) amplitude of carrier as well as signal varies

(c) amplitude of carrier changes with the amplitude of the signal

(d) none of the above.

13. The modulation index in AM

(a) Varies between 0 and 1 (b) has maximum value 1

(c) has maximum value of infinity (d) (a) and (b)

14. A low pass filter.

(a) allows low frequency signal to appear as output

(b) uses small capacitor

(c) is a differentiator (d) none of the above

(e) (a) and (b).

15. A bandpass filter

(a) is the arrangement of high and low pass filters

(*b*) acts like a high pass filters

(*c*) none of (*c*) and (*b*)

(*d*) both (*a*) and (*b*)

16. Characteristic impedance of low pass LC filter is given as

(*a*) $Z_0 = 1/\sqrt{L/C}$ (*b*) $Z_0 = \sqrt{L/C}$

(*c*) $\sqrt{X_L^2 + X_R^2}$ (*d*) both (*a*) and (*b*).

17. A multiplexer

(*a*) gives one output for many inputs (*b*) multiplies various inputs

(*c*) rejects some inputs (*d*) none of the above.

18. A demultiplexer

(*a*) does the function opposite to that of multiplexer

(*b*) gives many outputs for single input

(*c*) demultiplies the outputs

(*d*) both (*a*) and (*b*)

ANSWERS

1. (*c*)	2. (*b*)	3. (*b*)	4. (*a*)	5. (*c*)	6. (*b*)
7. (*b*)	8. (*b*)	9. (*b*)	10. (*b*)	11. (*a*)	12. (*c*)
13. (*d*)	14. (*e*)	15. (*a*)	16. (*b*)	17. (*a*)	18. (*d*)

❑❑❑

12
Output devices and displays

12.0. INTRODUCTION

Data available from any source can be used for several purposes such as for display recording, etc. The data can be recorded by several means such as computer printer, graphic recorder, tape recorder etc. In the present chapter we shall study few recording devices :

12.1. RECORDING DEVICES

Recording preserves the data in usable and measurable form. A recording device preserves the information obtained at any instant from any indicating instrument and it gives information about waveforms and transient behaviour or phase relationship in different parts of circuit under consideration.

Recording devices are of two types namely (1) analog recorders and (2) digital recorders. Analog recorders are divided into *(a)* graphic recorders *(b)* Oscillographic recorders and *(c)* magnetic tape recorders. Graphic recorders are described as under.

12.2. GRAPHIC RECORDERS

Graphic recorder is defined as an instrument which draws the graph relating two or more variable to each other. It is essentially an analog representation of discrete information. The graphic recorder is divided into *(a)* strip chart recorders and *(b)* circular chart recorder. The recorders are described below :

(a) **Strip chart recorder :** It is called x-t recorder as it plots one or more variable with respect to time. The recorder is shown in Fig. 12.1 Various parts of it are described below :

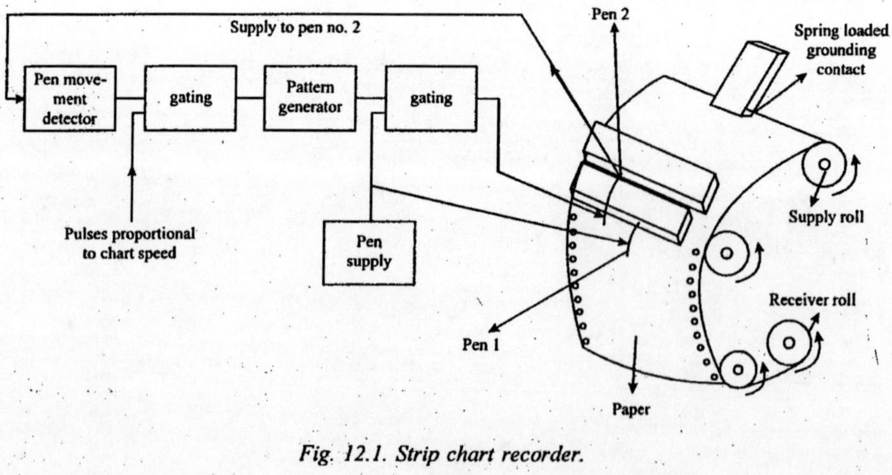

Fig. 12.1. Strip chart recorder.

1. Paper drive system : It consists of a long roll of graph paper called chart, moving vertically and driven by synchronous motor. The speed is selected by a speed selector in fixed increments.

2. Marking methods : Various marking methods used in the recorder are :

(a) **Pen and ink :** This is most familiar. Various types of pen such as V-pen, ball point pen and fiber tipped pen are used. Red ink is generally used. However other ink also can be used.

(b) **Impact Printing :** In this system a pointer is clamped in position for a longer time and a press bar above the pointer machine presses down on a carbon ribbon located between the pointer and chart. The carbon marking serves as the record.

(c) **Thermal Writing :** In this method an electric current flows through the tip of moving stylus. The current heats the stylus which causes a thin line to appear on the heat sensitive paper. This method overcomes the difficulties encountered in the ink-pen system.

(d) **Electric Writing :** In electric writing, an electric writing paper is used which is a black substrate coated with Aluminium. For writing a tungston system is used which remains in light contact with Aluminium coating. When the current passes through the stylus Aluminium is etched and the black surface is exposed, which is the trace. The cost of the paper used is very high. So it is rarely used.

3. Tracing system : In graphic recorder, two types of tracing systems are employed.

(a) **Curvilinear tracing system :** A galvanometer recorder is of this type. As the current proportional to the measurand flows through pointer arm of the galvanometer coil, angular movement of the coil takes place. A pen attached to the pointer arm of the galvanometer moves on the strip chart as the current through the coil flows proportional to the input quantity shown in Fig. 12.2. In curvilinear system non-linear relationship exists between angular deflection and displacement. The main drawback of curvilinear system is that it is very difficult to analyse the trace between input and the time. This difficulty is overcome, to some extent, by using long pointers of galvanometers.

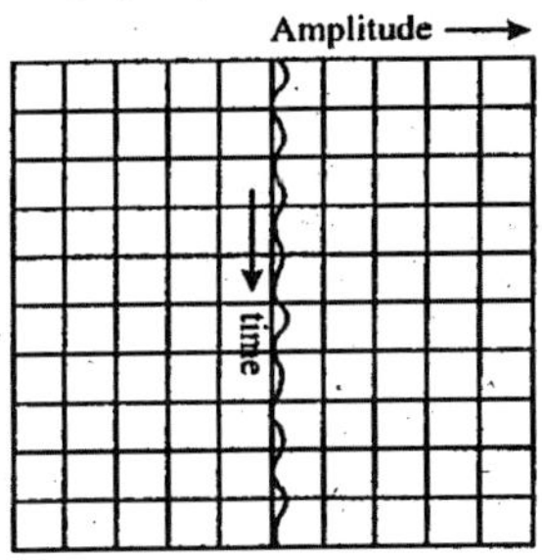

Fig. 12.2. Curviliner tracing system.

(b) **Rectilinear tracing system :** As the strip chart moves under pen or marker at a uniform speed, the graph on this paper is a square or rectangular co-ordinate matrix The direction of motion of the chart is perpendicular to the motion of assembly. Thus the lines that are ruled perpendicular to the pen direction function of chart are time lines. The pen assembly produces a straight line perpendicular to the motion of chart as shown in Fig. 12.3. This is called rectilinear trace. By using rectilinear trace uniform resolution can be obtained.

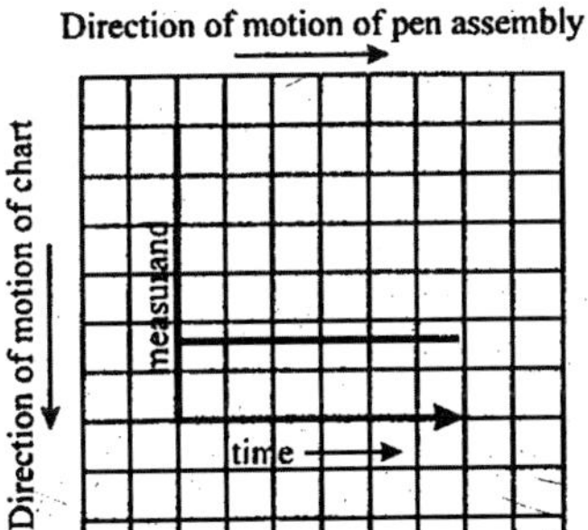

Fig. 12.3. Rectilinear tracing system.

Strip chart recorders are mainly of two types viz galvanometer recorder and null balance type as described below.

***(a)* Galvanometer recorder :** The constructional diagram of the recorder is shown in Fig. 12.4. The pen assembly is attached to the pointer of D' Arsonval movement. Measurand is fed as current to the coil of D' Arsonval galvanometer between two pole pieces N and S. To the pointer of the galvanometer is attached a pen which moves on the strip chart moving vertically downward. The trace is obtained as shown in the Fig. 12.4.

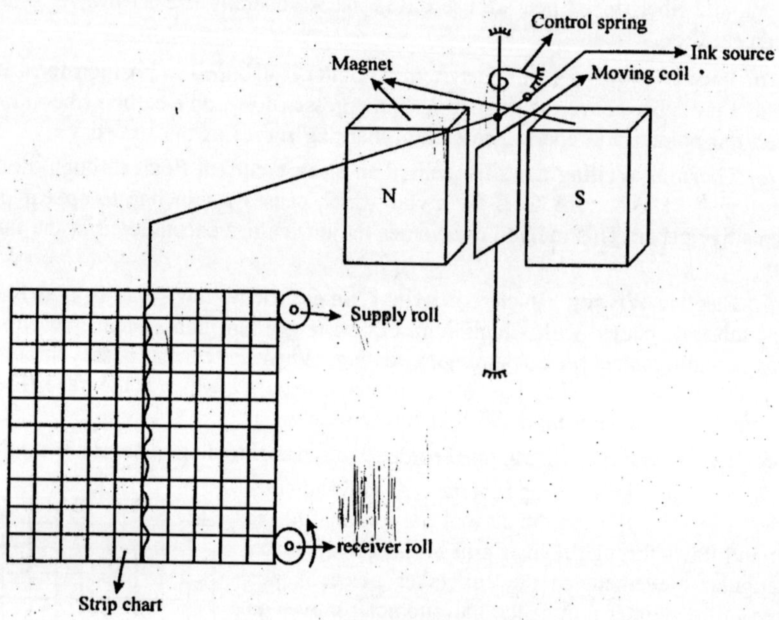

Fig. 12.4. Galvanometer recorder.

As the torque on the coil is applied the balancing torque is provided by the control spring. The arrangement in this recorder is slightly different from ordinary D' Arsonval arrangement. The frictional torque in galvanometer recorder is higher due to larger weight of moving system and higher friction between pen and the moving chart. Therefore large torque is required which in turn requires large controlling torque. So in such galvanometers larger moving coil and strong magnetic field is required. The response time of such galvanometers is slow as they are made critically damped.

Maximum working frequency of galvanometer recorder is about 100 Hz and their input impdeance of the order of 100 kΩ. The working frequency can be increased upto 13 kHz by replacing pen and pointer arrangement by a light beam.

Analysis : The deflecting torque developed by the moving coil due to current i flowing through it is given by

$$T_d = NBAi$$

$$= NBLri \quad ...(1)$$

$$= Gi \quad ...(2)$$

In eqn. (1)

N = No. of turns in the coil

B = Magnetic flux density wb/m^2

l = Length of the coil.

r = width of the coil

G = NB$l$$r$ = NBA is called the displacement constant of the galvanometer.

Due to deflecting torque to coil it deflects by an angle θ, The eqn. of motion is given by

$$J\frac{d^2\theta}{dt^2} + D\frac{d\theta}{dt} + K\theta + F = Gi \quad ...(3)$$

where J = moment of inertia of the coil kg m^2

D = damping constant

K = spring constant or restoration constant

F = Coulomb friction, N-m.

In steady state $\frac{d\theta}{dt} = \frac{d^2\theta}{dt^2} = 0$, giving

$$K\theta + F = Gi$$

or

$$\theta = \frac{Gi - F}{K}$$

$$\theta = \frac{Gi}{K} - \frac{F}{k} \quad ...(4)$$

From eqn. (4) one can deduce that deflecting torque should be made large and coulomb friction should be reduced to minimum value which is done by increasing the value of turns N and B as is obvious from the expression G = NAB.

Example 12.1. In a galvanometer recorder the number of turns in the coil whose area is 13 cm^2 is 30. Restoration constant and coulomb friction are respectively, 3 N/m and 0.1 N-m. If the current in the coil is 2.0 A compute deflection of the galvanometer. Magnetic flux density 3.0 wb/m^2.

Solution : The galvanometer constant is given by

$$G = NAB$$

$$= 30 \times 15 \times 10^{-4} \times 3.0$$

$$= 0.135 \text{ weber}$$

Galvanometer deflection :

$$\theta = \frac{Gi}{K} - \frac{F}{K}$$

$$= \frac{0.135}{3} \times 2.0 - \frac{0.1}{3}$$

$$= 0.90 - 0.03$$

$$= 0.06 \text{ radian}$$

$$= 10.8 \text{ degree } \textbf{Ans.}$$

***(b)* Null balance recorder** : Null balance recorder is shown in Fig. 12.5. The self balancing action of all such types of recorders is provided by servo motor. The input voltage is compared with the reference voltage generated by a feedback element. The servo amplifier amplifies the difference between the two signals. If the difference is non zero, servo motor is actuated which positions the indicator on the output display scale for zero error or null voltage condition between the unknown and reference voltage. The indicator is used to read directly magnitude of input signal under measurement.

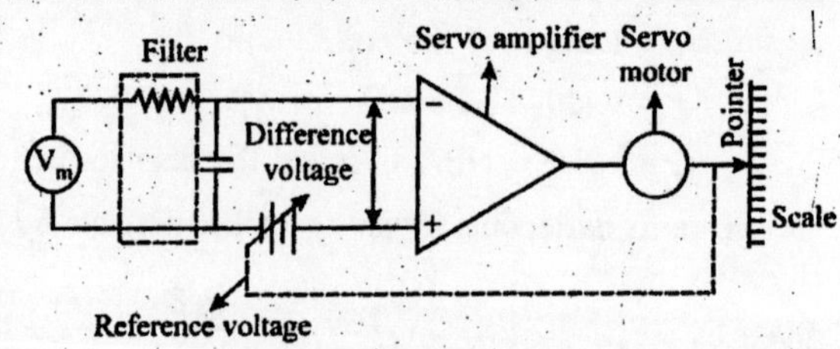

Fig. 12.5. Null balance recorder.

2. Circular chart recorder : Circular chart recorder is shown in Fig. 12.6. In this recorder a circular chart is used which has concentric circles ruled on it. Also there are arcs extending from the centre to the extreme end of the chart. The chart moves about an axis. The pen as it moves swings along these arcs. The separation between the arcs is the measure of time elapsed as the pen moves from one arc to another. These are called therefore as time arcs. The radial position of the pen indicates the instantaneous value of quantity under measurement

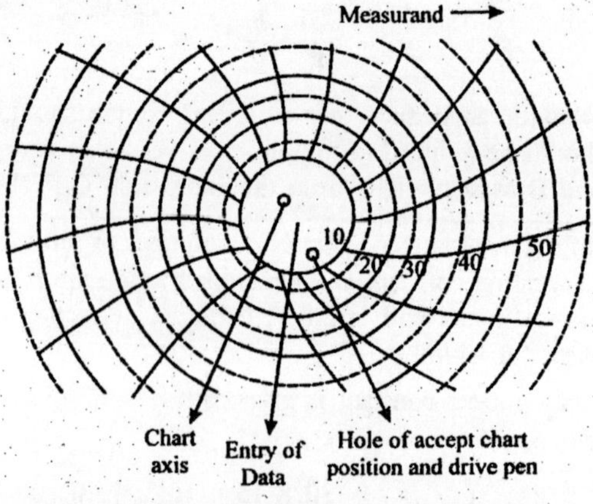

Fig. 12.6. Circular chart recorder.

Maximum diameter of the chart is 0.3 meter .The resolution of the chart is not uniform and these do not run for long periods. The speed of rotation of the chart is usually one revolution per 24 hours or more or several days.

12.3. X-Y RECORDERS

X-Y recorders are used where graphical relationship is to be obtained between two variables say X and Y. One variable is fed to X input and the other is fed to Y input. The recorder plots their variation against each other. In such recorders chart remains fixed and the pen moves simultaneously in both X and Y directions as the signals are applied to two input terminals. Most of the X-Y recorders have time base also hence these can be used like strip chart recorders by applying signal to one input terminal.

For the movement of the pen simultaneously in X and Y direction servo mechanism as shown in Fig. 12.7 is used. For any input given to X and Y terminals the corresponding change in the input to servomotor takes place. As a consequence the horizontal and vertical movement of the pen takes place. The pen used is simple ink pen and simple paper can be used. This is the advantages of X-Y recorder over strip chart recorder. However its use is limited to very low frequencies.

An X-Y recorder has sensitivity of 10 μV/mm and a frequency response of 6 Hz in both the X and Y directions. Input indepdance is from 100 ohms to about 50 M ohms.

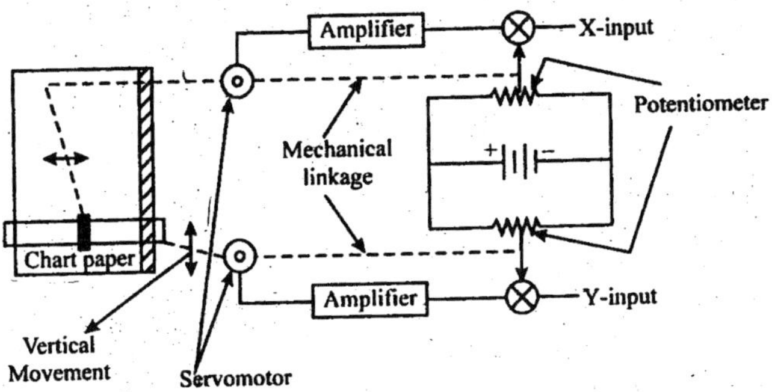

Fig. 12.7. Schematic of X-Y recorder.

By using X-Y recorders the graphical relationship is obtained between two variables such as voltage vs current curve of diodes, B-H curves for the magnetic materials simultaneously as shown in Fig. 12.7.

12.4. DIRECT RECORDER

This is the method of analog recording in which amplitude and frequency of the input signals are recorded as variation of the magnetization amplitude and wavelength on the tape

The arrangement is shown in Fig. 12.8. It has a reproducing and a recording head. Input voltage is converted into current which passes through the coil of recording head. The magnetic flux produced by this current in non-magnetic gap is $\phi = Ki$. As the tape passes through the non magnetic gap, the oxide particles on the tape are magnetized by the flux in the gap. If the input current is

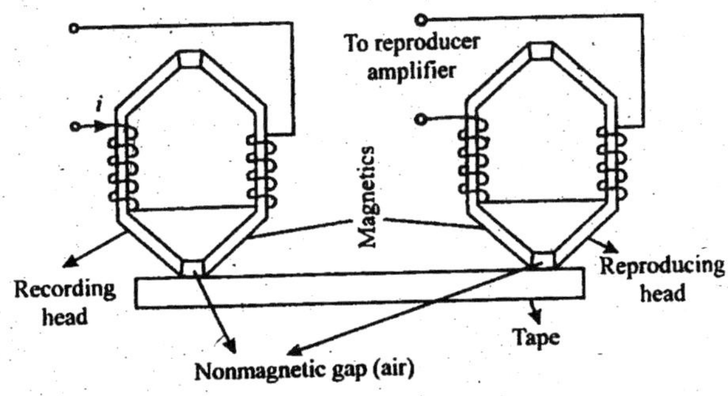

Fig. 12.8. Direct recorder.

$$I = I_0 \sin \omega t$$

Then the intensity of the magnetization along the tape varies with the displacement as

$$m = K_{in} K I_0 \sin \left(2\pi f \frac{x}{V}\right)$$

Thus resulting magnetic field in the gap enables the magnetic recording of the input information on a tape that passes under the gap.

For reproduction the tape is passed over a reproducing head thereby producing output voltage proportional to magnetic flux in the tape, across the coil of reproducing head. The output voltage is proportional to the rate of change of flux. If dc input is used the magnetization on the tape remains fixed which produces zero output of the reproducing head. The recording, therefore, should be done for varying input signal at a frequency not less than 50 Hz. There is an upper limit also of the frequency, as at high frequency, for a given speed and reproducing gap, the wavelength of magnetization becomes less than the gap. Under such condition average magnetization on the tape will be zero and there will be no output voltage. Direct method of recording suffers from the drawback of poor signal to noise ratio which is typically 25 db. For good working it should be 50 db.

Example 12.2. In a galvanometer recorder a coil of length 16 cm, width 9 cm and 40 turns is used. Damping constant and restoration constant are respectively 15 and 2 units. Neglecting coulomb friction calculate current through the coil. Angle of deflection is 15°. Assume magnetic flux density B = 2.4 wb/m².

Solution : The galvanometer constant is given by

$$G = NAB$$

$$= 40 \times 16 \times 10^{-2} \times 9 \times 10^{-2} \times 2.4 \text{ wb}$$

$$= 1.38 \text{ wb}$$

$$\theta = \frac{G}{k} i \text{ neglecting coulomb friction}$$

$$\theta = 15^{\circ} = \frac{\pi}{180} \times 15 \text{ radian} = \frac{\pi}{12} \text{ radian}$$

Putting the values, we obtain

$$\frac{\pi}{12} = \frac{1.38}{2} \times i$$

or

$$i = \frac{\pi}{12} \times \frac{2}{1.38} \text{ amp} = 1.45 \text{ amp} \quad \textbf{Ans.}$$

Example 12.3. In a direct recorder a sinusoidal current of 1.2 amp at 100 Hz flows through the coil of recording head. The tape moves with a constant velocity of 1 m/sec. Calculate magnetization of the tape having air gap of 1.6 cm, flux in the non magnetic gap. Given K = 1 unit, K_m = 3.2 units × 10^{-2} units.

Solution.

$$\phi = Ki$$

$$= KI_o \sin \omega t$$

$$I_o = \sqrt{2} \times 1.2 = 1.7 \text{ Amp.}$$

$$\phi = 1.7 \times 1 \sin 200 \pi t$$

$$\phi = 1.7 \sin 200 \pi t$$

$$M = K_m K I_o \sin\left(2\pi f \times \frac{x}{V}\right)$$

$$= 3.2 \times 10^{-2} \times 1 \times 1.7 \sin \left(200\pi \times \frac{1.6 \times 10^{-2}}{1}\right)$$

$$= 5.4 \times 10^{-2} \sin (3.2\ \pi) = 3.17 \times 10^{-2} \textbf{ Ans.}$$

12.5. OSCILLOGRAPHIC RECORDER

Although the recorders discussed in previous sections come under the category of oscillographic recorder, yet by oscillographic is meant galvanometer or CRT recorder. In what follows these two recorders are discussed separately.

***(a)* Galvanometer recorder.** This recorder is shown in Fig. 12.9. A light source (generally mercury bulb) is used which sends light to a reflector attached with the moving part of the galvanometer. The light after reflection falls on recording mirror which is sensed by light sensitive device through a lens.

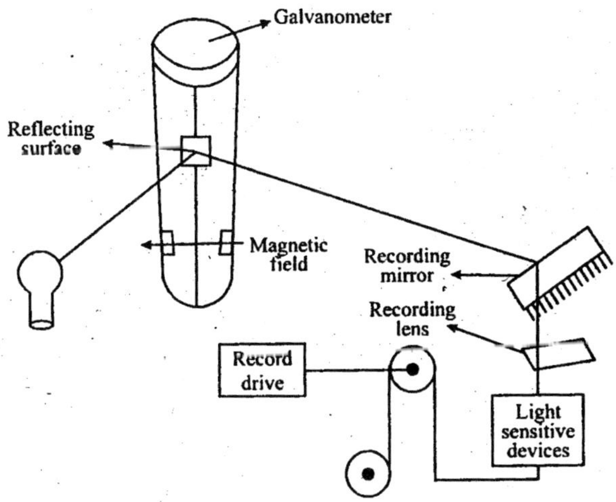

Fig. 12.9. Galvanometer recorder.

The input to be recorded is supplied to the galvanometer coils which deflects under the influence of magnetic field as shown in the figure. The direction of deflection of the coil and hence the reflecting surface depends upon the sense in which the current flows through the galvanometer coil. As a result of it the intensity of light passing through the recording mirror changes.

Corresponding to it the movement of pen system driven by light sensitive devices on the trace paper which moves as shown in the figure, takes place. Reference timings are recorded by a timing circuit. Thus a trace of input data is obtained. The working frequency of this recorder is about 20 kHz.

***(b)* CRT recorders.** A CRT recorder is a four dimensional device. It consists of four axes X, Y, Y′ and Z. The X axis is horizontal or time axis, input signal is fed through Y axis, Y′ axis is the movement of recording medium and Z axis is defined by the spot intensity on the recording surface as shown in Fig. 12.10.

As shown in the figure the output of cathode ray tube is focussed on the record by optical lens system or by using a fiber optic face plate and placing the record in contact with fiber optic. The reference grid lines on the record in horizontal and vertical directions are recorded by passing light from a flash tube through an aperture. The timing of the flash tube can be controlled and synchronized with the record drive system. The working frequency of CRT recorder may be as high as 20 MHz.

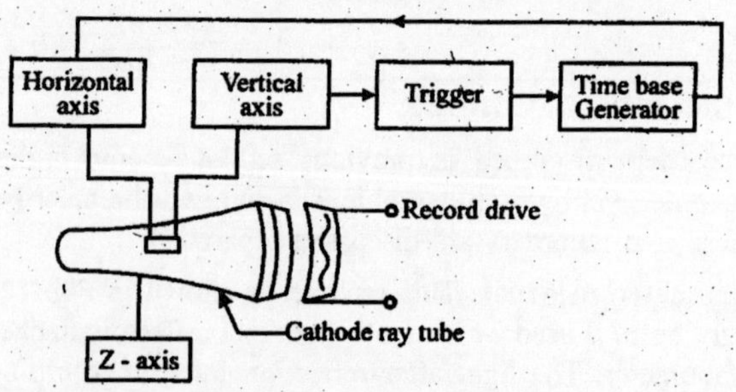

Fig. 12.10. CRT recorder.

12.6. MAGNETIC TAPE RECORDERS

Unlike the recorders discussed previously, in a magnetic tape recorder the data can be recorded and retrieved in electrical form whenever required. The data on magnetic tape recorder once recorded, can be removed to record new data whenever required.

The construction of a magnetic tape recorder is shown in Fig. 12.11. It consists of magnetic tape recording and reproducing heads and transport mechanism.

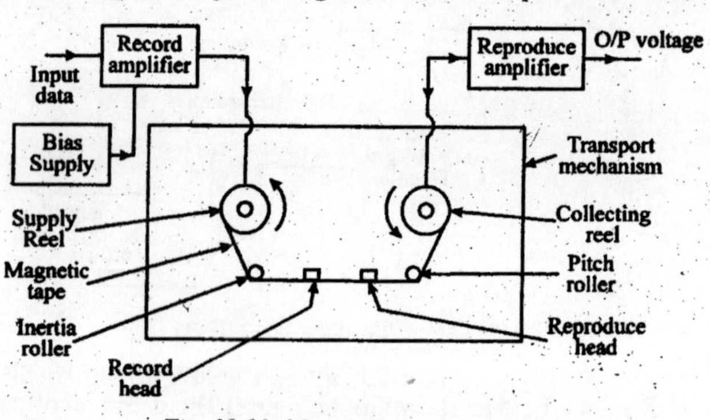

Fig. 12.11. Magnetic tape recorder.

The magnetic tape is made up of a flat plastic like PVC or Mylar which is coated on one surface by a magnetic material. The magnetic material is generally needle shaped ferric oxide (Fe_2O_3). The typical size of needles is 0.6 μm long 0.1 μm in diameter. These are spread on the tape with their lengths along the tape, at regular intervals. The typical value of tape size is 12.7 mm wide and 25.4 μm thick. As the tape moves the input data is recorded by magnetizing the magnetic material on it by the record head

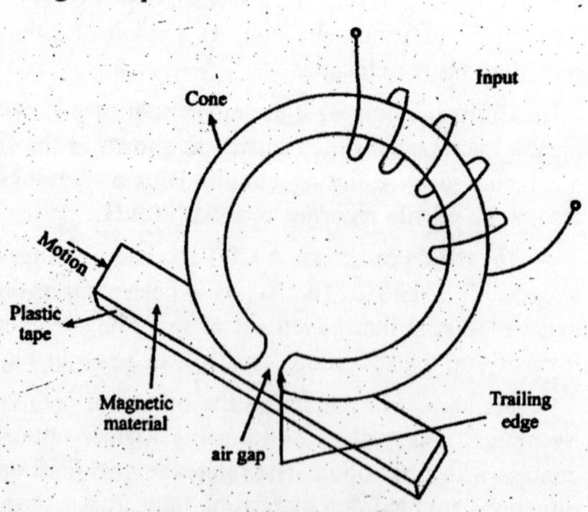

Fig. 12.12. Record head.

which is retrieved by the reproducing head. The construction of reproduce and record head are described in Fig. 12.12. As shown in the Figure the flux produced by the toroid which remains distributed in the core as well as air gap uniformly, depends upon the input current in the coil. When the tape moves under the air gap, the magnetic material on it gets magnetized at the trailing edge. Thus the information contained in the input is recorded on the tape.

The construction of reproducing head is similar to that of recording head except that in the former the induced output voltage is collected through the coil.

The tape transport mechanism moves the tape with constant speed without any strain or distortion. For this purpose the mechanism is set so as to maintain proper tension in the tape. The arrangements for other purposes like fast winding and reversing are also made.

Magnetic tape recorders which are available in various sizes are used for various purposes. Some of their applications are:

1. In communication surveillance and spying.
2. In industrial research and production monitoring and control.
3. Medical research and patient monitoring.
4. Data recording and then analysis of aircrafts, missiles and satellites.

12.7. Disk Files and Floppy Disk.

A disk file, as the name implies, is in the form of the circular sheet of metal having diameter between 0.25 to 0.5m and thickness of 1.5 to 6 mm. It is coated, on both the surfaces, with magnetic material. It is divided into a number of circular tracks which range from 100 to 1000. A read/Write head as shown in Fig. 12.13 can be adjusted by movable arm, to any track to read or write the data. For write operation the current through the coil of the head produces a magnetic flux which magnetizes the magnetic material on the track. On the contrary for read operation, the magnetic flux on the disk induces an e.m.f in the coil of the head which is detected as output.

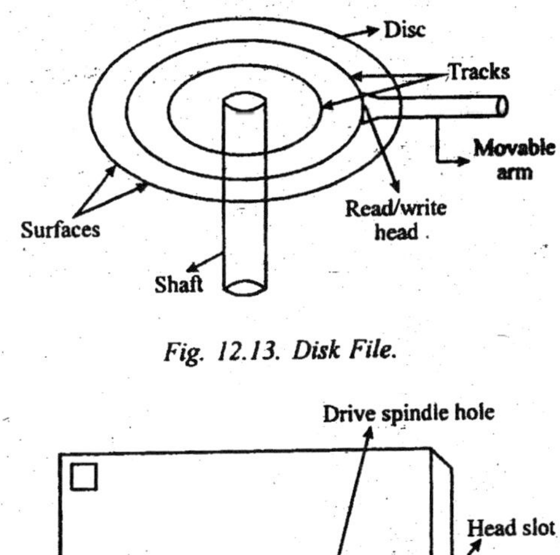

Fig. 12.13. Disk File.

The floppy disk is shown in Fig. 12.14. It is a mylar layer which is in the form of disc coated with oxide magnetic material. The disk has a thickness of 0.125 m and 195 mm diameter with 27.5 mm diameter hole in the centre. It is packaged in a flexible plastic envelope to protect

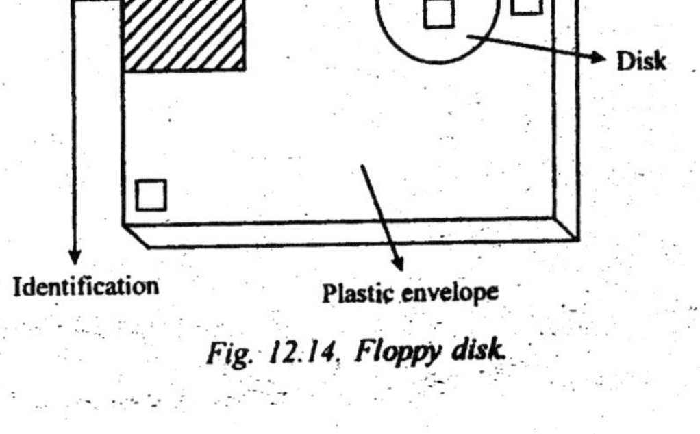

Fig. 12.14. Floppy disk.

the oxide layer. The disk with the envelopes is inserted into the device (generally personal computer is used) for read/write operation. The read/write operation of floppy disk is the same as that of the disk file. A disk has a storage capacity of 246000 bytes.

12.8. Digital Display Methods:

The output of an instrument can be displayed digitally. For this purpose, the analog input are first converted into digital forms which are then displayed in the form of numerals and alphanumeric characters. There is a provision for selecting the range selected-sometimes the range is automatically selected through the provision of autoranging.

For the display of inputs in digital forms, various units are available. These can be classified in various ways. On the basis of formats used, these can be classified as planar and nonplanar.. In former case the entire read out characters remain in same plane while in later case the characters remain in the different planes.

LEDs and LCDs are example of planar displays while in NIXIE tube the display is non linear. The LEDs and LCDs drive the segments of 7 segment or 14 segment displays. These are described in what follows:

12.8.1. LEDs

The light emitting diodes (LEDs) are most common devices used in digital displays. An LED consists of a P-N junction formed between a P type and N type material as shown in Fig. 12.15. As the electrical energy is supplied to the device, the electrons moving from *n* side to P side recombine with the holes. As a consequence, the energy is generated in the form of heat and light. This is the working principle of an LED. It is obvious from the figure that the rate of recombination of charge carriers depends upon the current flowing through the device. The typical values of current flowing through an LED various from few mA to 20 mA with a voltage drop of 1-2 volts.

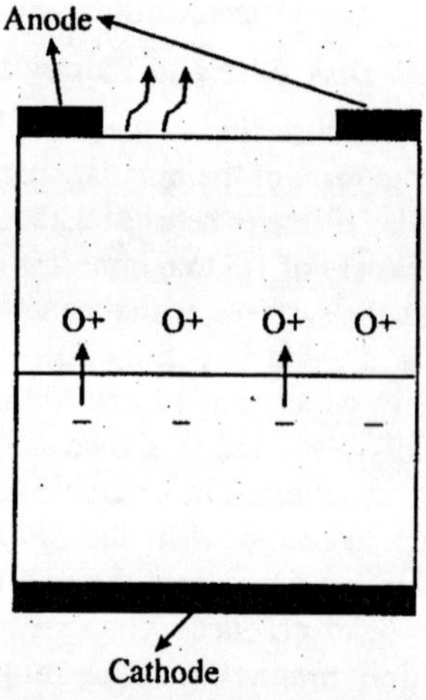

Fig. 12.15. LED construction

Semiconductor materials used for manufacturing of LEDs are galium arsenide phosphide (GaAsP) which emits red or yellow light. galium arsenide (GaAs) is another material used which emits green or red light. Thus light of different colours such as red, yellow, green and amber are emitted by the LEDs. These are extensively used materials presently for displaying numeric and alphanumeric characters in segmented and dot matrix displays. Some of the advantages of LEDs in display are:

1. They are miniature in size and can be stacked together to display numeric and alphanumeric characters.
2. They require moderate power. These are useful where low d.c. power is available.
3. Switching times of LEDs are small which is of the order of 1 ns.
4. These are reggued and can sustain shocks, vibrations etc.
5. Area occupied by them is small.

6. LEDs can be operated over wide range of temperature from 0-70°.

LEDs, however, have the drawbacks of large power consumption and are not suitable for large area displays.

12.8.2. Liquid Crystal Diodes (LCDs)

LEDs are active device which generate the light whose intensity depends upon the current flowing through them. The LCDs, on the other hand, are passive devices which modify the light falling on them.

An LCD is an organic compound which modifies the light passing through it. It is in the form of a liquid layer between two glass plates as shown in Fig. 12.16. Light from a separate light source passes through the crystal which is modified by it. The LCD can modify the light in two ways and accordingly it is of following types (a) Dynamic scattering type and (b) Field effect type.

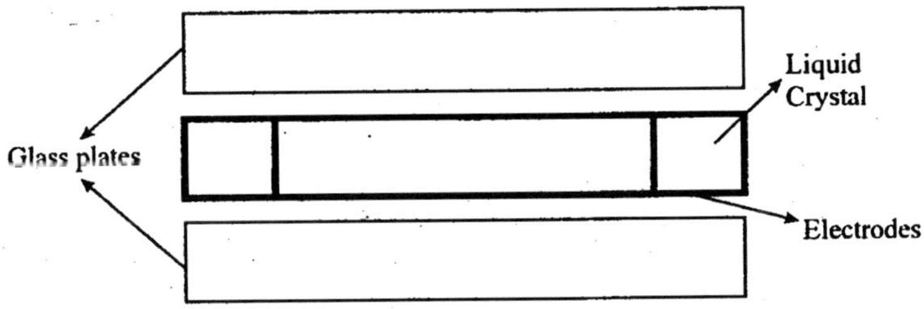

Fig. 12.16. LCD structure.

In dynamic scattering type LCD cell the light transmitting through the crystal is scattered when electric field is applied between the electrodes. The reason for this is that when electric potential is applied across crystal the molecular alignment in it is disturbed. These molecules scatter the light in all directions. Hence the cell appears bright.

In field effect type LCD, a polarizers and an analyzer is used one each at the inside surfaces of the glass plates. The crystal material used is of twisted nematic (TN) type. The light from the source is polarized by the polarizer which is twisted by 90° by the crystal material in the absence of the field. If polarizer and analyser are properly aligned, this light is transmitted by the analyser. As a result, the crystal appears bright. If the field is applied, the molecules in the liquid crystal get aligned in the direction of the field. As a result, the light is untwisted by the crystal and it does not transmit through the analyser. As a result, the crystal appears dark.

Twisted nematic (TN) structure is used in most of the applications because it requires very small voltage, typically 10 volts for the alignment of the molecules. The power consumed by an LCD is of the order of few microwatts (10 μW per segment). An ac is generally used for its excitation as dc produces flickering and the device failure is more likely.

An Important disadvantages of LCD is its high switching times. Its turn ON and turn OFF times are typically 10 ns and 100 ns respectively. Another disadvantage is that its use limited within a small temperature range (0 to 60°C). It reliability and life span are poor.

12.8.3. Nixie Tube

Nixie Tube is used for non-planar display unlike seven-segment display. It consists of a glass envelope with an inert gas such as Neon gas filled in it. Ten electrodes are fitted in the form of different numerals from 0 to 9 in the tube. These are called cathodes as shown in Fig. 12.17. A gauze electrode in the tube performs the function of anode.

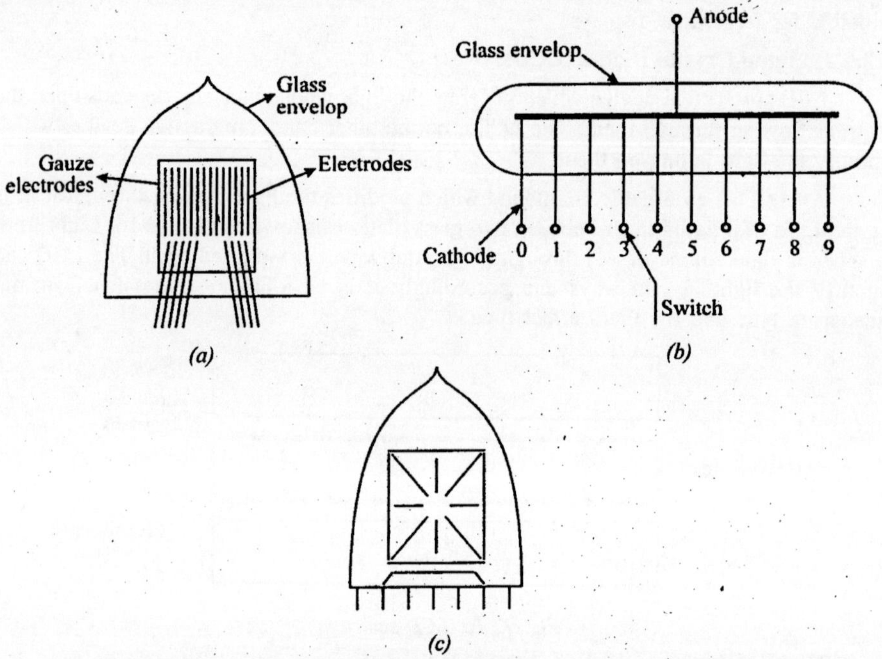

Fig. 12.17. Nixie tube (a) construction, (b) schematic symbol, (c) modern mixie.

The working principle of nixie tube is based on the fact that when electric current flows through a gas, it breaks down and produces a glow at that location. When negative voltage is applied to any cathode, the glow is produced and corresponding numeral is displayed. The colour of display depends upon the gas used.

Nixie tubes required a voltage of the range from 150-220 volts to produce glow discharge. A current from 1 to 5 mA is required. These are bulkier in size as compared to seven segment displays.

In modern nixie tubes more than 10 cathodes are used to display additional characters. A nixie tube having 15 cathodes is shown in Fig. 12.17 (c). It is used to display numeric and alphanumeric characters.

SUMMARY

In the present chapter a recorder has been clearly explained. Various types of recorders such as strip chart recorder, circular recorder, X-Y recorder and direct recorders have been described. Various types of tracing systems such as curvilinear tracing system, rectilinear tracing system etc. have also been described. In addition emphasis has been laid on oscillographic recorder, magnetic tape recorder, disk files and floppy disk.

EXERCISE

1. What are recording devices ? Describe circular chart recorder.
2. What are the types of graphic recorders. Describe construction and working of strip chart recorder.

3. Write short note on tracing systems in recorders.
4. Draw constructional diagram of galvanometer recorder. Derive expression for deflection.
5. What is the working principle of null balance recorder ? Explain
6. What is time axis in circular chart recorder ?
7. What is X–Y recorder ? How can it be used as X–t recorder?
8. Explain working of direct recorder.
9. Describe oscillographic recorders.
10. Explain the working of magnetic tape recorder. What is its advantage over conventional recorders.
11. Write a note on disk files.
12. You are familiar with floppy disk used in computer. Describe its construction.

NUMERICAL PROBLEMS

1. The coil of a moving coil galvanometer has 3,000 turns and it is suspended in a uniform magnetic flux density of 0.12 Tesla. The restoration constant is 0.12 Newton/meter and coulomb friction is 0.15×10^{-6} N-m. The Coil is 25 mm high and 50 mm wide. Determine the current required in the coil to produce a deflection of 60^0.
2. In the above problem what will be the deflection if number of turns are reduced to 300.
3. In problem 1 if damping constant is 0.016 N-m and moment of inertia of the coil is 0.006 kg m^2 determine whether the instrument is over damped, under damped or critically damped.

OBJECTIVE TYPE QUESTIONS

1. Graphic recorder is called
   (a) X–Y recorder (b) X–t recorder
   (c) *t–t* recorder (d) none of the above
2. The frequency range of strip chart recorder is
   (a) 50 Hz (b) 100 Hz
   (c) 20 kHz (d) 100 kHz
3. Uniform resolution in obtained in
   (a) Curvilinear tracing system (b) rectilinear tracing system
   (c) None of the above (d) both of the above.
4. Drawback of strip chart recorder is
   (a) Paper used is expensive (b) Paper used is aluminium coated
   (c) only *(a)* (d) both *(a)* and *(b)*
5. In circular chart recorder
   (a) Concentric circles are ruled on the paper
   (b) The chart moves about an axis
   (c) Both *(a)* and *(b)* (d) none of the above.
6. Galvanometer constant is given as
   (a) KBL (b) NBL
   (c) NAB (d) B/r.

7. In galvanometer recorder
   (a) coil size should be large (b) coulomb friction should be large
   (c) Coulomb friction should be minimum
   (d) (a) and (c) (e) (a) and (b)
8. X–Y recorder
   (a) has maximum operating frequency of 6 Hz
   (b) Works for both ac and dc
   (c) has high operating frequency (d) none of the above.
9. Minimum working frequency of direct recorder is
   (a) 50 Hz (b) 100 Hz
   (c) dc (d) 1 kHz
10. Direct recorder works
    (a) On the principle of electromagnetic induction
    (b) requires low frequency (c) requires high frequency
    (d) both (a) and (c) (e) both (a) and (b)
11. The example of oscillographic recorder is
    (a) direct recorder (b) magnetic tape recorder
    (c) X-Y recorder (d) galvanometer recorder.
12. The working frequency of galvanometer recorder is about (a) 100 KHz (b) very low (c) 20 KHz (d) 1 MHz .
13. CRT recorder uses
    (a) Two axes (b) Four axes
    (c) three axes (d) none of the above.
14. In magnetic tape recorder the tape is made of
    (a) PVC only (b) Mylar only
    (c) any of (a) and (b) (d) none is correct.
15. The magnetic material used in magnetic tape recorder is
    (a) oxide layer (b) Fe_2O_3
    (c) Nickel (d) both (a) and (b).
16. Disk file and floppy disk use
    (a) Magnetic material for recording (b) paper for recording
    (c) oxide layer (d) (a) and (c) both.

ANSWERS

1. (b)	2. (d)	3. (b)	4. (d)	5. (c)	6. (c)
7. (d)	8. (a)	9. (a)	10. (d)	11. (d)	12. (c)
13. (b)	14. (c)	15. (d)	16. (d).		

□□□

SUBJECT REVIEW QUESTIONS

1. SCR is a .... layer .... terminal .... junction device.
2. The terminals of an SCR are called .... and ....
3. For switching operation anode is .... and gate is .... with respect to cathode.
4. The voltage drop across SCR in the ON state is very ....
5. The resistance of SCR in ON state is about .... ohm.
6. The resistance of SCR in OFF state is about about ....
7. In ON state the current through SCR is controlled by ....
8. For turning ON the SCR a .... voltage is applied at the gate terminal.
9. Thyristor current below which the SCR is turned OFF from ON state is called ....
10. Minimum current required to keep the SCR in ON state after the removal of gate current is called ....
11. Maximum forward voltage across SCR at which it is turned ON is called ....
12. Pulse triggering of the gate is preferred over continuous triggering as later leads to wastage of .... power.
13. DIAC is a .... terminal device.
14. DIAC can work for ....
15. Triac is a .... terminal .... layer .... junction device.
16. Triac can be triggered by applying .... voltage to the gate terminal.
17. Triac can work for ....
18. SCR can be turned OFF by .... methods.
19. For ac supply the thyristor is turned OFF by ....
20. For dc supply the SCR is turned OFF by ....
21. Snubber circuit is used to control ....
22. Heat sink is used to ....
23. Thermal resistance of heat sink has unit ....
24. Heat sink efficiency ....

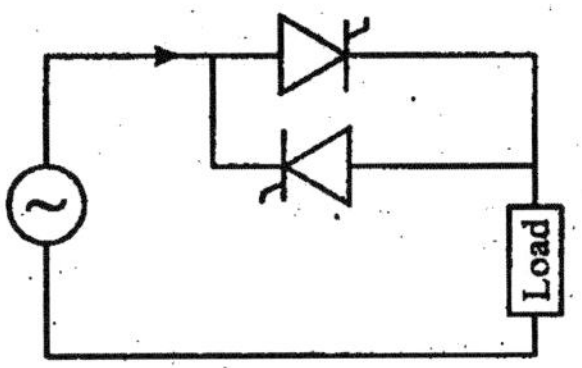

Fig. B.1.

25. AC phase control is used for .... inputs.
26. Voltage regulator is used to convert constant ac into .... ac.
27. In half wave voltage regulator .... thyristor and .... diode is used.
28. Circuit shown in Fig. B.1 is of ....
29. In full wave voltage regulator .... thyristors are used.
30. For charging a battery it is connected in .... with the charger.
31. When input voltage to an electric device reaches over voltage the SCR is turned ....
32. Output of controlled rectifier can be controlled by .... SCR.
33. In half wave controlled rectifier .... SCR is used.
34. Output of half wave controlled rectifier is given as ....
35. In full wave controlled rectifier .... SCRs are used.
36. In half-wave rectifier peak reverse voltage across diode is .... where V_m is peak value of input voltage.

37. In full wave centre tapped rectifier peak reverse voltage across diode .... where V_m is peak value of input voltage.
38. In full wave bridge controlled rectifier .... thyristors are used.
39. In full wave bridge controlled rectifier peak reverse voltage is .... where V_m is peak value of input voltage.
40. The average output voltage of a full wave controlled rectifier is .... where V_m is peak value of input voltage and α is firing angle.
41. In three phase half wave rectifier each diode conducts for .... degrees.
42. In three phase full wave rectifier each diode conducts in .... degrees.
43. In three phase half controlled bridge converter each thyristor conducts for .... degrees.
44. In three phase full wave controlled bridge rectifier the average load voltage is given by .... where V_m is peak value of input phase voltage and α is firing angle.
45. In ac phase control using R-firing circuit the firing angle can be changed from .... to .... degrees.
46. In ac phase control method .... input is required.
47. In ac phase control using R-C firing angle the firing angle can be varied from .... degree to .... degree.
48. UJT firing circuit is an example of .... triggering.
49. Triac can not be used for ....
50. Illumination of any lamp can be controlled by varying .... angle.
51. AC regulators can not be used to control the speed of .... motors.
52. Rectifiers can not be used to control the speed of .... motors.
53. Output of a step down chopper is given by V_o = ... where α is called duty cycle and V_{dc} is dc source voltage.
54. Voltage eqn. of dc motor is given by V = ... where E_b is called .... and R_a is called .... I_a is armature current.
55. Back e.m.f. of dc motor is given by .... where Z is ...., A is .... and *p* is ...., N is the speed of motor in r.p.m.
56. Synchronous speed in ac motor is given by N_s = ... where *f* is .... and *p* is no of ....
57. In induction motor the rotor speed is given by *n* = ...
58. In parallel inverter circuit commutating elements are in parallel ....
59. In series inverter circuit commutating elements are .... load.
60. Mc Murray bridge inverter uses .... commutation.
61. Mc Murray Bedford half bride inverter uses .... commutation.
62. 180° conduction scheme in three phase inverter is used particularly for .... loads.
63. 120° conduction scheme in three phase inverter is used for .... load.
64. In current source inverter current source is constructed by connecting a large inductor in .... with ....
65. For induction heating .... input at .... frequency is required.
66. Induction heating of magnetic material is due to .... and ....
67. Induction heating is same for both magnetic and non magnetic material about .... point.
68. For induction heating at higher power .... tubes are used.

69. Dielectric heating occurs at .... frequency.
70. Dielectric heating is more for material having .... dielectric constant.
71. Strain gauge is an example of .... transducer.
72. LVDT is an example of .... transducer.
73. Piezoelectric crystal is an example of .... transducer.
74. Solar cell is .... transducer.
75. Thermocouple is .... transducer.
76. When LVDT is used for pressure measurement it acts like a .... transducer.
77. Thermistor has .... temperature coefficient.
78. When capacitive transducer is used for displacement measurement, it acts like .... transducer.
79. Gauge factor for semiconductor strain gauge is ....
80. As the liquid level increases the capacitance of capacitive transducer ....
81. The output of LVDT is taken in .... mode.
82. In the middle position the output of an LVDT is ....
83. In solid state sensor as a result of rise in temperature the current ....
84. Thermocouple works on the basis of .... effect.
85. Thermocouple has two .... metals.
86. Certain materials when subjected mechanical stress develop electric potential between the two faces. This effect is called ....
87. Piezoelectric effect is observed in the crystal having ....
88. An inverse transducer is one which converts electrical energy into other energy. An example of inverse transducer is ....
89. A photodiode to measure light intensity is used in .... mode.
90. As the intensity of light increases, ... current through the photodiode increases.
91. Piezoelectric transducer can be used for .... as well as .... inputs.
92. The output of a digital transducer is in the form of pulses. The height of the pulse is proportional to ....
93. Temperature coefficient of resistance of a thermistor is given by ....
94. Voltage between input terminals of an ideal operational amplifier is .... and its input impedance is ....
95. CMRR of ideal operational amplifier is ....
96. DC output offset voltage of an ideal operational amplifier is ....
97. Drift is given in the rate of change of output voltage w.r.t. ....
98. In strip chart recorder the graph is plotted between measurand and ....
99. For better resolution .... tracing system should be used.
100. In X-Y recorder the graph is plotted between .... measurands.
101. Galvanometer recorder uses the principle of ....
102. In a CRT recorder the output of the tube is focussed on the recorder by ....
103. Magnetic tape recorder one .... head and one .... head.
104. A disk file is divided into .... to .... tracks.
105. A CRT recorder uses .... axes.

ANSWERS

1. four, three, three, 2. Anode, cathode, gate, 3. positive, positive, 4. small, 5. few ohms, 6. few magaohms, 7. load impedance, 8. positive, 9. holding current. 10. latching current, 11. breakover voltage, 12. more, 13. two, 14. dc and ac both, 15. three, four, three, 16. positive, 17. dc and ac both, 18. natural commutation and forced commutation, 19. natural commutation, 20. forced commutation, 21. high dv/dt rating, 22. dissipate heat to surrounding. 23. °C/watt, 24. $\frac{\text{output power}}{\text{power transferred to heat sink}}$, 25. ac 26. variable, 27. one, one, 28. full wave ac regulator, 29. two, 30. opposition, 31. ON, 32. changing firing angle, 33. one, 34. $V_{oav} = \frac{V_m}{2\pi}(1+\cos\alpha)$, 35. two, 36. V_m, 37. $2V_m$, 38. four, 39. V_m, 40. $\frac{2V_m}{\pi}\cos\alpha$. 41. 120, 42. 120, 43. 120, 44. $\frac{3\sqrt{3}V_m}{\pi}\cos\alpha$, 45. 3°, 77°, 46. ac, 47. 0°, 180°, 48. pulse, 49. rectification, 50. firing, 51. dc, 52. ac, 53. αV_{dc}, 54. $E_b + I_a R_a$, back e.m.f., armature resistance, 55. $\frac{N\phi PZ}{60A}$, no. of conductors, no. of parallel paths, no. of poles, 56. $\frac{120f}{p}$, frequency, no. of poles, 57. $n_s(1-s)$, 58. with transformer primary, 59. in series, 60. auxiliary, 61. complimentary, 62. star connected, 63. star as well as delta connected, 64. series, voltage source, 65. ac, high, 66. eddy current loss, hysteresis loss, 67. curie, 68. vacuum, 69. very high, 70. higher, 71. passive, 72. passive, 73. inverse, 74. active, 75. an active, 76. secondary, 77. negative, 78. primary, 79. 140, 80. increases, 81. differential, 82. non zero, 83. rises, 84. seeback, 85. dissimilar, 86. piezoelectric effect, 87. assymetry, 88. piezoelectric crystal, 89. reverse, 90. reverse saturation, 91. dc, ac, 92. measurand, 93. $-b/T^2$, 94. zero, ∞, 95. ∞, 96. zero, 97. time, 98. time, 99. rectangular, 100. two. 101. electromagnetic induction, 102. optical fiber plate, 103. record, reproduce. 104, 100, 1000, 105. four.

Solutions to Unsolved Numerical Problems

Chapter 1

Solution 1. The voltage safety factors for a thyristor is given by

$$V_f = \frac{V_{RSM} \text{ (or PIV)}}{\sqrt{2} \times \text{RMS value of input voltage}}$$

Under given conditions

$$V_f = 2.1,\ V_m = 70 = \sqrt{2} \times V_{rms}$$

Putting the values, we get

$$2.1 = \frac{\text{PIV}}{70}$$

or $$\text{PIV} = 70 \times 2.1 = 147.0 \text{ volts} \quad \textbf{Ans.}$$

Solution 2. Assuming linear variation of voltage and current, the average power dissipated has the value

$$P_{av} = \frac{1}{T}\int_0^T VI\,dt$$

$$= \frac{1}{T} VIT = VI$$

$$= 50 \times 3 = 150 \text{ watt.}$$

The curve showing the switching loss in shown below:

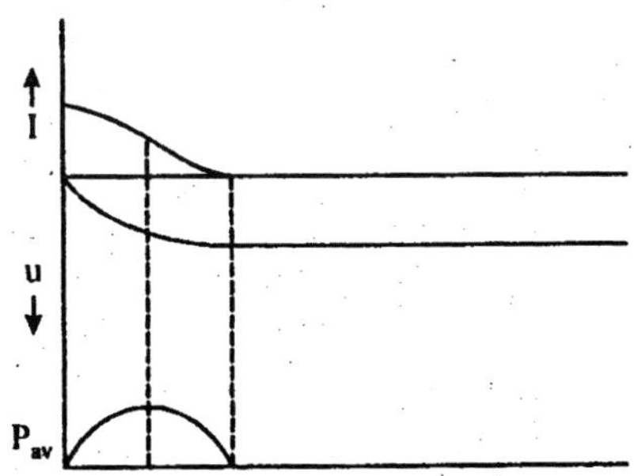

Solution 3. Using the relationship

$$RC \geq \frac{4}{\omega}, \text{ we get}$$

$$\omega \leq \frac{RC}{4}$$

The maximum value of ω is

$$\omega = \frac{RC}{4}; \; R = 6\Omega \text{ and } C = 10 \times 10^{-6} \text{ F; giving}$$

$$\omega = \frac{6 \times 10 \times 10^{-6}}{4} \text{ rad/sec}$$

Time period,

$$T = \frac{2\pi}{\omega}$$

$$= \frac{2\pi \times 4}{6 \times 10^{-5}} \text{ sec}$$

$$= 4.19 \times 10^{5} \text{ sec.} \quad \textbf{Ans.}$$

Solution 4. It is given that

$$v_s = 2 \text{ volts}$$

$$v_g = 0.7 \text{ volts}$$

$$v_{01} = 0.7 \text{ volts}$$

$$I_{g(min)} = 6 \text{ mA}$$

$$R = ?$$

Using the relation

$$R = \frac{v_s - v_g - v_{01}}{I_g \text{ (min)}}; \text{ we get}$$

$$R = \frac{2 - 1.4}{6 \times 10^{-3}} \Omega$$

$$= 0.1 \times 10^{+3} \ \Omega$$

$$= 100 \ \Omega \quad \textbf{Ans.}$$

Solution 5. We are given:

$$V = 100 \text{ volts}$$

$$L = 20 \text{ mH}$$

$$C = 47 \ \mu F$$

The peak value of capacitor current is given by

$$I_{cm} = V\sqrt{\frac{C}{L}} \text{ Amp.}$$

$$= 100\sqrt{\frac{47 \times 10^{-6}}{20 \times 10^{-3}}} \text{ Amp}$$

$$= 4.84 \text{ Amp.} \quad \textbf{Ans.}$$

Chapter 2

Solution 1. The heat dispated by the junction to the ambient is given by the equation

$$P_{av} = \frac{T_j - T_a}{\theta_{ja}}$$

$$= \frac{T_s - T_a}{\theta_{sa}} = \frac{T_j - T_s}{\theta_{js}}$$

From the last relationship, we get

$$3 = \frac{100 - 50}{\theta_{js}}; \text{ giving}$$

$$\theta_{js} = \frac{50}{3} = 16.67 \text{ °C/watt} \quad \textbf{Ans.}$$

Solution 2. Assuming ambient temperature of 23°C, we can write the total resistance as

$$\theta_{ja} = \frac{T_j - T_a}{P_{av}}$$

$$= \frac{100 - 23}{3} \text{ °C/watt}$$

$$= \frac{77}{3} \approx 26° \text{C/watt} \quad \textbf{Ans.}$$

Solution 3. The heat sink efficiency is given as

$$\eta H = \frac{T_s - T_A}{\theta_{sA} P_{av}}$$

In the present problem

$$T_s = 40° \text{ C}, T_A = 23° \text{ C}, \theta_{sA} = 35° \text{ C/watt}$$

P_{av} = 10 watt, ηH = ? Putting the values ; we get

$$\eta H = \frac{40 - 23}{35 \times 10} \times 100\%$$

$$= \frac{170}{35}\% \approx 5\% \quad \textbf{Ans.}$$

Solution 4. The heat transfer coefficient is given as

$$h = 0.15 \text{ (V/L)}^{0.5} \text{ w/m}^2$$

Here V = 5 m/sec, L = 22 cm giving;

$$h = 0.15\left(\frac{5}{22 \times 10.2}\right)^{0.5} \text{ w/m}^2$$

$$= 0.72 \text{ w/m}^2 \quad \textbf{Ans.}$$

Chapter 3

Solution 1. The phase angle between the load voltage and load current for R-L load is given as

$$\tan\phi = \frac{\omega L}{R}$$

Here $\omega = 2\pi \times 50 = 100\pi$ rad/sec

$L = 20$ mH; $R = 10\Omega$. Hence

$$\tan\phi = \frac{100\pi \times 20 \times 10^{-3}}{10} = 200\,\pi \times 10^{-3}$$

or $\phi = \tan^{-1}(200\,\pi \times 10^{-3}) = 32.13°$ **Ans.**

Solution 2. Refer to Fig. 3.6. The 6 volt battery supplies the power to the bulb through the path battery-bulb-SCR-transformer secondary-battery. Assuming no loss through transformer secondary and SCR, the current supplied by the battery is divided into two parts-one through the bulb and two through 500 Ω resistance. The total resistance in the circuit is:

$$R = R_b \parallel (R_g + 500);\ R_b \text{ is bulb resistance}$$

$$= 5 \parallel (500 + R_g)$$

$$= \frac{5 \times (500 + R_g)\Omega}{505 + R_g} \approx 5\Omega$$

Thus the total current $I = I_b + I_g = \frac{V}{R} = \frac{6}{5} = 1.2$ Amp.

The current through the bulb

$$I_b = I - I_g$$

$$= 1.2 - 0.003 \text{ amp} = 1.197 \text{ amp}$$

(*a*) Heat dissipated by the bulb

$$= (1.197)^2 \times 5 \text{ watt} = 7.16 \text{ watts} \quad \textbf{Ans.}$$

(*b*) Total resistance in the gate circuit, by definition is:

$$R_g' = \frac{V}{I_g} = 500 + R_g$$

$$R_g' = \frac{6}{3 \times 10^{-3}} = 2 \times 10^3\,\Omega \quad \textbf{Ans.}$$

Solution 3. The maximum current flows when the firing angle has zero value.

The average current is given by

$$I_{av} = \frac{V_{av}}{R} = \frac{V_m}{2\pi R}(1 + \cos\alpha),\ \alpha = 0$$

$$= \frac{V_m}{\pi R}$$

Here
$$V_m = 15\sqrt{2} \text{ volts}$$
$$R = 3\Omega$$

Hence
$$I_{av} = \frac{15\sqrt{2}}{3} = 3\sqrt{2} \text{ amp}$$
$$= 4.2 \text{ amp} \quad \textbf{Ans.}$$

Solution 4. The time delay is given as

$$T = RC\ln\left(\frac{1}{1-\eta}\right)$$

Here
$$R = R_1 + R_2 = 5 + 10 = 15\Omega$$
$$C = 0.47 \ \mu F$$
$$\eta = 0.6 \text{ giving}$$

$$T = 15 \times 0.47 \times 10^{-6} \ln\left(\frac{1}{1-0.6}\right)$$
$$= 7.05 \times 10^{-6} \ln(2.5)$$
$$= 6.46 \times 10^{-6} \text{ sec.} \quad \textbf{Ans.}$$

Chapter 4

Solution 1. (*a*) The average load voltage is given as

$$V_{oav} = \frac{V_m}{\pi}(1 + \cos\alpha)$$
$$V_m = 220\sqrt{2} \text{ volts, giving}$$
$$V_{oav} = \frac{220\sqrt{2}}{\pi}(1 + \cos 20°)$$
$$= 191.62 \text{ volts} \quad \textbf{Ans.}$$

(*b*) Average load power $= V_{oav} \times I_{oav}$
$$= V_{oav}^2/R, \; R = 50 \ \Omega$$
$$= 734.36 \text{ watts} \quad \textbf{Ans.}$$

Solution. 2. (*a*) The average load voltage across R-L load in continuous conduction mode is given by

$$V_{oav} = \frac{2V_m}{\pi}\cos\alpha$$
$$= \frac{2 \times 220\sqrt{2}}{\pi} \cos 20°$$
$$= 185.66 \text{ volts} \quad \textbf{Ans.}$$

(*b*) The average output power to the load has the value

$$P_{oav} = V_{oav} \cdot I_{oav}$$
$$= V_{oav} \times V_{oav}/z$$
$$= \frac{V_{oav}^2}{z}; z = \sqrt{R^2 + (\omega L)^2}$$
$$z = \sqrt{(40)^2 + (100\pi \times 20 \times 10^{-3})^2}$$
$$= 74.46$$

Hence $P_{oav} = 493.13$ watts.

Solution 3. The average load voltage for R-L load in discontinuous conduction mode is given by

$$V_{oav} = \frac{V_m}{\pi}(\cos\alpha - \cos\beta)$$
$$= \frac{220\sqrt{2}}{\pi}[\cos 30^\circ - \cos(180 + 20^\circ)]$$
$$= \frac{220\sqrt{2}}{\pi}\left[\frac{\sqrt{3}}{2} + \cos 20^\circ\right]$$
$$= 178.29 \text{ volts.} \quad \textbf{Ans.}$$

Average load current

$$I_{oav} = \frac{V_{oav}}{z}; z = \sqrt{R^2 + \omega^2 L^2}$$
$$z = \sqrt{50^2 + (10 \times 10^{-3} \times 100\pi)^2} = 50.09$$

Hence $I_{oav} = 3.56$ **Ans.**

The average output power

$$P_{oav} = V_{oav} \cdot I_{oav}$$
$$= 178.29 \cdot 3.56$$
$$= 634.71 \quad \textbf{Ans.}$$

Solution 4. Given $v = V_m \sin \omega t$

$$= 200 \sin 100\, t$$

Here $\omega = 100$ rad/sec

The average load voltage is given as

$$V_{oav} = \frac{V_m}{\pi}(\cos\alpha - \cos\beta)$$
$$= \frac{2V_m}{\pi}\cos\alpha; \ \alpha = 35^\circ$$
$$= 104.35 \text{ volts}$$

The average load current is given as

$$I_{oav} = \frac{V_{oav}}{z} = \frac{V_{oav}}{\sqrt{R^2 + 100^2 L^2}}$$

$$L^2 = \frac{\left(\frac{V_{oav}^2}{I_{oav}^2} - R^2\right)}{100}; \; I_{oav} = 9 \text{ amp}$$

$$= 0.34$$

$$L = (0.34)^{0.5} = 0.59 \text{ henry} \quad \textbf{Ans.}$$

Solution 5. The rms load voltage in given by

$$V_{orms} = \frac{V_m}{\sqrt{2\pi}}\left[\beta - \alpha - \frac{1}{2}\{\sin 2\beta - \sin 2\alpha\}\right]^{0.5}$$

For purely resistive load $\beta = \pi$, giving

$$V_{orms} = \frac{V_m}{\sqrt{2\pi}}\left[\pi - \alpha + \frac{1}{2}\sin 2\alpha\right]^{0.5}$$

Here $\alpha = 45°$

$V_m = 200$ volts

Therefore

$$V_{orms} = \frac{200}{\sqrt{2\pi}}\left[\frac{3\pi}{4} + \frac{1}{2}\sin 90°\right]^{0.5}$$

$$= \frac{200}{\sqrt{2\pi}}\left[\frac{3\pi}{4} + \frac{1}{2}\right]^{0.5} = 134.85 \text{ volts} \quad \textbf{Ans.}$$

$$V_{oav} = \frac{V_m}{\pi}(1 + \cos\alpha)$$

$$= \frac{200}{\pi}\left(1 + \frac{1}{\sqrt{2}}\right) = 108.73 \text{ volts} \quad \textbf{Ans.}$$

$$\text{Form factor} = \frac{V_{orms}}{V_{oav}}$$

$$= \frac{134.85}{108.73} = 1.24 \quad \textbf{Ans.}$$

Solution 6. (*a*) From equation (16)

$$V_{oav} = \frac{V_m}{\pi}(1 + \cos\alpha_1) + \frac{V_m}{\pi}(1 + \cos\alpha_2)$$

where $\alpha_1 = 30°$; $\alpha_2 = 45°$. Therefore,

$$V_{oav} = \frac{V_m}{\pi}[(1 + \cos 30°) + (1 + \cos 45°)]$$

Given $V_m = 300$ volts. Hence

$$V_{oav} = \frac{300}{\pi}\left[2+\frac{\sqrt{3}}{2}+\frac{1}{\sqrt{2}}\right]$$

$$= 341.27 \quad \textbf{Ans.}$$

(*b*) Load current $I_{oav} = \frac{V_{oav}}{R}; \quad R = 40\ \Omega$

$$= \frac{341.27}{40} = 8.53 \text{ Amp} \quad \textbf{Ans.}$$

(*c*) Average O/P power

$$P_{oav} = V_{oav} \cdot I_{oav}$$

$$= 2911.6 \text{ watts} \quad \textbf{Ans.}$$

Solution 7. (*a*) The average load voltage for R-L load has the value

$$V_{oav} = \frac{V_m}{\pi}[(\cos\alpha_1 + 1) + (\cos\alpha_2 + 1)]$$

$$= \frac{V_m}{\pi}[2 + \cos\alpha_1 + \cos\alpha_2]$$

In the present case

$$v = 300 \sin 200\, t$$

Hence

$$V_{oav} = \frac{V_m}{\pi}[\cos 30^\circ + 1^\circ + \cos 45^\circ + 1]$$

$$= \frac{300}{\pi}\left[2+\frac{\sqrt{3}}{2}+\frac{1}{\sqrt{2}}\right] \text{ volts}$$

$$= 341.27 \text{ volts} \quad \textbf{Ans.}$$

(*b*) Average load current

$$I_{oav} = \frac{V_{oav}}{z}$$

Here $z = \sqrt{R^2 + (\omega L)^2}$

$$= \sqrt{(50)^2 + (25\times 10^{-3})^2 \times (200)^2}$$

$$= 50.25$$

Hence $I_{oav} = 6.79$ amp **Ans.**

(*c*) Average O/P power

$$P_{oav} = V_{oav} \times I_{oav}$$

$$= 2317.22 \text{ watts} \quad \textbf{Ans.}$$

Solution 8. The average load voltage three phase rectifier is given by

$$V_{oav} = \frac{3\sqrt{3}V_m}{\pi}\cos\alpha$$

Given $v = 230 \sin 100\, t;\ \alpha = 34.6°$

Therefore $$V_{oav} = \frac{3\sqrt{3} \times 230}{\pi} \cos 34.6°$$
$$= 312.92 \text{ volts} \quad \textbf{Ans.}$$

Solution 9. The average load voltage for R-L load is

$$V_{oav} = \frac{3\sqrt{3}V_m}{\pi} \cos\alpha\,;\ \alpha = 34.6°$$
$$= 312.92$$

Average load current

$$I_{oav} = \frac{V_{oav}}{z}$$
$$z = \sqrt{(100)^2 + (100 \times 10 \times 10^{-3})^2}$$
$$= 100\ \Omega$$

Therefore

$$I_{oav} = \frac{312.92}{100} = 3.13 \text{ amp} \quad \textbf{Ans.}$$

Average output power

$$P_{oav} = V_{oav} \cdot I_{oav}$$
$$= 312.92 \times 3.13$$
$$= 979.44 \text{ watts} \quad \textbf{Ans.}$$

Solution 10. The line to line voltage = 415 volts

$\therefore$ The phase voltage $= \dfrac{415}{\sqrt{3}} = 239.88$ volts

Peak value of phase voltage $= \sqrt{2} \times \dfrac{415}{\sqrt{3}}$ volts $= 338.24$ volts

The average value of load current $I_{oav} = 100$ amp.

(*a*) Average output power

$$P_{oav} = V_{oav} \times I_{oav}$$

Here $$V_{oav} = \frac{3\sqrt{3}V_m}{\pi} \cos\alpha\,;\ \alpha = 35°$$
$$= 457.96 \text{ volts.}$$

Hence $$P_{oav} = V_{oav} \times I_{oav}$$
$$= 45796 \text{ watts} \quad \textbf{Ans.}$$

(*b*) Since the current through a thyristor flows for $\frac{1}{3}$ period of cycle. Hence average current through the SCR

$$I_{TAV} = \frac{I_{oav}}{3}$$

$$= \frac{100}{3} = 33.3 \text{ amp} \quad \textbf{Ans.}$$

The rms current-through the thyristor

$$= \frac{\text{rms load current}}{3}$$

$$= \frac{100\sqrt{2}}{\sqrt{6}} = \frac{100}{\sqrt{3}} = 57.80 \text{ amp.} \quad \textbf{Ans.}$$

(*c*) Peak reverse voltage through the thyristor in the same as peak line so line voltage. Here

$$V_{P\,(line)} = 415\sqrt{2}$$

Hence $$V_{T\,(peak)} = 415\sqrt{2} = 585.15 \text{ volts.} \quad \textbf{Ans.}$$

Solution 11. For 3ϕ rectifier $\alpha = 0°$. By putting $\alpha = 0$ in above question, the quantities can be computed.

Solution 12. Average load voltage for a full wave (diode) rectifier is given as

$$V_{oav} = \frac{2V_m}{\pi}$$

The input voltage in

$$v = 100 \sin \omega t; \; V_m = 100 \text{ volts}$$

$$\therefore \quad V_{oav} = \frac{100 \times 2}{\pi} \text{ volts} = 63.7 \text{ volts}$$

$$z = \sqrt{R^2 + \omega^2 L^2}$$

$$= \sqrt{10^2 + (25 \times 10^{-3} \times \omega)^2}$$

Average load current

$$I_{oav} = \frac{V_{oav}}{z} = 6 \text{ amp (given)}$$

or $$z = \frac{V_{oav}}{I_{oav}} = \frac{63.7}{6} = 10.6\ \Omega$$

or $$100 + (2.5 \times 10^{-3} \times \omega)^2 = (10.6)^2$$

giving;

$$\omega^2 v^2 = 12.36$$

$$\omega = \sqrt{\frac{12.36}{L^2}}$$

$$= 1.4 \times 10^5 \text{ rad/sec} \quad \textbf{Ans.}$$

Chapter 5

Solution 1. In R-C firing circuit, the device turns on when the capacitor voltage attains the value of 1.4 volts. Thus

$$1.4 = V_m e^{-\frac{t}{RC}}$$

or $$= 100\sqrt{2}e^{-\frac{t}{RC}}$$

Taking natural logarithm on both sides

$$\ln(1.4) = \ln(100\sqrt{2}) - \frac{t}{RC}$$

or $$\frac{t}{RC} = \ln(100\sqrt{2}) - \ln(1.4)$$

or $$t = RC[\ln(100\sqrt{2}) - \ln(1.4)]$$

Given $$R = 100\ \Omega;\ C = 47\ \mu F$$

$\therefore$ $$t = 100 \times 15 \times 10^{-6}\ [\ln(100\sqrt{2}) - \ln(1.4)]$$

$$= 6.9 \times 10^{-3}\ \text{sec}$$

Hence the firing angle $$\alpha = \omega t$$

$$= 100 \times 180 \times 6.9 \times 10^{-3} - \frac{\pi}{2}\ \text{degree}$$

$$= 124.2^\circ - 90^\circ = 34.2^\circ.$$

$\therefore$ The average load voltage $$= \frac{V_m}{2\pi}(1 + \cos\alpha)$$

$$= \frac{100\sqrt{2}}{2\pi}(1 + \cos\alpha)$$

$$= \frac{141}{6.28}(1 + \cos 34.2^\circ) = 41.02\ \text{volts.}$$ **Ans.**

Solution 2. From the figure the expression for the duration of pulse is given by

$$T = RC\ln\left(\frac{1}{1-\eta}\right)$$

$$R = 100\ \Omega;\ C = 10\ \mu F = 10 \times 10^{-6}\ F$$

$\therefore$ $$T = 100 \times 10 \times 10^{-6} \ln\left(\frac{1}{1-0.5}\right)$$

$$= 10^{-3} \ln 2 = 0.693 \times 10^{-3}$$

$\therefore$ The firing angle $$= \omega T$$

$$= 100\ \pi T = 100 \times 180 \times 0.693 \times 10^{-3}\ \text{degree}$$

$$= 12.48^\circ$$

The rms value of load voltage is given as

$$V_{orms} = V_m\left[\frac{\pi-\alpha}{2\pi}+\frac{\sin 2\alpha}{4\pi}\right]^{0.5}$$

$$V_m = 70\sqrt{2} = 98.99 \text{ volts}$$

$$V_{orms} = 98.99\left[\frac{180-12.48}{2\times 180}+\frac{\sin 2\times 12.48}{4\times 3.14}\right]^{0.5}$$

$$= 98.99 \times [0.4989]^{0.5}$$

$$= 69.92 \text{ volts} \quad \textbf{Ans.}$$

Solution 3. The average output power supplied by a half wave AC regulator to the resistive load is given by

$$P_{oav} = V_{orms} \times I_{orms}$$

Here

$$V_{orms} = \frac{V_m}{2\sqrt{\pi}}\left[2\pi-\alpha+\frac{1}{2}\sin 2\alpha\right]^{0.5}$$

$$= \frac{100\sqrt{2}}{2\sqrt{\pi}}\left[2\pi-\frac{\pi}{6}+\frac{1}{2}\sin 60^{\circ}\right]^{0.5}$$

$$= \frac{107}{\sqrt{\pi}}\left[\frac{11\pi}{6}+\frac{\sqrt{3}}{4}\right]^{0.5}$$

$$= 99.26 \text{ volts.}$$

$$P_{oav} = V^2_{orms}/R \;;\; R = 10\,\Omega$$

$$= 985.25 \text{ watts} \quad \textbf{Ans.}$$

Solution 4. The rotor speed $\alpha\ (V_{orms})^2$

$$V_{orms}\ \frac{V_m}{\sqrt{2\pi}}\left[\pi-\alpha+\frac{\sin\alpha}{2}\right]^{0.5}$$

First case

$$N_1\alpha\frac{V_m^2}{2\pi}\left[\pi-\alpha_1+\frac{\sin 2\alpha_1}{2}\right]$$

Second case

$$N_2\alpha\frac{V_m^2}{2\pi}\left[\pi-\alpha_2+\frac{\sin 2\alpha_2}{2}\right]$$

$$\frac{N_1}{N_2} = \left[\frac{\pi-\alpha_1+\sin 2\alpha_1/2}{\pi-\alpha_2+\sin 2\alpha_2/2}\right] \quad \textbf{Ans.}$$

Chapter 6

Solution 1. (*a*) The speed of the motor is given as

$$N = \frac{V - I_a R_a}{K}$$

or $$N \; \alpha \; V - I_a R_a$$

Here $V = \alpha V_{dc}$; α is duty cycle

In first case

$$1000 \; \alpha \; 250 - 40 \times 0.5 \quad \text{...}(i)$$

In IInd case $$700 \; \alpha \; \alpha \times \; 250 - 40 \times 0.5 \quad \text{...}(ii)$$

dividing (*ii*) by (*i*); we get;

$$\frac{700}{1000} = \frac{250\alpha - 20}{230}$$

or $$250\alpha - 20 = 161$$

or $$\alpha = \frac{181}{250} = 0.724 \quad \textbf{Ans.}$$

(*b*) The motor torque is given by

$$T = K\phi I_a$$

At 1000 rpm the rator torque has the relationship with speed as

$$N = \frac{(V - I_a R_a)I_a}{K'T}$$

or $$T = \frac{(V - I_a R_a)I_a}{N}$$

Under given condition

$$T = \frac{(250 - 40 \times 0.5)40}{1000} \quad \text{...}(i)$$

$$\frac{1}{2}T = \frac{(250\alpha - 40 \times 0.5)40}{700} \quad \text{...}(ii)$$

Dividing equation (*i*) by (*ii*) ; we get

$$2 = \frac{230/1000}{(250\alpha - 30)/700}$$

$$= \frac{230 \times 0.7}{250\alpha - 30}$$

or $$500\alpha - 60 = 161.0$$

or $$\alpha = \frac{221}{500} = 0.442 \quad \textbf{Ans.}$$

Solution 2. Duty cycle is $\dfrac{8}{8+16} = \dfrac{1}{3}$

$$V_{orms} = \sqrt{\alpha}\ V_{dc}$$

$$= \sqrt{\frac{1}{3}} \times 230$$

$$= 132.79 \text{ volts} \quad \textbf{Ans.}$$

Solution 3. (*a*) The rms load voltage has the value

$$V_{orms} = \frac{V_m}{2\sqrt{\pi}}\left[2\pi - \alpha + \frac{\sin 2\alpha}{2}\right]^{0.5}$$

$$V_{orms} = \frac{230\sqrt{2}}{2\sqrt{\pi}}\left[2\pi - \frac{\pi}{6} + \frac{\sin 60}{2}\right]^{0.5}$$

$$= 129.409\left[\frac{11 \times 3.14}{6} + \frac{\sqrt{3}}{4}\right]^{0.5}$$

$$= 228.33 \text{ volts}$$

$$P_{oav} = \frac{V_{orms}^2}{R}; \ R = 40\ \Omega$$

$$= 1303.32 \text{ watts} \quad \textbf{Ans.}$$

(*b*)

$$V_{orms} = 129.409\left[\frac{5\pi}{3} + \frac{\sin 120}{2}\right]^{0.5}$$

$$= 218.47$$

$$P_{oav} = \frac{V_{orms}^2}{R}; \ R = 40\ \Omega$$

$$= 1193.25 \text{ watts} \quad \textbf{Ans.}$$

Solution 4. In full wave ac regulator

$$V_{orms} = \frac{V_m}{\sqrt{2\pi}}\left[\pi - \alpha + \frac{\sin \alpha}{2}\right]^{0.5}$$

Putting the values we can compute the values of power dissipated.

Solution 5. The temp depends upon duty cycle as

$$T \ \alpha \ \sqrt{\alpha}$$

In the question

$$T_1 \ \alpha \ \sqrt{0.70}$$

$$T_2 \ \alpha \ \sqrt{0.80}$$

$$\frac{T_1}{T_2} = \sqrt{\frac{0.70}{0.80}} = 0.935 \quad \textbf{Ans.}$$

Solution 6. The O/P of half wave converter is given as

$$V_{oav} = \frac{V_m}{2\pi}(1+\cos\alpha_a)$$

$$= \frac{230}{2\pi}(1+\cos 30^\circ)$$

The motor speed

$$N = V - I_a R_a$$

or

$$700 = \frac{\frac{230}{2\pi}(1+\cos 30^\circ) - 10\times 0.4}{K}$$

$$N = \frac{\frac{230}{2\pi}(1+\cos 150^\circ) - 10\times 0.4}{K}$$

$$\frac{N}{700} = \frac{\frac{230}{2\pi}(1+\cos 150^\circ) - 4}{\frac{230}{2\pi}(1+\cos 30^\circ) - 4} = \frac{1.493}{64.34}$$

or

$$N = 16.24 \text{ rpm} \quad \textbf{Ans.}$$

Solution 7. We are given

No. of conductors, $Z = 36$

No.of poles, $P = 4$

No. of parallel paths, $A = 2$ for wave winding

Flux $\phi = 100$ mWb

Speed $N = 600$ rpm.

The back e.m.f of the motor

$$E_b = \frac{N\phi Pz}{60A}$$

$$= \frac{600\times 4\times 36\times 100\times 10^{-3}}{60\times 2} = 72 \text{ volts}$$

The converter output voltage

$$V_{oav} = \frac{3\sqrt{3}\,V_m}{\pi}\cos\alpha$$

$$= \frac{3\sqrt{3}\times 230}{\pi}\cos 30^\circ$$

$$= 329.2 \text{ volts.}$$

Using the relation

$$E_b = V - I_a R_a$$

or $$I_a = \frac{V - E_b}{R_a} \text{ gives}$$

$$I_a = \frac{329.2 - 72}{5} \;;\; R_a = 0.5\ \Omega$$

$$= \frac{257.2}{5}$$

$$= 51.44 \text{ amp.} \quad \textbf{Ans.}$$

Solution 8. The rms O/P voltage of half wave AC regulator is given by

$$V_{orms} = \frac{V_m}{2\sqrt{\pi}}\left[2\pi - \alpha + \frac{\sin 2\alpha}{2}\right]^{0.5} \quad ...(i)$$

The speed of the motor

$$N \; \alpha \; V_{orms}^2$$

In first case

$$N_1 = 600 \text{ rpm}, \; \alpha = 30°$$

In second case

$$N_2 = ? \; \alpha = 75°$$

From equation (*i*)

$$\frac{N_1}{N_2} = \frac{2\pi - \alpha_1 + \frac{\sin 2\alpha_1}{2}}{2\pi - \alpha_2 + \frac{\sin 2\alpha_2}{2}}$$

or $$\frac{600}{N_2} = \frac{2\pi - \frac{\pi}{6} + \frac{\sin 60}{2}}{2\pi - \frac{5}{12}\pi + \frac{\sin 150}{2}}$$

or $$N_2 = \frac{600\left(\frac{19}{12}\pi + \frac{\sin 150}{2}\right)}{\left(\frac{11}{6}\pi + \frac{\sin 60°}{2}\right)}$$

$$= \frac{600(5.22)}{6.18}$$

$$= 506 \text{ rpm} \quad \textbf{Ans.}$$

Chapter 7

Solution 1. (*a*) The O/P voltage of half bridge inverter is given by

$$V_0 = \sum_{i=1,3,5...}^{\infty} \frac{2V_s}{n\pi} \sin n\omega t$$

In the present case $V_s = 50$ volts

The fundamental component of O/P voltage is

$$V_{01} = \frac{2V_s}{\pi} \sin \omega t$$

$$V_{01\ rms} = \frac{2V_s}{\pi\sqrt{2}} = \frac{2 \times 50}{3.14 \times \sqrt{2}}$$

$$= 22.45 \text{ volts} \quad \textbf{Ans.}$$

(*b*) The rms value of output voltage is given as

$$V_{orms} = \sqrt{\frac{1}{T}\int_0^T V_0^2\, dt}$$

$$= \sqrt{\frac{2}{T}\int_0^{T/2} V_0^2\, dt}$$

$$= \sqrt{\frac{V_0^2 \frac{T}{2}}{\frac{2}{T}}} = \sqrt{V_0^2} = V_0 = \frac{V_s}{2}$$

$$= \frac{50}{2} = 25 \text{ volts.}$$

Hence average O/P power

$$P_{oav} = \frac{V_{orms}^2}{R};\ R = 5\Omega$$

$$= \frac{25^2}{5} = 125 \text{ watts} \quad \textbf{Ans.}$$

(*c*) Peak value of thyristor current

$$I_{T\ (peak)} = \text{average current}$$

$$= \frac{V_{orms}}{R} = \frac{25}{5} = 5 \text{ amp} \quad \textbf{Ans.}$$

The average value of thyristor current. The SCR conducts for 50% of the duration. Hence its average value

$$I_{T(av)} = \frac{I_{T(peak)}}{2} = \frac{5}{2} = 2.5 \text{ amp} \quad \textbf{Ans.}$$

Solution 2. (*a*) The O/P voltage of a full bridge inverter is given as

$$V_0 = \sum_{n=1,3,5...}^{\infty} \frac{4V_s}{n\pi} \sin n\omega t$$

The rms value of O/P voltage is

$$V_{orms} = \sqrt{\frac{1}{T}\int_0^T V_0^2\,dt}$$

$$= \sqrt{\frac{2}{T}\int_0^{T/2} V_s^2\,dt}$$

$$= V_s = 50 \text{ volts}$$

rms value of load current

$$I_{orms} = V_{orms}/R;\ R = 5\Omega$$

$$= \frac{50}{5} = 10 \text{ Amp.} \quad \textbf{Ans.}$$

(*b*) The output power

$$P_{oav} = V_{oav} \times I_{oav}$$

$$= 50 \times 10 = 500 \text{ watts} \quad \textbf{Ans.}$$

(*c*) Peak thyristor current = rms load current

$$I_{T(peak)} = 10 \text{ amp}$$

An SCR conducts for 50% duration. Hence

$$I_{T(rms)} = \frac{10}{2} = 5 \text{ amp} \quad \textbf{Ans.}$$

Solution 3. (*a*) The output voltage of full bridge invertor is given as

$$V_0 = \sum_{n=1,3,5...}^{\infty} \frac{4V_s}{n\pi} \sin n\omega t$$

The output current will have the value then

$$I_0 = \sum_{n=1,3,5...}^{\infty} \frac{4V_s}{n\pi[R^2 + (n\omega L)^2]^{1/2}} \sin n\omega t$$

The fundamental current component

$$I_{01} = \frac{4V_s}{\pi(R^2 + \omega^2 L^2)^{0.5}} \sin \omega t$$

The peak value of fundament current

$$I_{01(peak)} = \frac{4V_s}{\pi(R^2 + \omega^2 L^2)^{0.5}}$$

Here

$$V_s = 200 \text{ volts}, \omega = 2\pi \times 60, R = 10\Omega, L = 20\text{mH}$$

Hence

$$I_{01(peak)} = \frac{4 \times 200}{3.14[10^2 + (120\pi \times 20 \times 10^{-3})^2]^{0.5}}$$

$$= 20.35 \text{ amp}$$

rms value

$$I_{01(rms)} = \frac{20.35}{\sqrt{2}}$$

$$= 14.39 \text{ amp.} \quad \textbf{Ans.}$$

(*b*) The total load current

$$I_0 = \sqrt{I_{01}^2 + I_{03}^2 + I_{05}^2 + \ldots}$$

$$I_{03(peak)} = \frac{4V_s}{\pi\sqrt{(R^2 + 9\omega^2 L^2)}}$$

$$= \frac{4 \times 200}{3.14[10^2 + 9(120\pi + 20 \times 10^{-3})^2]^{0.5}}$$

$$= 0.23 \text{ amp}$$

$$I_{05(peak)} = \frac{4 \times 200}{3.14[10^2 + 25(120\pi + 20 \times 10^{-3})^2]^{0.5}}$$

$$= 0.135 \simeq 0.14$$

$$I_{0(peak)} = \sqrt{I_{01(peak)}^2 + I_{03(peak)}^2 + I_{05(peak)}^2}$$

$$= 20.35 \text{ amp.} \quad \textbf{Ans.}$$

Solution 4. In 120° conduction scheme no phase current flows through the load during $\frac{1}{3}$ of the cycle. The average current is given as

$$I_0 = \frac{V}{2R}; \text{ since two phases remain connected in series}$$

$$= \frac{150}{2 \times 10} = 7.5 \text{ amp}$$

The rms value of the current is given by

$$I_{orms} = \sqrt{\frac{2}{2\pi}\int_0^{2\pi/3} I_0^2 d(\omega t)}$$

$$= \left[\frac{1}{\pi}(7.5)^2\left(\frac{2\pi}{3}\right)\right]^{0.5} = 6.12 \text{ amp} \quad \textbf{Ans.}$$

The output power has the value

$$P_{oav} = I^2_{orms} \times R \times 3 \text{ (for three phase)}$$
$$= (6.12)^2 \times 10 \times 3$$
$$= 1123.63 \text{ watts} \quad \textbf{Ans.}$$

Peak thyristor current has the same value as average load current. Hence

$$I_T(\text{peak}) = 7.5 \text{ amps} \quad \textbf{Ans.}$$

Each thyristor conduct for 120°. Hence its average current will be given as

$$I_T(av) = \frac{1}{2\pi}\int_0^{\frac{2\pi}{3}} I_0 d(\omega t)$$

$$= \frac{1}{2\pi}(7.5) \times \frac{2\pi}{3}$$

$$= 2.5 \text{ amps}$$

r.m.s. thristor current

$$I_{T(rms)} = \sqrt{\frac{1}{2\pi}\int_0^{\frac{2\pi}{3}} I_0^2 d(\omega t)}$$

$$= \left[\frac{1}{2\pi}(7.5)^2 \frac{2\pi}{3}\right]^{0.5}$$

$$= 4.33 \text{ amp} \quad \textbf{Ans.}$$

Solution 5. *Do this problem yourself.*

Solution 6. Given R = 60 Ω, L = 6mH, C = 1μF. The output frequency is maximum when turn off time between the thyristors is zero. It is given by

$$W_d = \sqrt{\frac{1}{LC} - \frac{R^2}{4L^2}}$$

$$= \sqrt{\frac{1}{6\times10^{-3}\times1\times10^{-6}} - \frac{60^2}{4\times(6\times10^{-3})^2}}$$

$$= 11982.89 \text{ rad/sec}$$

$$\approx 12 \text{ k rad/sec} \quad \textbf{Ans.}$$

Solution 7. We are given R = 100 Ω, L = 10mH, C = 1.3μF and $t_{off} = 6 \times 10^{-6}$ sec.

Duration of half pulse is given by

$$t = \frac{\pi}{\omega d}$$

Here $$\omega d = \sqrt{\frac{1}{LC} - \frac{R^2}{4L^2}}$$

$$= \sqrt{\frac{1}{1.3 \times 10^{-6} \times 10 \times 10^{-3}} - \frac{100^2}{4 \times (10 \times 10^{-3})^2}}$$

$$= 7205.77 \text{ rad/sec}$$

$\therefore$ $$t = 4.36 \times 10^{-4} \text{ sec}$$

$\therefore$ Frequency of output

$$f = \frac{1}{2(t + t_{off})} = \frac{1}{2(4.36 \times 10^{-4} + 6 \times 10^{6})}$$

$$= 1.13 \times 10^3 \text{ Hz} \quad \textbf{Ans.}$$

Solution 8. (*a*) The load voltage of full bridge inverter is given as

$$V_0 = \sum_{n=1,3,5}^{\infty} \frac{4V_s}{n\pi} \sin n\omega t$$

The rms value of load voltage

$$V_{orms} = \sqrt{\frac{1}{T}\int_0^T V_0^2 \, d(\omega t)}$$

$$= V_s = 50 \text{ volts}$$

The output power

$$P_{oav} = \frac{V_{orms}^2}{R}; \; R = 5\ \Omega$$

$$= \frac{50^2}{5} = 500 \text{ watts} \quad \textbf{Ans.}$$

(*b*) Average output current

$$I_{orms} = \frac{V_{orms}}{R} = 10 \text{ amp}$$

Peak thyristor current = average load current

= 10 amp

Average thyristor current – The thyristor conducts for half cycle. Hence its average current

$$= \frac{\text{average load current}}{2}$$

$$= \frac{10}{2} = 5 \text{ amp.} \quad \textbf{Ans.}$$

(*c*) The source voltage appears across the device in the OFF state. Hence peak inverse voltage = source voltage = 50 volts **Ans.**

Chapter 8

Solution 1. We are given

$$K = 1.00$$
$$B_m = 6\text{Wb/m}^2$$
$$V = (3 \times 2 \times 15)\ \text{cm}^3$$
$$f = 20\ \text{kHz}$$
$$W_e = ?$$

Using the formula for eddy current losses

$$W_e = kf^2 B_m^2 V$$
$$= 1.00 \times (20 \times 10^3)^2 \times (6)^2 \times (3 \times 2 \times 15) \times 10^{-6}$$
$$= 1296 \times 104 \text{ watts}$$

Electrical energy required for two hours

$$= 1296 \times 10^4 \times 2 \text{ kwh}$$
$$= 2592 \times 10^4 \text{ kwh}$$
$$\text{Heat produced} = (860 \times 2592 \times 10^4) \text{ kCal}$$
$$= 2229120 \times 10^4 \text{ kCal}$$
$$= 2.23 \times 10^{10} \text{ kCal} \quad \textbf{Ans.}$$

Solution 2. The hysteresis loss is given as

$$W_h = k'f B_m^{1.6} V$$
$$= 3.68 \times 20 \times 10^3 \times 6^{1.6} \times (3 \times 2 \times 1.5) \times 10^{-6} \text{ watt}$$
$$= 3.68 \times 20 \times 10^3 \times 1.758 \times 9 \times 10^{-6} \text{ watt}$$
$$= 11.64 \text{ watt}$$
$$\text{Electrical energy} = 11.64 \times 2 \text{ kwh}$$
$$= 23.28 \text{ kwh}$$
$$\text{Heat produced} = (860 \times 23.28) \text{ k Cal} = 2 \times 10^{10} \text{ Cal} \quad \textbf{Ans.}$$
$$\text{The total loss } W = W_e + W_n$$
$$= (2230 \times 10^{10} + 2 \times 10^{10}) \text{ Cal}$$
$$= 2232 \times 10^{10} \text{ Cal} \quad \textbf{Ans.}$$

Solution 3. It is given

$$R = 2\ \Omega$$
$$L = 30 \text{ mH}$$
$$V = 240 \text{ volts}$$
$$f = 20 \text{ kHz}$$
$$V = 25 \text{ cm}^3$$

$$\text{The impedance of the coil} = \sqrt{R^2 + W^2 L^2}$$
$$= \sqrt{4 + (20 \times 10^3 \times 30 \times 10^{-3})^2} = 600\ \Omega.$$

$$\text{Current through the coil} = \frac{V}{Z} = \frac{240}{600} = 0.40 \text{ amp}$$

The flux through the coil, is given by

$$V = N\frac{d\phi}{dt}$$

$$= N\phi f$$

or $$\phi = \frac{V}{N \times f} = \frac{240}{10 \times 20 \times 10^3} = 1.2 \times 10^{-3} \text{ Wb}$$

Assuming unit area the flux density

$$B_m = \text{Max flux/area}$$

$$\text{Max flux } \phi_m = 1.2 \times \sqrt{2} \times 10^{-3} \text{ Wb}$$

$$= 1.68 \times 10^{-3} \text{ Wb}$$

$$B_m = 1.68 \times 10^{-3} \text{ Wb/m}^2.$$

$$\text{Induction loss } W_e = kf^2B_m^2 V; \text{ Let } K = 1$$

$$= f^2B_m^2V \text{ watts}$$

$$= (20 \times 10^3)^2 (1.68 \times 10^{-3})^2 \times 25 \times 10^{-6}$$

$$= 28.2 \text{ mWatt}$$

$$\text{The resistance heating } W_R = I^2R$$

$$= (0.40)^2 \times 2$$

$$= 0.32 \text{ watt}$$

Hence $$\frac{W_R}{W_e} = \frac{0.32}{28.2 \times 10^{-3}} = 11.35 \quad \textbf{Ans.}$$

Solution 4. The eddy current loss

$$W_e = kf^2B_m^2 V$$

$$W_e = 1.2 \times 10^5 \text{ kW}$$

$$K = ?$$

$$f = 100 \text{ kHz}$$

$$B_m = 2.5 \text{ Wb/m}^2$$

$$V = 10 \times 10^{-6} \text{ m}^3$$

From above equation

$$K = \frac{W_e}{f^2B_m^2V} = \frac{1.2 \times 10^5 \times 10^3}{(100 \times 10^3)^2 (2.5)^2 \times 10 \times 10^{-6}}$$

$$= 0.192 \times 10^8 \times 10^{-5}$$

$$= 0.192 \times 10^{83}$$

$$= 192 \quad \textbf{Ans.}$$

Solution 5. The dielectric loss is given by

$$P = V^2 \omega\, c \tan \delta$$

$$\delta = 90 - \theta;\ \theta \text{ is phase angle}$$

$$V = 300\ V;\ \omega = 2\pi \times 100 \times 10^3 \text{ rad/sec}$$

$$C = 0.01\ \mu F,\ \theta = 45°$$

Putting the values, we get

$$P = (300)^2 \times (2\pi \times 100 \times 10^3)(0.01 \times 10^{-6}) \tan (90 - 45°)$$

$$= 9 \times 10^4 \times (2\pi) \times 10^5 (0.01 \times 10^{-6})$$

$$= 56.52 \times 10^3$$

$$= 565.2 \text{ watts}$$ **Ans.**

Solution 6. It is given that

$$A = 6 \text{ cm}^2 = 6 \times 10^{-4} \text{ m}^2$$

$$d = 10 \text{ cm} = 10 \times 10^{-2} \text{ m}$$

$$\varepsilon_0 = 8.85 \times 10^{-11}$$

$$\varepsilon_r = 100$$

$$\omega = 2\pi \times 1 \times 10^6 \text{ rad/sec}$$

$$V = 200 \text{ volts}$$

$$\delta = 90 - \theta = 90 - 30° = 60°$$

$$P = ?$$

$$P = V^2 \omega c \tan \delta.$$

Here

$$c = \frac{\varepsilon_0 \varepsilon_r A}{d} = \frac{8.85 \times 10^{-11} \times 100 \times 6 \times 10^{-4}}{10 \times 10^{-2}}$$

$$= 53.2 \times 10^{-15} \times 10^3$$

$$= 53.2 \times 10^{-12} \text{ farad}$$

$$P = 200^2 \times 2\pi \times 10^6 \times 53.2 \times 10^{-12} \tan 60°$$

$$= 2311.94 \times 10^4 \times 10^6 \times 10^{-12}$$

$$= 2311.94 \times 10^{-2} \text{ watts} = 23.12 \text{ watts}$$ **Ans.**

Chapter 10

Solution 1. The output obtained across potentiometer terminal is given as

$$V_0 = \frac{x}{y} V_i$$

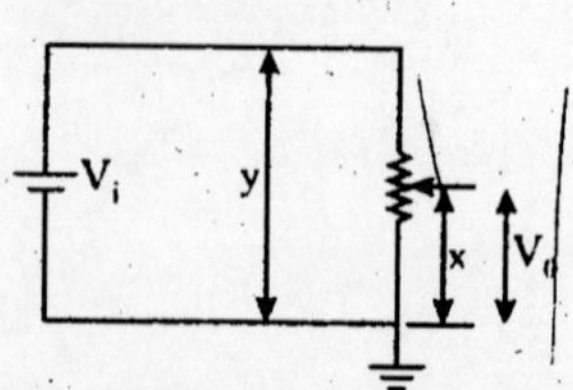

Here

$$x = 7 \text{ cm}$$

$$y = 10 \text{ cm}$$

$$V_i = 30 \text{ volts}$$

$$V_0 = \frac{7}{10} \times 30 = 21 \text{ volts}$$ **Ans.**

Solution 2. Assuming that the potentiometer turns are in tight closeness to each other the total no of turns of wire accomodated in the entire length

$$N = \frac{30\text{cm}}{0.8\text{mm}} = \frac{30\text{cm}}{0.08\text{mm}} = 375$$

$$\begin{aligned}
\text{Hence total length of the wire} &= \text{length of one turn} \times \text{total turns} \\
&= 9 \times 375 = 3375 \text{ cm} \\
\text{Cross sectional area of the wire} &= \pi r^2 \\
&= \pi (0.4)^2 \text{ mm}^2 \\
&= 0.5024 \text{ mm}^2 \\
&= 0.5024 \times 10^{-6} \text{ m}^2 \\
\text{wire resistance R} &= \frac{\rho l}{A}; \rho = 8 \times 10^{-8} \text{ ohm meter} \\
&= \frac{8 \times 3375 \times 10^{-2} \times 10^{-8}}{0.5024 \times 10^{-6}} \\
&= 5.37\ \Omega \quad \textbf{Ans.}
\end{aligned}$$

Solution 3. The O/P of a wheatstone bridge using a strain gauge is given by

$$\Delta V_0 = \frac{V}{(1+9)^2}[G\varepsilon]$$

Here

$$\Delta V_0 = 1.2 \text{ volts}$$
$$V = 10 \text{ volts}$$
$$a = 1$$
$$G = 2.1$$
$$\varepsilon = ?$$

Hence

$$1.2 = \frac{10}{2^2}[2.1\varepsilon]$$

or

$$\varepsilon = \frac{4 \times 1.2}{10 \times 2.1} \text{siemens}$$
$$= 0.23 \text{ siemens} \quad \textbf{Ans.}$$

Solution 4. (*a*)

$$R = ae^{b/T}$$

Here

$$a = -0.045$$
$$b = 3985$$
$$T = 273 + 40 = 313° \text{ K}$$
$$R = -0.045\exp\left[\frac{3985}{313}\right]$$
$$= 15222.28\ \Omega \quad \textbf{Ans.}$$

(*b*) Here. $T = 333$

$$\therefore \quad R = -0.045 \exp\left[\frac{3985}{333}\right]$$

$$= 7085.98 = 7086\ ^\circ C \quad \textbf{Ans.}$$

Solution 5. The capacitance is given by

$$C = \frac{\varepsilon_0 A}{d}$$

Under first condition $C = \dfrac{\varepsilon_0 A}{d}$

$$C = \frac{8.85 \times 10^{-11} \times 2 \times 10^{-6}}{2.1 \times 10^{-3}} \text{ Farad}$$

$$= 8.85 \times 10^{-14} \text{ Farad}$$

Under 2nd condition $C' = C + \Delta C = \dfrac{\varepsilon_0 A}{d \mp \Delta d}$

If separation is increased $C' = \dfrac{\varepsilon_0 A}{d + \Delta d} = \dfrac{885 \times 10^{-11} \times 2 \times 10^{-6}}{2.1 \times 10^{-3}}$

$$= 8.43 \times 10^{-14} \text{ Farad}$$

$$\therefore \quad \Delta C = C' - C = -0.42 \times 10^{-14}$$

i.e. capacitance decreases.

If separation decreases

$$C' = \frac{8.85 \times 10^{-11} \times 2 \times 10^{-6}}{1.9 \times 10^{-3}}$$

$$= 9.32 \times 10^{-14}$$

$$\therefore \quad \Delta C = C' - C = 0.47 \times 10^{-14} \text{ Farad}$$

Thus the capacitance increases.

Solution 6. For dc input whole voltage appears across the capacitor in steady state. **Ans.**

Solution 7. When a dielectric material is inserted between the capacitor plates, its capacitance has the value

$$C = \frac{\epsilon_0 \epsilon_r A}{d}$$

$$= \frac{8.85 \times 10^{-11} \times 16 \times 2 \times 15^{-6}}{2 \times 10^{-3}}$$

$$= 141.6 \times 10^{-14} \text{ Farad}$$

Hence the change in capacitance

$$\Delta C = (16 - 1)\ 8.85 \times 10^{-14} \text{ Farad}$$

$$= 132.75 \times 10^{-14} \text{ Farad}$$
$$= 1.33 \text{ PF} \quad \textbf{Ans.}$$

Solution 8. In the linear range of operation of the LVDT

$$V_0 \; \alpha \; x; \; x \text{ is displacement of the core}$$

$\therefore$ $$V_{01} \; \alpha \; x_1$$

and $$V_{02} \; \alpha \; x_2$$

$$\frac{V_{01}}{V_{02}} = \frac{x_1}{x_2}$$

Here $$V_{01} = 1.2 \text{ mV}, \; x_1 = 2 \text{ mm}$$
$$V_{02} = ? \; x_2 = 2.08 \text{ mm}$$

Putting the values, we get

$$\frac{1.2 \times 10^{-3}}{V_{02}} = \frac{2 \times 10^{-3}}{2.08 \times 10^{-3}}$$

or $$V_{02} = \frac{1.2 \times 10^{-3} \times 2.08}{2}$$
$$= 1.248 \text{ mV} \quad \textbf{Ans.}$$

In opposite direction $$x^2 = 2 - 0.08 = 1.92 \text{ cm}$$

$\therefore$ $$V_{02} = \frac{1.2 \times 10^{-3} \times 1.92}{2}$$
$$= 1.152 \text{ mV} \quad \textbf{Ans.}$$

Solution 9. $$V_{BE} = \frac{KT}{q} \ln \left[\frac{I_E + I_{EO}}{I_{EO}} \right]$$

neglecting leakage current

$$V_{BE} = \frac{KT}{q} \ln I_E$$

Here $$K = 1.38 \times 10^{-23} \text{ Joules/°K}$$
$$T = 40 + 273 = 313° \text{ K}$$
$$q = 1.6 \times 10^{-19} \text{ coulomb}$$
$$I_E = 5 \text{ mA}$$

$\therefore$ $$V_{BE} = \frac{1.38 \times 10^{-23} \times (313)}{1.6 \times 10^{-19}} \ln(5 \times 10^{-3})$$
$$= 269.96 \times 10^{-4} [\ln(5) - \ln 1000]$$
$$= -0.14 \text{ volts}$$

$\therefore$ $$V_{EB} = 0.14 \text{ volts} \quad \textbf{Ans.}$$

Solution 10. In first case

$$e_1 = a\Delta T_1 + b\Delta T_1^2$$

$$1.2 \times 10^{-3} = a(298) + b(298)^2$$

In IInd case

$$2.6 \times 10^{-3} = a(318) + b(318)^2$$

Solving above two equations

$$a = -61.8 \times 10^{-3} \text{ volts/}^\circ\text{K}$$

$$b = 2.07 \times 10^{-4} \text{ volts/}(^\circ\text{K})^2 \quad \textbf{Ans.}$$

Solution 11. The output of a piezo-electric crystal is given as

$$V_0 = g\frac{F}{A}t$$

$$= g\,Pt$$

Here g = voltage sensitivity

P = pressure

and t = crystal thickness

In the present case

$$g = 3 \times 10^{-6} \text{ mV/N} = 3 \times 10^{-9} \text{ V/N}$$

$$P = 2 \times 10^{6} \text{ N/m}^2$$

and $t = 2.1 \text{ mm} = 2.1 \times 10^{-3}$ meter

$$\therefore \quad V_0 = 2 \times 10^{-9} \times 2 \times 10^{6} \times 2.1 \times 10^{-3} \text{ volts}$$

$$= 8.4 \times 10^{-6} \text{ volts}$$

$$= 8.4\ \mu\text{V} \quad \textbf{Ans.}$$

Solution 12. The charge sensitivity d is related with voltage sensitivity by the expression

$$g = \frac{d}{\epsilon_0 \epsilon_r}$$

$$\therefore \quad d = g\epsilon_0\epsilon_r$$

Here $\epsilon_r = 25$

$$g = 3 \times 10^{-6} \text{ mV/N} = 3 \times 10^{-9} \text{ V/N}$$

$$\epsilon_0 = 8.85 \times 10^{-11}$$

$$\therefore \quad d = 3 \times 10^{-9} \times 8.85 \times 10^{-11} \times 25 \text{ Coulomb/N}$$

$$= 663.75 \times 10^{-20} \text{ Coulomb/Newton} \quad \textbf{Ans.}$$

Solution 13. The speed in r.p.s is given by

$$n = \frac{60f}{P}$$

f = no. of counts/sec = 115

P = no. of holes = 28

$$\therefore \quad n = \frac{60 \times 115}{28}$$

$$= 246 \text{ rps.} \quad \textbf{Ans.}$$

Solution 14.

$$n = \frac{60f}{t}$$

f = frequency of rotation (no. of pulses)

t = no. of teeth.

or

$$f = \frac{nt}{60}$$

$$x = 700 \text{ rpm} = \frac{700}{60} \text{ r.p.s.}$$

$$t = 8$$

$$\therefore \quad f = \frac{700 \times 8}{60 \times 60} = 1.55 \text{ counts/sec} \quad \textbf{Ans.}$$

Chapter 11

Solution 1. The O/P of OPAMP in inverting mode is given by

$$V_0 = -\frac{R_f}{R_i} V_i$$

$$R_f = 10 \text{ K}\Omega,\ R_i = 50\ \Omega,\ V_i = 10 \text{ mV}$$

$$\therefore \quad V_0 = -\frac{10 \times 10^3}{50} \times 10 \times 10^{-3}$$

$$= -2 \text{ volts} \quad \textbf{Ans.}$$

Solution 2. For non inverting mode

$$V_0 = \left(1 + \frac{R_f}{R_i}\right) V_i$$

$$= \left(1 + \frac{10 \times 10^3}{50}\right) 10 \times 10^{-3}$$

$$= 2.01 \text{ volts} \quad \textbf{Ans.}$$

Solution 3. The O/P of a voltage follows has the value

$$V_0 = \frac{A_{OL}}{1 + A_{OL}} V_i$$

$$A_{OL} = 99985$$

$$V_i = 12 \text{ mV}$$

$$\therefore \quad V_0 = \frac{99985}{1+99985} \times 12 \times 10^{-3}\text{ V} = -11.9\text{ mV} \quad \textbf{Ans.}$$

Solution 4. The common mode voltage is

$$V_c = \frac{V_1 + V_2}{2}$$

$$= \frac{1 + 10 \times 10^{-3}}{2}\text{ volts}$$

$$= 0.505\text{ volts.} \quad \textbf{Ans.}$$

$$\text{Differential voltage} = V_2 - V_1 = 0.99\text{ volts} \quad \textbf{Ans.}$$

Solution 5. Using the relation

$$\text{CMRR} = \frac{A_d}{A_c},\ A_d = 10^5,\ A_c = 1.5$$

$$= \frac{10^5}{1.5} = 6.7 \times 10^4 \quad \textbf{Ans.}$$

Solution 6. The O/P of instrumentation amplifier is given as

$$V_0 = \left(1 + \frac{2R_2}{R_g}\right)(V_2 - V_1)$$

Here $R_2 = 100\text{ k}\Omega$, $Rg = 100\ \Omega$, $V_2 = 10\text{ mV}$, $V_1 = 9.2\text{ mV}$

$$\therefore \quad V_0 = \left(1 + \frac{2 \times 100 \times 10^3}{100}\right)(10 - 9.2) \times 10^{-3}\text{ volts.}$$

$$= 1.6\text{ volts} \quad \textbf{Ans.}$$

Solution 7. The O/P voltage is given by

$$V_0 = (2^0 a + 2^{-1} b + 2^{-2} c + \ldots)\, V_R$$

Here input data is = 1001 and $V_R = 8$ volts

$$\therefore \quad V_0 = \left(2^0 \times 1 + \frac{1}{2} \times 0 + \frac{1}{4} \times 0 + \frac{1}{8} \times 1\right) \times 8$$

$$= \left(1 + \frac{1}{8}\right) \times 8 = 9\text{ volts} \quad \textbf{Ans.}$$

Solution 8. The O/P will be maximum when input data has value 1111. In this case

$$V_0 = \left(2^0 \times 1 + \frac{1}{2} \times 1 + \frac{1}{4} \times 1 + \frac{1}{8} \times 1\right)8$$

$$= \left(1 + \frac{1}{2} + \frac{1}{4} + \frac{1}{8}\right)8 = 15\text{ volts} \quad \textbf{Ans.}$$

Chapter 12

Solution 1. We are given

$$\begin{aligned}
\text{Number of turns N} &= 3000 \\
\text{Magnetic flux density B} &= 0.12 \text{ Tesla} \\
\text{Restoration constant K} &= 0.12 \text{ Newton/meter} \\
\text{Coulomb friction F} &= 0.15 \times 10^{-6} \text{ Nm} \\
\text{Area of the coil A} &= 25 \text{ mm} + 50 \text{ mm} = 1250 \times 10^{-6} \text{ m}^2 \\
\text{Galvanometer deflection } \theta &= 60^\circ = \frac{\pi}{3} \text{ radian} \\
\text{Coil current } i &= ?
\end{aligned}$$

The galvanometer constant has the value

$$\begin{aligned}
G &= NAB \\
&= 3000 \times 1250 \times 10^{-6} \times 0.12 \\
&= 0.45 \text{ m}^2\text{T}
\end{aligned}$$

The coil current is given by

$$\theta = \frac{G_i}{K} - \frac{F}{K}$$

or

$$\frac{\pi}{3} = \frac{0.45i}{0.12} - \frac{0.15 \times 10^{-6}}{0.12}$$

or

$$i = \frac{\left(\frac{\pi \times 0.12}{3} + 0.15 \times 10^{-6}\right)}{0.45} \text{ amp}$$

$$= 0.28 \text{ amp} \quad \textbf{Ans.}$$

Solution 2. When the number of turns are reduced to 300 then

$$G = 0.045 \text{ Tm}^2$$

$$\therefore \quad \theta = \frac{G_i}{K} - \frac{F}{K}$$

$$= \frac{0.045 \times 0.28}{0.12} - \frac{0.15 \times 10^{-6}}{0.12}$$

$$= 6.02^\circ \quad \textbf{Ans.}$$

Solution 3. Writing the quadratic equation

$$J\frac{d^2\theta}{dt^2} + D\frac{d\theta}{dt} + K\theta + F - G_i = 0$$

or

$$\frac{d^2\theta}{dt^2} + \frac{D}{J}\frac{d\theta}{dt} + \frac{K\theta}{J} = \frac{G_i - F}{J}$$

L.H.S. of above equation gives the characteristic equation

$$m^2 + \frac{D}{J}m + \frac{K}{J} = 0$$

This is quadratic equation. The system is underdamped, critically damped or overdamped accordingly as

$$\left(\frac{D}{J}\right)^2 - 4\frac{K}{J} < = > 0$$

It is given that

$$D = 0.016, \ K = 0.12 \text{ and } J = 0.006$$

Putting the values

$$\left(\frac{0.016}{0.006}\right)^2 - 4\left(\frac{0.12}{0.006}\right) = -72.8 < 0$$

Hence the system is underdamped. **Ans.**

EXPERIMENTS

EXPERIMENT 1

Aim : Observation of waveshape and measurement of voltage at relevant points of an SCR based single phase half wave controlled rectifier circuit using resistive triggering circuit.

Apparatus required : Half wave controlled rectifier kit 1, cathode ray oscilloscope 1, connecting leads.

Theory : Rectification is conversion of ac in dc power. Input ac signal can be converted into dc signal by using simple diodes, SCR and other devices. By using simple diode whole of positive half or negative half of input signal can be allowed to appear across the load. Sometimes the requirement of power across the load is different from that is fed by using diodes. These requirements of power are met by using SCR. In controlled rectifier the SCR is triggered at different phases by changing the potentiometer resistance across the gate circuit. This is the principle of phase controlled rectification. The current is shown in Fig. 1.1. In the Fig., we observe that the current does not flow through the circuit unless SCR. This turned on.

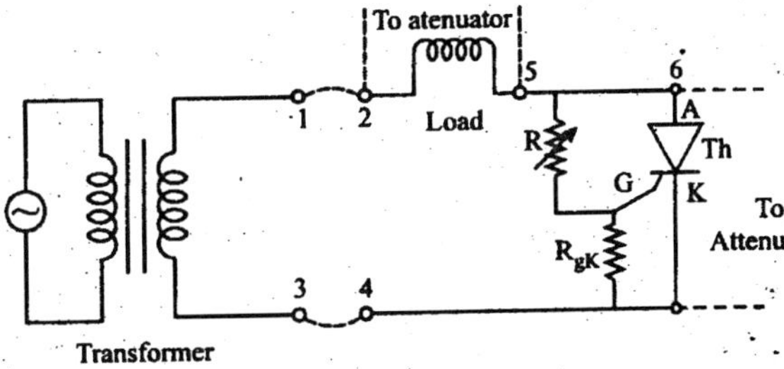

Fig. 1.1.

In the off condition input voltage appears across the thyristor. By changing the value of potentiometer R the voltage required to turn on the thyristor may be changed which is equivalent to saying that the phase at which thyristor is turned on, can be changed. In the negative half cycle the SCR is turned off. This is the reason it is called half wave phase controlled rectification. Average voltage across the load is given by

$$V_{oav} = \frac{V_m}{2\pi}(1+\cos\alpha)$$

where α is called the firing angle.

Waveforms are shown in Fig. 1.2.

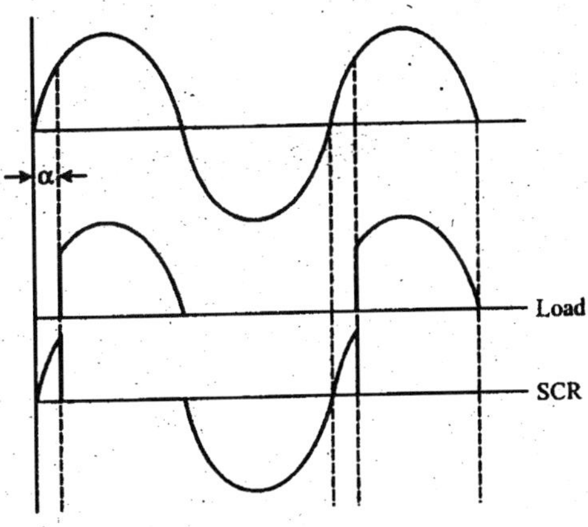

Fig. 1.2.

Procedure :

1. Make the connections as shown in Fig. 1.1.
2. Switch on the power supply.
3. Connect the CRO between pair of points 2, 5 and 6, 7.
4. Change the resistance until the SCR is turned on.
5. Observe the waveforms across load and SCR.

Calculation : Firing angle α^o

$$= 180 \times \frac{\text{no. of divisions on CRO screen covered by firing angle}}{\text{Total no. of divisions on CRO screen covered by half cycle on time axis}}$$

Result : Waveforms have been observed at different firing angles and firing angle and average load voltage have been computed.

Precautions :

1. Mains supply should be switche on only after all the connections are complete.
2. Do not touch any point with bare hand.
3. Signal should be applied to CRO through attenuator only.

EXPERIMENT 2

Aim : Observation of waveshapes and measurements of voltages at relevant points of an SCR based single phase half wave controlled rectifier circuit using R-C phase shift gate triggering circuit.

Apparatus : Half wave controlled rectifier kit 1, CRO 1, connecting leads.

Theory : R-C phase triggering employed for controlled rectification is also like R-phase triggering with the difference that firing angle in this case ranges from 0° to 180°. Thus the range of power fed to the load is large by using R-C phase triggering. The circuit is shown in Fig. 2.1.

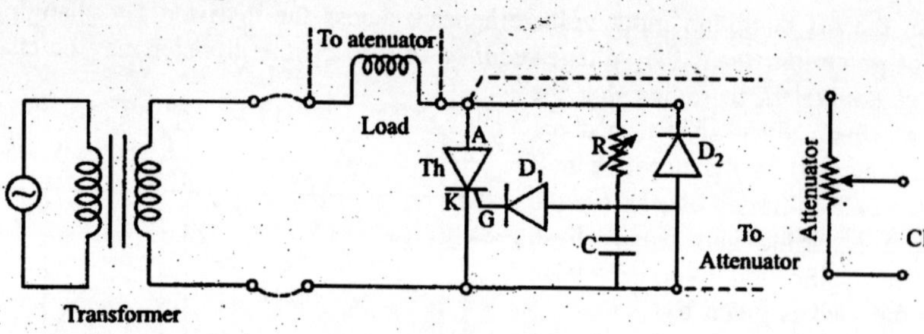

Fig. 2.1.

During negative half cycle the capacitor C is charged to negative peak value with upper plate negative. After negative peak the negative cycle starts decresing towards. As a result the capacitor starts through resistance R and source. After it reaches zero value it starts charging in opposite direction, *i.e.*, with upper plate positive and attain a value of 0.7 + 0.7 = 1.4 volt the thyristor is triggered and the current starts flowing through the load. In negative half cycle the SCR is turned off.

The discharge time of the capacitor is governed by the product RC. By increasing the product RC the rate of discharge can be decreased which means it will attain zero voltage at higher ϕ angle. Thus the plate angle at which SCR is turned on can be changed by changing the product RC. This is the working principle of RC firing circuit. The waveforms are shown in Fig. 2.2. At angle α, the capacitor voltage reaches the value which is sufficient to trigger the SCR. The voltage across the SCR is reduced to zero while that across the load follows the input voltage. The angle α can be changed by changing the rate of discharge of capacitor as is obvious from the waveform. Thus controlled power can be fed to the load.

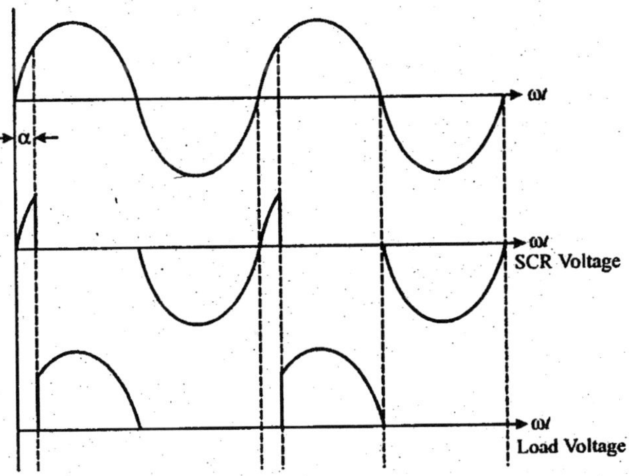

Fig. 2.2.

Procedure :

1. Make the connections as shown in Fig. 2.1.
2. Switch on the main supply.
3. Connect the CRO across load, capacitor and SCR.
4. Observe the waveforms.
5. Change the potentiometer resistance.
6. Observe the waveforms.
7. **Repeat Steps 5 and 6.**

Calculation : Firing angle α

$$= 180 \times \frac{\text{no. of divisions covered by firing angle on time axis}}{\text{Total no. of divisions covered by half wave on time axis}}$$

$$\text{Conduction angle} = 180^\circ - \alpha$$

$$\text{Output voltage } V_{oav} = \frac{V_m}{2\pi}(1+\cos\alpha)$$

Result : The firing angles and output voltage have been calculated from given waveforms.

Precautions :

1. Switch on the supply only after the connections are complete.
2. Do not tough any terminal.

EXPERIMENT 3

Aim : Observation of waveshapes and measurement of voltages at relevant points of an SCR based single phase half wave controlled rectifier circuit using UJT relaxation oscillator for gate triggering.

Apparatus required : Half wave controlled rectifier kit with UJT relaxation oscillator 1, CRO 1, Connecting leads.

Theory : In pulse triggering methods of SCR the wastage of power in gate circuit can be avoided. Amongst various methods used UJT relaxation oscillator is an important

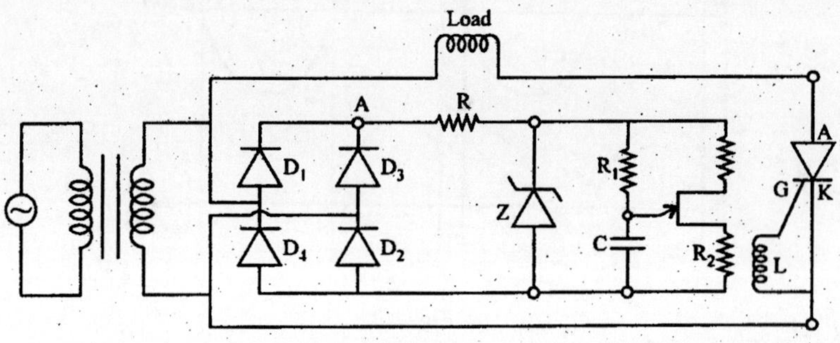

Fig. 3.1.

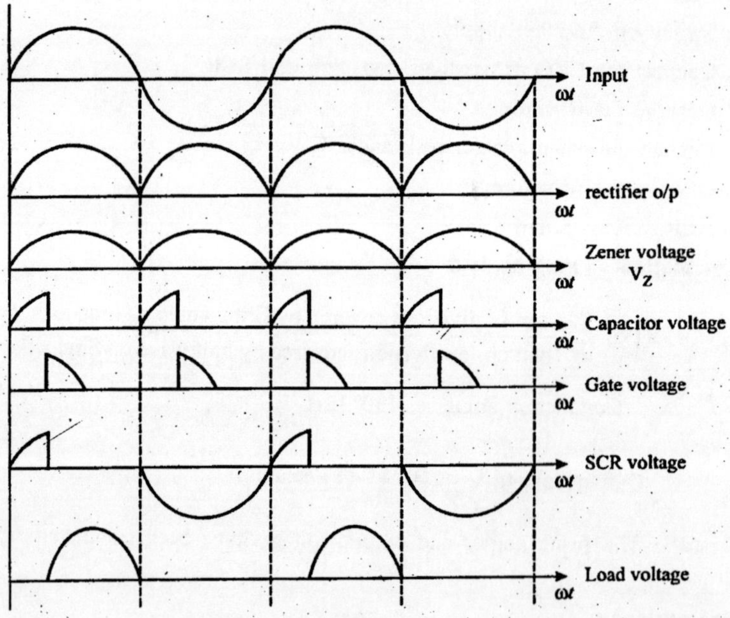

Fig. 3.2.

method. The circuit used is shown in Fig. 3.1. Diodes $D_1 - D_4$ form bridge rectifier whose output is made constant dc by zener diode Z except at zero crossing. The capacitor C is charged through resistance R_1 and finally when the voltage across it reaches the value of ηV_z where η is called intrinsic stand off ratio of UJT. At this value of the voltage the UJT is turned on and the capacitor starts discharging through the path containing UJT and resistor R_2. The current in R_2 induces an e.m.f. in a coil L which triggers the SCR. Only first pulse to gate in half cycle is responsible to turn on, other pulses are of no use. We assume that resistance R_2 is very small, then pulse frequency is given by

$$f = \frac{1}{T} \text{ where } T = R_C \ln\frac{1}{1-\eta}$$

The waveforms at different points are shown in Fig. 3.2.

Procedure :

1. Make connections as shown in Fig. 3.1.
2. Switch on the supply.
3. Observe the waveforms across load, SCR, capacitor C, point A, zener diode and gate.

Result : Waveforms at different points have been observed.

Precautions :

1. Connections should not be loose.
2. Before switching on ensure that connections are complete.
3. Any point should not be touched by bare hand.

EXPERIMENT 4

Aim : Observation of waveshapes and measurement of voltages at relevant points of an SCR based single phase full wave controlled rectifier circuit.

Apparatus : Full wave controlled rectifier kit 1, CRO 1, connecting leads.

Theory : The working of full wave controlled rectifier is similar to that of half wave controlled rectifier. Both positive and negative half cycle of input are switched to the load at desired firing angles. The circuit is shown in Fig. 4.1. The potentiometer R_1 and R_2 are generally the same or sometimes they are derived from one potentiometer only so that firing angle in positive and negative half cycle may remain same. The output voltage is given by

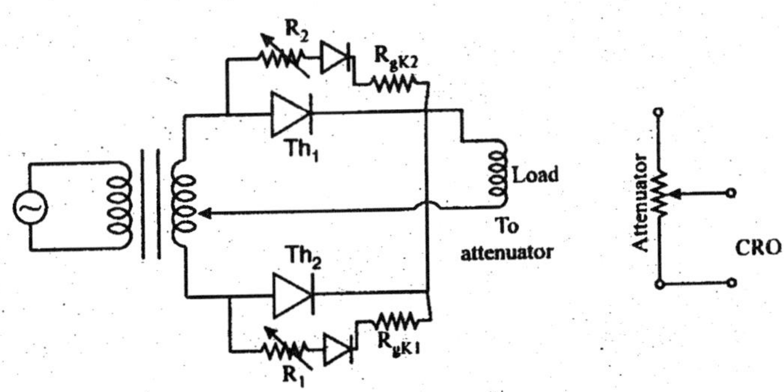

Fig. 4.1.

$$V_{av} = \frac{V_m}{2\pi}(2+\cos\alpha_1+\cos\alpha_2)$$

Here α_1 and α_2 are, respectively, the firing angle of SCRs, Th_1 and Th_2. The waveforms are as shown in Fig. 4.2.

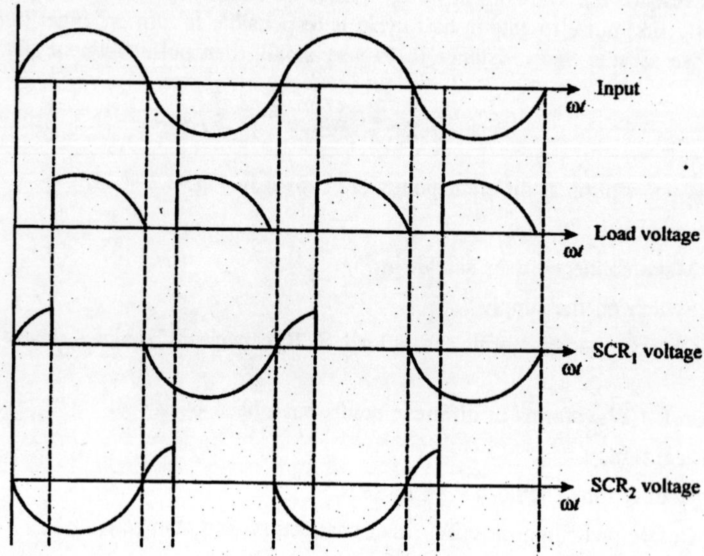

Fig. 4.2.

Procedure :

1. Make connections as shown in Fig. 4.1.
2. Switch on power supply.
3. Observe waveforms and measure voltages across SCRs and load.
4. Change potentiometer resistance R_1 and R_2.
5. Repeat Step 3.

Calculations : Firing angle $= 100 \times \dfrac{\text{No. of divisions under firing angle}}{\text{Total no. of divisions under half cycle}}$

Conduction angle = 180° – firing angle

Result : The waveforms and voltages have been measured at different points.

Precautions :

1. Do not touch any terminal when supply is on.
2. Connect the CRO through attenuator only.
3. Supply should not be turned on for prolonged period.

EXPERIMENT 5

Aim : Observation of waveshapes and measurement of voltage at relevant points of an SCR based single phase controlled bridge rectifier circuit.

Apparatus : Single phase controlled bridge rectifier kit 1, CRO 1, connecting leads.

Theory : Theory of all types of controlled rectifier is same. Bridge rectifier using four SCRs is shown in Fig. 5.1. Firing pulse FP_1 is used trigger thyristor pair Th_1, Th_2 simultaneously while firing pulse FP_2 is used for thyristor pair Th_3 Th_4. During positive half cycle thyristors Th_1 and Th_2 conduct while during negative half cycle thyristors Th_3 and Th_4 conduct. The load voltage is given by

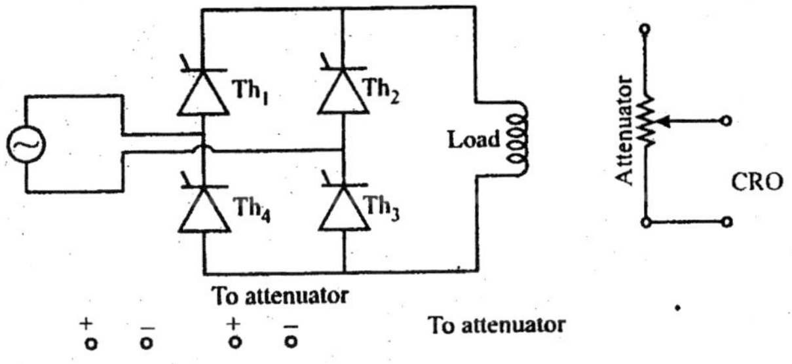

Fig. 5.1.

$$V_{oav} = \frac{V_m}{\pi}(1+\cos\alpha_1), \text{ for } Th_1\ Th_2 \text{ pair}$$

$$= \frac{V_m}{\pi}(1+\cos\alpha_2), \text{ for } Th_3\ Th_4 \text{ pair}$$

The waveforms are as shown in Fig. 5.2.

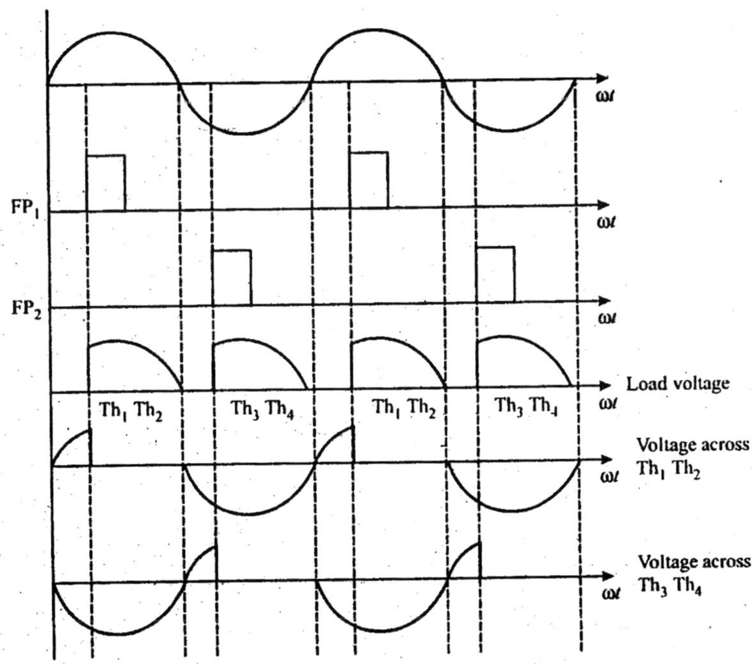

Fig. 5.2.

Procedure :

1. Make connections as shown in Fig. 5.1.

2. Turn on power supply.
3. Observe waveforms across thyristors and load.
4. Measure the voltage across SCRs and load by using CRO.

Calculations : From the waveforms

$$\text{Firing angles, } \alpha_1, \alpha_2 = 180 \times \frac{\text{no. of divisions covered between rising edge of pulse and } 0^\circ}{\text{Total divisions covered by half cycle}}$$

Result : The waveforms have been observed and voltages across different points measured are .....

Precautions :

1. Loose connections should be avoided.
2. Damage to the CRO should be avoided.
3. Observation on CRO screen should be taken carefully.

EXPERIMENT 6

Aim : Observation of waveshapes and measurement of voltage at relevant points in a triac based ac phase control circuit used for lamp intensity and / r ac fan speed control.

Apparatus : AC regulator using triac, electric bulb, small ac fan, speedometer (techiometer), CRO, connecting leads etc.

Theory : AC regulator is a circuit which converts ac into variable ac. Thyristor family such as thyristor, triac etc. can be used as ac regulator. For regular ac input desired part of it at different firing angles can be switched to the load. The circuit of ac regulator is shown in Fig. 6.1. We observe that unless triac is off no current will flow through lamp or the fan. As the potentiometer resistance R is varied, the gate current reaches to a value sufficient to turn it on. Remaining part of the half cycle will appear across the load. Thus by changing potentiometer resistance desired current can be supplied which in turn changes the speed of fan or intensity of lamp.

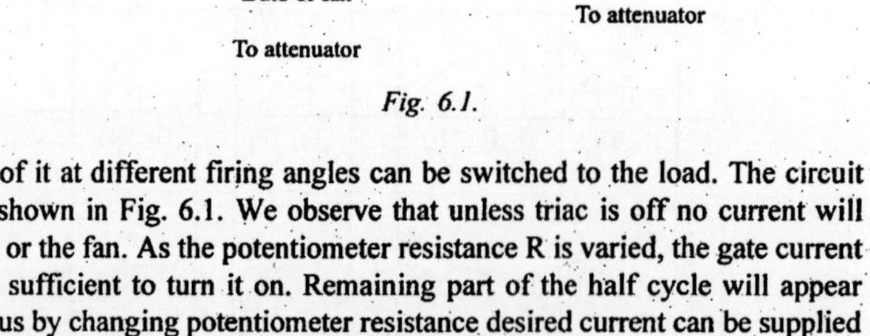

Fig. 6.1.

The waveforms are shown in Fig. 6.2 and graph between fan speed and current is a straight line as shown in Fig. 6.3. Load voltage is given by

$$V_L = V_m\left[\frac{1}{2\pi}\left(\pi - \alpha + \frac{\sin 2\alpha}{2}\right)\right]^{0.5}$$

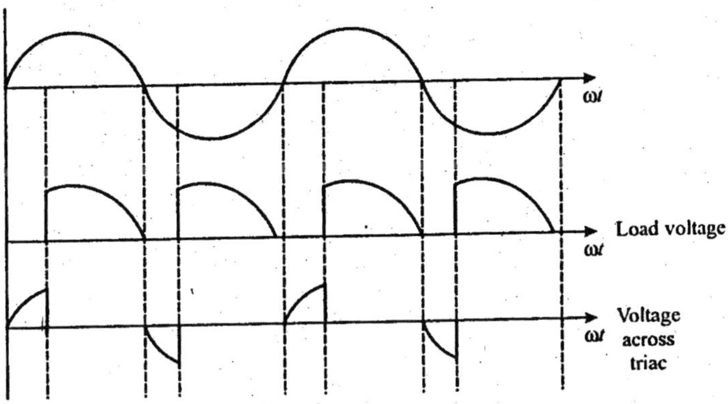

Fig. 6.2.

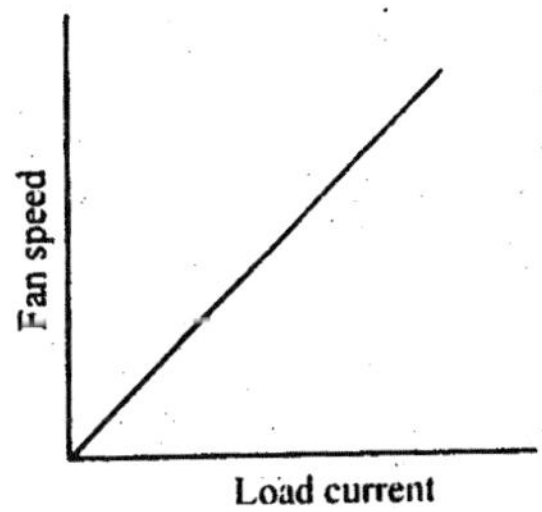

Fig. 6.3.

Procedure :

1. Make the connections as shown in Fig. 6.1.
2. Turn on the power supply.
3. Observe the waveforms across load and triac.
4. Vary potentiometer resistance R.
5. Repeat step 3 for different values of R.
6. Measure the speed of fan by potentiometer.
7. Plot the graph between fan speed and armature current.

Calulations : Firing angle $= 180^o \times \dfrac{\text{no. of divisions covered by firing angle}}{\text{no. of divisions covered by half cycle}}$

Result : The waveforms at all relevant points have been observed and the measured voltages at these points is ....

EXPERIMENT 7

Aim : To observe waveforms and measurement of voltages at various points of a circuit for overvoltage protection using SCR.

Apparatus : Kit for overvoltage protection - 1, CRO - 1, connecting leads.

Theory : The working of overvoltage protection circuit is based on the control of firing angle. The firing angle of the thyristor is set at the overvoltage of the electrical device. The circuit is shown in Fig. 7.1. Thyristors Th_2 and Th_1 are used for overvoltage protection in positive and negative half cycles respectively. Breakdown voltage of zener diode Z is equal to overvoltage for the load. At overvoltage is positive half cycle, the firing of thyristor Th_2 takes place through R - D_1 - Z and D_4 and the current now flows through R - Th_2 - R' - Z' and source. In the negative half cycle thyristor Th_1 is triggered through D_2, Z and D_3. The waveforms have been shown in Fig. 7.2.

Procedure :

1. Make the connections as shown in Fig. 7.1.
2. Turn on the supply.

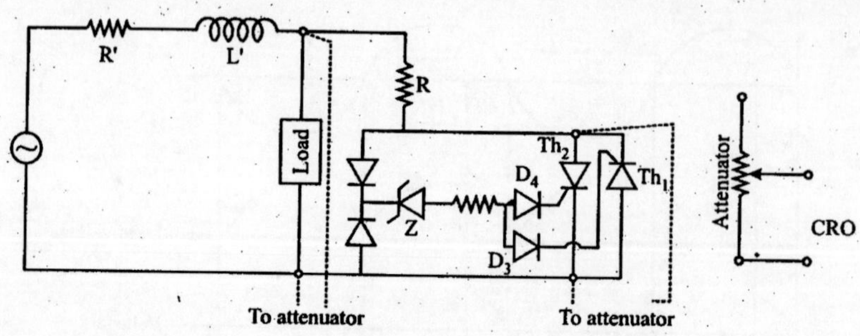

Fig. 7.1.

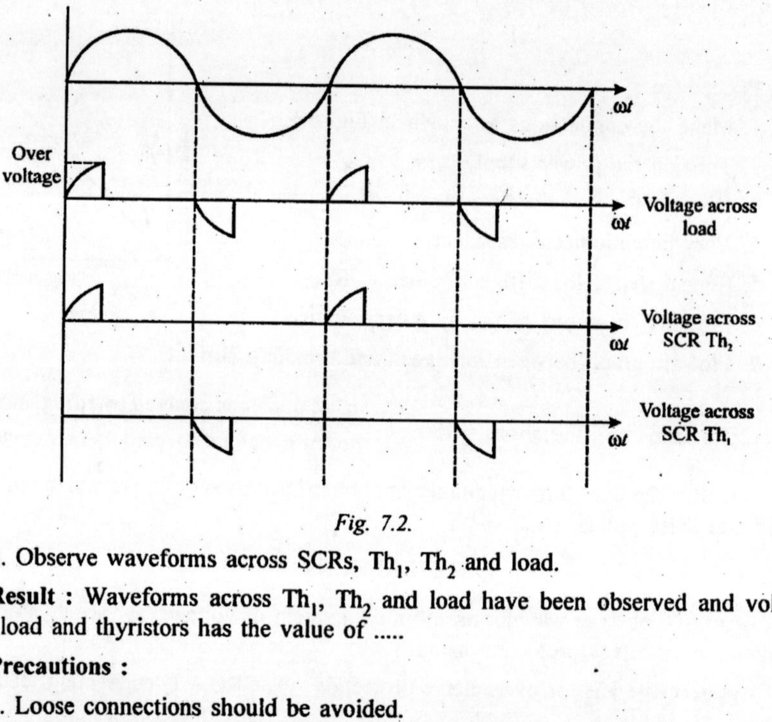

Fig. 7.2.

3. Observe waveforms across SCRs, Th_1, Th_2 and load.

Result : Waveforms across Th_1, Th_2 and load have been observed and voltage across load and thyristors has the value of .....

Precautions :

1. Loose connections should be avoided.
2. CRO should be connected between proper points.
3. Terminals should not be touched with bare hands.

EXPERIMENT 8

Aim : Study of various transducers like strain gauge, thermistor photodiode, phototransistor etc.

Apparatus required : Instrumentation trainer, strain gauge, thermistor photodiode, phototransistor-1 each, voltmetr (mV range)-1, Ammeter (mA range)-1, connecting wires.

Theory : Strain gauge : It is a resistive transducer which when expands or contracts under the application of free observses a change in its resistance. This change in resistance is detected and a number of physical quantities can be measured. The output of the strain gauge is measured in terms of what is called gauge factor given by

$$G = \frac{dR/R}{dl/l} = 1 + 2\mu + \frac{d\rho/\rho}{dl/l}$$

where $\frac{dl}{l}$ = strain of the material of strain gauge designated ε

μ = Poisson's ratio having value of 0.3 for most of metals

$\frac{d\rho}{\rho}$ = change in resistance.

Let us consider Fig. 8.1. A strain gauge is bonded to a specimen which is subjected to the force at one end in terms of the weight mounted on it. As the weight on the cantilever increases the cantilever (specimen) bends downwards further. As a result, the strain gauge expands in length and its resistance changes correspondingly. If initially the bridge is balanced, *f* will be unbalanced due to change in strain gauge resistance.

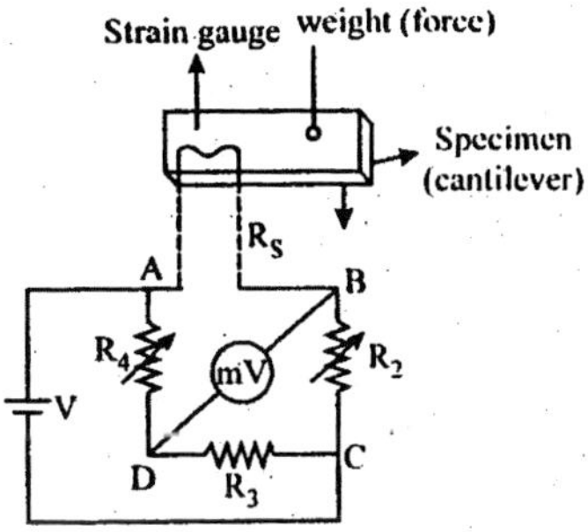

Fig. 8.1.

Let the resistance R_s of strain gauge and all other resistances have equal values-R say, then output voltage as a result of change of strain gauge resistance is given by

$$V_o = \frac{V}{(1+a)^2}[G\varepsilon]$$

where $a = \frac{R_s}{R_2} = \frac{R_4}{R_3} = 1$ in this case

hence $$V_o = \frac{V}{4}\varepsilon G$$

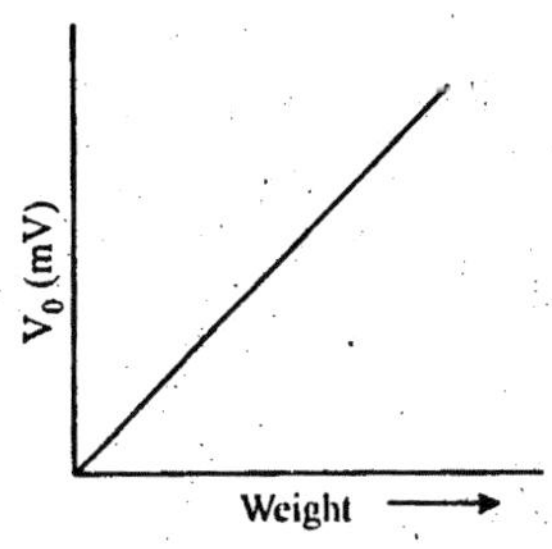

Fig. 8.2.

Here ε = strain in the strain gauge.

As the force is applied the strain in developed in the gauges which lead to a change in the output voltage. Thus output voltage is directly proportional to the applied force, *i.e.*, the graph between weight and output voltage of the bridge should be a straight line as shown in Fig. 8.2.

Procedure :

1. Turn on power supply to the strain gauge bridge.

2. Adjust resistance so that no deflection is observed in the voltmeter connected between points B and D.

3. Mount the weight on the specimen.

4. Observe output voltage.

5. Repeat step 3 and 4.

Observation :

S. No.	Weight in gms	Output voltage in mV
1		
2		
3		
4		
5		

Result : The graph between load and output voltage has been drawn which is a straight line.

Thermistor : Thermistor is a temperature sensing device which has negative temperature coefficient, *i.e.*, with an increase in temperature the resistance of the device decreases. The change in resistance is detected which is the measure of the temperature of any medium in which thermistor is placed. The relationship between resistance and temperature is given by expression

$$R = ae^{b/T}$$

where a and b are constants.

The resistance decreases exponentially with temperature. The circuit using a thermistor to measure temperature is shown in Fig. 8.3. As the temperature changes the resistance R_{Th} of the thermistor for changes which is detected by a change in ammeter current. The graph between resistivity and temperature is shown in Fig. 8.4, while in Fig. 8.5 voltage across thyristor Vs. current is shown.

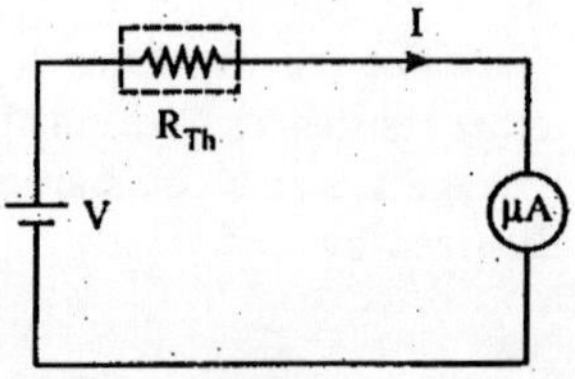

Fig. 8.3.

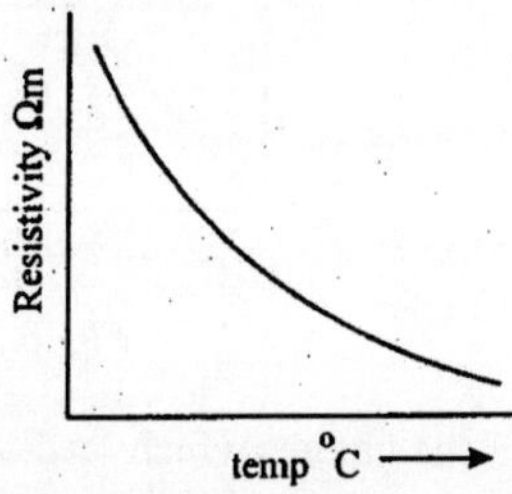

Fig. 8.4.

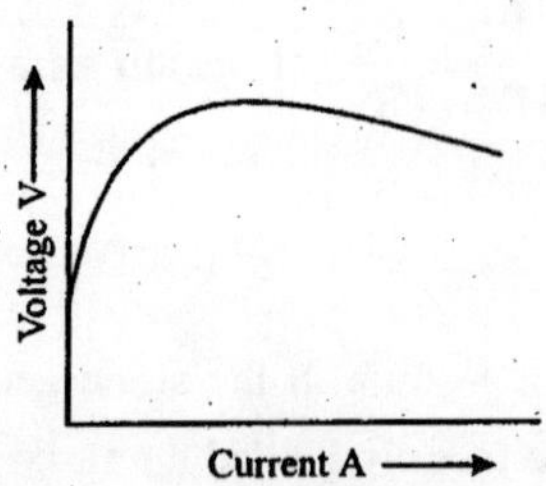

Fig. 8.5.

Procedure :

1. Maintain thermistor at °C using standard thermometer.
2. Switch on power supply.
3. Measure the current in ammeter.
4. Decrease the temperature of the medium.
5. Measure the current in ammeter.
6. Repeat steps 4 and 5.

Observation : Source voltage = V volt

S. No.	Temperature °C	R_{th} = V/I Ω
1		
2		
3		
4		
5		

Result : The graph between resistance and temperature of the thermistor has been drawn.

Photodiode : Theory : Photodiode is a p-n junction diode. When intensity of light is directed in the viscinity of the junction electron pairs are generated and a current flows. Photodiode is conneced in reverse mode. As a result of increase in light intensity the reverse saturation current increases. Fig. 8.6 shows the circuit photodiode while its characteristics are shown in Fig. 8.7.

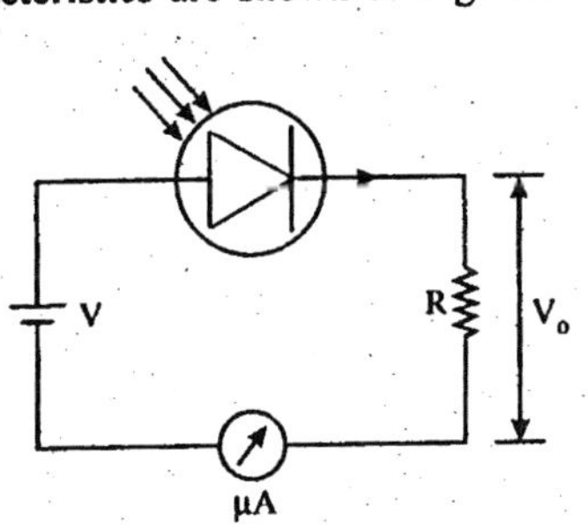

Fig. 8.6.

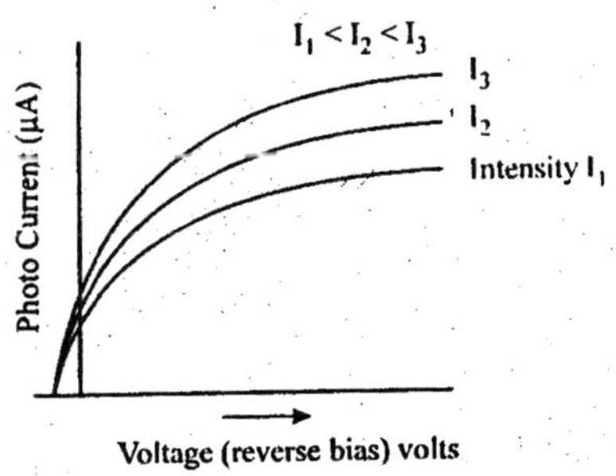

Fig. 8.7.

It is clear from the curve that even in the drak there is a current through the device which is called the dark current of the device.

Procedure :

1. Place the photodiode in complete darkness.
2. Switch on the power supply.
3. Measure output voltage.
4. Expose the device to visible light.
5. Measure output voltage.
6. Increase the light intensity in steps and measure output voltage.
7. Repeat step 6.

Observation : For different intensities of light following table is used.

S.No.	Bias voltage (V)	Photocurrent (μA)
1		
2		
3		
4		
5		

Phototransistor : Theory : Phototransistor is like simple BJT except that envelop containing junction is made transparent. This allows the light to impinge on base emitter junction of the transistor. As a result electron hole pairs are generated which leads to an increase in the current. The circuit diagram is shown in Fig. 8.8 and characteristics are plotted in Fig. 8.9 for different values of resistance R.

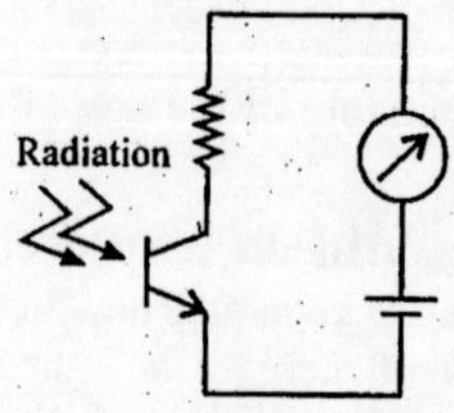

Fig. 8.8.

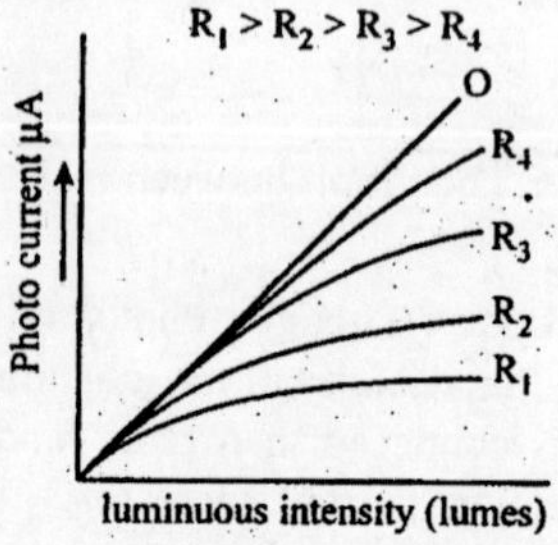

Fig. 8.9.

Procedure :

1. Make the connections as shown in the Fig. 8.8.
2. Observe the current with transistor in complete darkness.
3. Increase the light intensity falling in the device.
4. Measure current.
5. Repeat steps 3 and 4.
6. Plot the graph.

Observation : For different values of resistance following table is used.

S. No.	Light intensity (lumen)	Collector current (μA)
1		
2		
3		
4		
5		

Result : The characteristics have been studied.

EXPERIMENET 9

Aim : To assemble and test an instrumentation amplifier, measure its gain, input and output impedance.

Apparatus Required : Three operational amplifiers IC 741, Resistance 100 kΩ - 2, resistance, variable resistance 10 kΩ - 1, resistance 1 kΩ - 1, voltmeter (mV) - 2, voltmeter - 1, Ammeter (mV) - 3, connecting wires.

Theory : Instrumentation amplifier is a high adjustable gain amplifier whose gain can be adjusted by simply having one variable resistance. It consists of two or three operational amplifiers as shown in Fig. 9.1. For the sake of simplicity one of the input terminals has been grounded. Output voltage may have maximum value of supply voltage.

Input voltage and input current are small and measured respectively in mV and μA. Following are the parameters described.

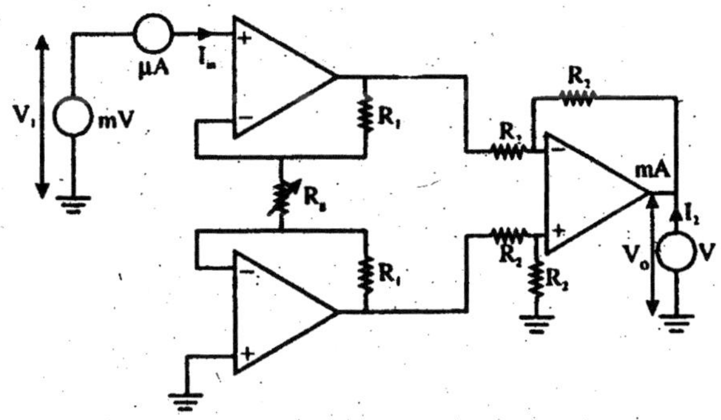

Fig. 9.1.

1. Gain : Gain is the ratio of output voltage to input voltage. In case of instrumentation amplifier

$$\text{gain} = \frac{V_0}{V_1} = -\left(1 + \frac{2R_1}{R_g}\right)V_1$$

Input impedance : It is the ratio of input voltage to input current, *i.e.*,

$$Z_{in} = \frac{V_1}{I_{in}}\Omega$$

Output impedance : It is the ratio of output voltage to output current, *i.e.*,

$$Z_{out} = \frac{V_2}{I_{out}}\Omega$$

Procedure :

1. Assemble the circuit as shown in Fig. 9.1.
2. Switch on power supply.
3. Apply the signal to the amplifier.
4. Measure output voltage, output current and input voltage input current.
5. Vary the potentiometer resistance R_g.
6. Repeat step 4.

Observation : R_1 = 100 kΩ, R_g = 10 kΩ variable, R_2 = 10 kΩ.

Observation Table

S. No.	Input voltage	Input current	Output voltage	Output current	Gain	Input impedance	Output impedance
	mV V_1	I_{in} = mA	V_0 (V)	(mA)	$\frac{V_0}{V_1}$	$\frac{V_1}{I_n}$	$\frac{V_0}{I_2}$
1							
2							
3							
4							
5							

Result : Gain of the amplifier =

Input impedance of the amplifier = ... Ω

Output impedance of the amplifier = ... Ω

Precautions : Instrumentation amplifier is very sensitive amplifier. Therefore the observations should be taken very carefully for good accuracy in the result.

EXPERIMENT 10

Aim : To study an X-Y recorder and graphic recorder.

Apparatus : X-Y recorder, graphic recorder, chart signal source.

Theory : X-Y recorder : This recorder draws the curve between two variables - one fed to X-input and the other fed to Y-input. For variable in X and Y-input the pen on the paper movers simultaneously. For example voltage vs current of a diode are plotted by using this. Voltage is one variable at X-input while current is fed at Y-input. The construction of X-Y recorder is shown in Fig. 10.1.

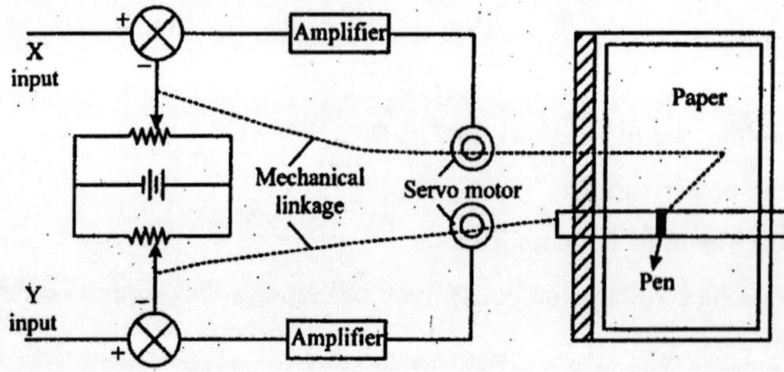

Fig. 10.1.

Procedure :

1. Apply the inputs at X and Y-inputs
2. Turn off the power supply to recorder.
3. Observe the graph.

Result : The graph between two variables (diode voltage and diode current) is obtained.

Graphic recorder : This type of recorder plots a variable with respect to time for this reason it is called X-t recorder. In this recorder measurand is fed to one terminal and time is the internal variable of the recorder. A chart paper moves vertically downward and open assembly moves perpendicular to the direction of motion of the chart. Thus a plot in a direction normal to the motion of paper is obtained. The direction of motion of chart is the time axis while the magnitude is along perpendicular axis.

Procedure :

1. Apply the measurand to the terminal of the recorder.
2. Turn on the power supply.
3. Observe waveforms.

Result : The maximum amplitude of the variable = ...

Minimum amplitude of the variable = ...

Previous Examination Question Papers

Previous Examination Question Paper
December 1999

Section-A

1. Fill in the blanks :

(i) The minimum anode current required to maintain the thyristor in the on-state immediately after a thyristor has been turned on and the gate signal has been removed is called _______.

(ii) Latching current is _______ than holding current.

(iii) dc/dt can be limited by adding a _______ in a thyristor circuit.

(iv) A semi-converter is _______ quadrant converter.

(v) A dual converter can work in quadrant.

(vi) Thermistor has a _______ temperature coefficient of resistance.

(vii) In a transducer the observed output deviates from correct value by a constant factor, the resultant error is called _______ error.

(viii) In a resistance potentiometer the non-linerity _______ with increase of load to potentiometer resistance.

(ix) Piezo electric crystals can act as an _______ transducer.

(x) For a single phase half wave rectifier V_{dc} = _______ Vm.

(xi) Heat sink thermal resistance is given by ΘsA = _______.

(xii) Ripple factor of half wave rectifier is RF = _______.

(xiii) Full wave rectifier has theoretical maximum efficiency of _______.

(xiv) The lowest limit of useful working range of transducer is determined by _______ & _______.

(xv) In LVDT two secondary windings are connected in _______.

Section-B

2. Attempt any six questions :

(a) What is the purpose of di/dt protection in SCR ?

(b) Explain the VI characteristics of SCR ?

(c) Discuss the principle of phase control.

(d) What do you understand by thermal resistance of heat sink ?

(e) Explain the working of Automatic Battery charger.

(f) Differentiate between active and passive transducer.

(g) Explain the principle of strain gauge.

(h) Describe the working of xy recorder.

Section-C

3. Attempt any three questions :

(i) Explain the triggering circuit of SCR.

(ii) Describe the principle of operation and working of piece electric transducer.

(iii) Explain the construction and working of Graphic recorder.

(iv) Describe the speed control of DC motor.

(v) Explain with wave forms working of three phase full wave bridge controlled rectifier.

Previous Examination Question Paper
May 2000

Section-A

1. Fill in the blanks :

(i) When junction J_2 is reverse biased and junction J_1 and J_3 are in forward biased condition the thyristor is in _______ state.

(ii) The minimum anode current required to maintain the thyristor circuit.

(iii) di/dt can be limited by adding a _______ in a thyristor circuit.

(iv) The period of oscillation of UJT relaxation oscillator is given by T = _______.

(v) A full converter is a _______ quadrant converter.

(vi) Strain gauge converts mechanical displacement into a change in _______.

(vii) Photovoltaic cell is _______ transducer.

(viii) Resistance potentiometer is a _______ order instrument.

(ix) Thermocouples are _______ transducers.

(x) Air cored inductive transducers are suitable for use at _______ frequencies.

(xi) In LVDT two secondary windings are connected in _______.

(xii) For a single phase full wave rectifier Vd * c = _______ Vm.

(xiii) For a three phase bridge rectifier the output is VDc = _______.

(xiv) The operating junction temperature in a device is given by Tj = _______.

(xv) Rectification efficiency of a rectifier is n = _______.

Section-B

2. Attempt any six questions :

(a) What is the purpose of dv/dt protection of a thyristor ?

(b) What are the means of turning on thyristor ?

(c) What is natural or line communication ?

(d) Describe the waveforms of 3 phase full converter ?

(e) What are various types of heat sink and their mounthing techniques ?

(f) Explain the circuit of fan control using phase control method ?

(g) Discuss the working of proximity probes ?

(h) Explain the chracteristics of an instrumentation amplifier.

Section-C

3. Attempt any three questions :

(i) Explain the forced turn off methods of SCR.

(ii) Describe the principle of operation and working of time delay relay circuit.

(iii) Explain the construction, principle of operation and working of LVDT.

(iv) Discuss with waveforms the working of parallel commutated inverter.

(v) Explain with waveforms working of single phase full wave bridge controlled rectifier.

Previous Examination Question Paper
December 2000

Section-A

1. Do as directed :

(i) The thyristor can be considered device. (T/F)

(ii) V-I characteristics of a SCR is alike in both first and third quadrants. (T/F)

(iii) A series inverter produces output distortion in the output waveform. (T/F)

(iv) In a parallel inverter the capacitor is connected in series with the load. (T/F)

(v) In a light operated SCR alarm circuit the relay is energised in presence of light. (T/F)

(vi) Induction heating is possible with both a.c. and d.c. power supplies. (T/F)

(vii) Efficiency of induction heating process is very _______.

(viii) In piezoelectric transducers the materials used are _______.

(ix) LVDT _______ type transducer.

(x) For a three phase bridge rectifier the output is Vdc = _______ Vm.

(xi) The thermocouple, cold junction is used as _______ junction.

(xii) The thermistor is a resistor which has _______ temperature coefficient.

(xiii) _______ technique is very convenient method of a.c. motor speed control.

(xiv) While conducing, the resistance of an SCR is very _______.

(xv) Response of a transistor is _______ in comparison to a thyristor.

Section-B

2. Attempt any six questions :

(a) Explain how the anode, cathode and the gate connections are made in a thyristor ?

(b) List the names of the methods adopted for triggering a thyristor.

(c) Explain the wave shapes of voltage and current for a single phase series inverter.

(d) Explain d.c. shunt motor for speed control.

(e) Explain with the help of a diagram the working of a light operated SCR alaram circuit.

(f) Explain how a capacitive transducer may be used for measurement of pressure or force.

(g) List the factors governing selection of a transducer for a particular measurement system.

(h) Explain graphic recorder.

Section-C

3. Attempt any three questions :

(i) Explain the principle working of strain gauge types of torque transducers. Explain its advantages and disadvantages.

(ii) Explain with diagram the salient feasures for designing the SCR controlled emergency light.

(iii) Discuss with waveforms, working of single phase controlled full wave bridge rectifier.

(iv) Explain in detail the gate triggering process for firing an SCR.

(v) Discuss in detail the working principle and applications of proximity process.

Previous Examination Question Paper
December 2001

Section-A

1. Fill in the blanks :

(a) Holding current is _______ than latching current.

(b) Thyristor is turned on by increasing the _______ current.

(c) The ratio of dc power to ac power is known as _______ of a rectifier.

(d) An inverter is called a _______ if input voltage remains constant, a _______ if input current is maintained constant and _______ if input voltage is controllable.

(e) _______ is the SI unit of luminance intensity.

(f) _______ is an example of active transducer.

(g) _______ determine temperature through radiation or optical sensors.

(h) Unit of seeback coefficient (α) is _______.

(i) _______ states that change in length is proportional to applied force.

(j) Additional strain without corresponding increase in stress is called_______.

(k) The instrumentation amplifier uses _______ inputs and defects _______.

(l) LVDTs monitors change in _______ to measure displacement.

Section-B

2. Attempt any six questions :

(a) Explain thyristor turn on using gate triggering ?

(b) What characteristics need to be considered when designing heat sinks ?

(c) Explain a basic measurement system ?

(d) Discuss performance parameters of rectifiers ?

(e) Explain principle of operation of inverters.

(f) What are the advantages of 3-phase rectifier over single phase rectifier ?

(g) What are the applications of dielectric heating and inductance ?

(h) Write short notes on any two :

(i) Graphic recorders

(ii) Working of LVDTs

(iii) Principle of SCR controlled battery charger.

Section-C

3. Attempt any three questions :

(a) Explain with circuit diagram, the working of DC and small AC motors.

(b) Discuss three phase half wave bridge rectifier with relevant waveforms.

(c) Discuss at least two applications of AC phase control rectifier with relevant diagrams.

(d) Explain the basic requirements for transducer selection.

(e) Explain working of instrumentation amplifier. A strain gauge with a nominal resistance of 350 Ω and a GF of 2.0; is connected into the quarter bridge circuit shown below. The excitation voltage for the bridge is set to 15 V. The other bridge resistors, including the dummy gage, are 350 Ω. Assume the bridge is initially balanced with no strain and has an output 45 μv under strain. *(i)* Compute the exact applied strain. *(ii)* Compute applied strain assuming linear bridge.

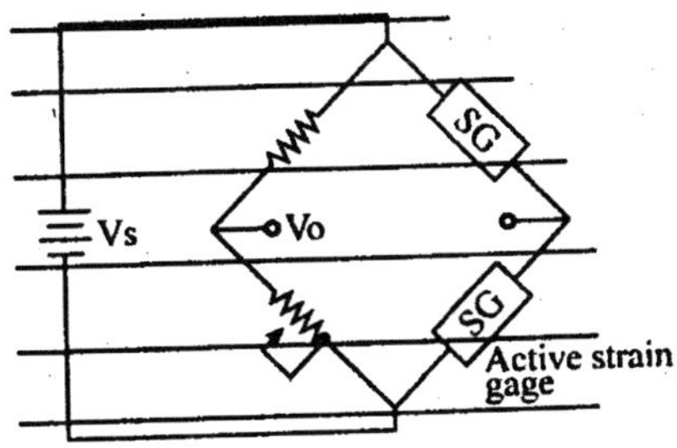

□□□

INDEX

□□□

Win Prizes

ATTENTION READERS

Win Prizes

We request for your frank assessment regarding various aspects of the book as given below :

Industrial Electronics And Instrumentation

By : SUNIL KUMAR

Please Fill up the Sheet in Neat and Clean Writing

(i) What topics of your syllabus, important for your examination point of view, are not covered in this book ?

....

....

....

....

(ii) What are the chapters wherein the treatment of the subject matter is not systematic or organised or updated ?

....

....

....

....

(iii) Have you come across misprints/mistakes/factual inaccuracies in the book ? Please specify the chapters and the page numbers.

....

....

....

....

(iv) What other books on the same subject have you found/heard better than the present book ? Please specify in terms of price and quality.

....

....

....

....

(v) Further suggestions and comments for the improvement of this book.

....

....

....

....

(PS : If need be please attach a separate sheet to write your views)

OTHER DETAILS

(i) Who recommended you this book ? (Please tick (✓) in the box near the option relevant to you.

☐ Teacher ☐ Friends ☐ Bookseller

....

....

....

....

(ii) Name of the recommending teacher, his designation and address:

....

....

....

....

(iii) Name and address of the bookseller you purchased the book from :

....

....

....

....

(iv) Name and address of your institution (Please mention the University or Board as the case may be)

....

....

....

....

(v) Your name and complete address :

....

....

....

....

(vi) Course you are enrolled in :

....

....

....

....

The best assessment will be awarded each month : The award will be in the form of our publications as decided by the Editorial Board.

Please mail the filled up coupon at your earliest to :

S.K. KATARIA & SONS®
4885/109, PrakashMahal, Dr. Subhash Bhargav Lane,
Daryaganj, New Delhi-110002
☎ +91-11-43551243
Mobile: +91-9871775858
Email: katariabook@gmail.com